Real-Time Systems Development with RTEMS and Multicore Processors

Embedded Systems
Series editor: Richard Zurawski

Time-Triggered Communication
Roman Obermaisser

Embedded Software Development
The Open-Source Approach
Ivan Cibrario Bertolotti, Tingting Hu

Real-Time Embedded Systems
Open-Source Operating Systems Perspective
Ivan Cibrario Bertolotti, Gabriele Manduchi

Communication Architectures for Systems-on-Chip
José L. Ayala

Event-Based Control and Signal Processing
Marek Miskowicz

**Real-Time Systems Development with RTEMS
and Multicore Processors**
Gedare Bloom, Joel Sherrill, Tingting Hu, Ivan Cibrario Bertolotti

For more information about this series, please visit: https://www.
routledge.com/Embedded-Systems/book-series/CRCEMBSYS

Real-Time Systems Development with RTEMS and Multicore Processors

Gedare Bloom

Joel Sherrill

Tingting Hu

Ivan Cibrario Bertolotti

CRC Press
Taylor & Francis Group
Boca Raton London New York

CRC Press is an imprint of the
Taylor & Francis Group, an **informa** business

First edition published 2021
by CRC Press
6000 Broken Sound Parkway NW, Suite 300, Boca Raton, FL 33487-2742

and by CRC Press
2 Park Square, Milton Park, Abingdon, Oxon, OX14 4RN

Library of Congress Cataloging-in-Publication Data

Names: Bloom, Gedare, author. | Sherrill, Joel, author. | Hu, Tingting,
   author. | Bertolotti, Ivan Cibrario, author.
Title: Real-time systems development with RTEMs and multicore processors /
   Gedare Bloom, Joel Sherrill, Tingting Hu, Ivan Cibrario Bertolotti.
Description: First edition. | [Boca Raton : CRC Press, 2020] | Series:
   Embedded systems | Includes bibliographical references and index.
Identifiers: LCCN 2020036898 (print) | LCCN 2020036899 (ebook) | ISBN
   9780815365976 (hardback) | ISBN 9781351255790 (ebook)
Subjects: LCSH: Embedded computer systems. | Real-time data processing. |
   Automatic control.
Classification: LCC TK7895.E42 B66 2020 (print) | LCC TK7895.E42 (ebook)
   | DDC 006.2/2--dc23
LC record available at https://lccn.loc.gov/2020036898
LC ebook record available at https://lccn.loc.gov/2020036899

ISBN: 978-0-8153-6597-6 (hbk)
ISBN: 978-1-351-25579-0 (ebk)

Contents

PART III Inter-Task Synchronization and Communication

PART IV   Network Communication

PART V   Multicores in Real-Time Embedded Systems

Preface

This book is the outcome of several decades of cumulated research, teaching, and consultancy experience in the field of real-time operating systems and communications applied to control systems and other classes of embedded applications, often carried out in strict cooperation with industrial and academic partners. During this time, we have been positively influenced by many other people we came in contact with. They are too numerous to mention individually, but we are nonetheless indebted to them for their contribution to our knowledge and professional growth.

A special thank you also goes to our university students, who first made use of the lecture notes this book is based upon. The suggestions and remarks that we collected along the years were helpful in making the book clearer and easier to read. We are also thankful to CRC Press publishing, editorial, and marketing staff, Nora Konopka and Prachi Mishra in particular. Without their valuable help, the book would have probably never seen the light of day.

The Authors

Gedare Bloom earned his PhD in computer science from The George Washington University, Washington, D.C., in 2013. He joined the Department of Computer Science at University of Colorado Colorado Springs as Assistant Professor in 2019. Previously, he was an Assistant Professor of Computer Science at Howard University from 2015–2019, Research Scientist at The George Washington University from 2014–2015, and Postdoctoral Fellow at The George Washington University from 2013–2014. His research expertise is computer system security with particular focus on real-time embedded systems used in automotive, avionics, industrial control, and critical infrastructure domains. The techniques he applies to solve problems along the hardware-software interface range from computer architecture, computer security, cryptography, operating systems, and real-time analysis. Prof. Bloom teaches undergraduate and graduate courses in computer architecture, operating systems, and system security. He brings novel content to these courses covering real-time embedded systems and giving practical exercises for students to gain exposure to the subtleties of working directly with hardware. He has received departmental and university recognition for excellence in teaching.

Since 2011, Prof. Bloom has been a maintainer for the RTEMS open-source hard real-time OS, which is used in robotics frameworks, unmanned vehicles, satellites and space probes, automotive, defense, building automation, medical devices, industrial controllers, and more. Some of his key contributions to RTEMS include the first 64-bit architectural port of RTEMS, design and implementation of a modern thread scheduling infrastructure, support for running RTEMS as a paravirtualized guest for avionics hypervisors, and implementation of POSIX services required for conformance with the FACE avionics technical standard. Additionally, he mentors and guides students around the world through learning about and developing with RTEMS.

Dr. Bloom is active in service to the professional community. He is a Senior Member of the ACM (M'09, SM'19) and a Senior Member of the IEEE (M'09, SM'19), with membership in the SIGARCH, SIGOPS, SIGBED, SIGSAC, and SIGCSE special interest groups of the ACM, and in the IEEE Computer Society and the IEEE Computer Society Technical Committee on Real-Time Systems. He has served as a program committee member, artifact evaluation committee member, and technical referee for flagship conferences and journals in the area of real-time and embedded systems.

Joel Sherrill earned his PhD in computer science from The University of Alabama in Huntsville in 1999. He joined On-Line Applications Research Corporation in 1989 and is currently the Director of Research and Development. He is one of the original developers of the free real-time operating system RTEMS and current project lead. Dr. Sherrill has ported RTEMS to multiple processor architectures and has been responsible for development of many RTEMS capabilities. He has also supported student activities through mentoring in programs such as the Google Summer of Code (GSoC) and ESA Summer of Code in Space (SOCIS) since 2008. One of his focuses with RTEMS is ensuring that RTEMS meets the stringent requirements required when deploying software in critical systems while ensuring conformance to open standards such as POSIX.

Since 2011, Dr. Sherrill has represented what is now the U.S. Army Combat Capabilities Development Command Aviation & Missile Center on The Open Group's FACE Consortium. He is a principal author of the FACE Technical Standard (https://www.opengroup.org/face) and multiple supporting documents. Additionally, Dr. Sherrill designed and co-authored the Basic Avionics Lightweight Sample Application (BALSA) examplar for the FACE Technical Standard. Dr. Sherrill led the effort to integrate a paravirtualized RTEMS with the ARINC 653 RTOS Deos to create the Deos+RTEMS product, which has achieved formal conformance to the FACE Technical Standard.

Dr. Sherrill is a senior member of the ACM having joined in 1997 as well as a senior member of the IEEE with membership since 1991. He has been a member of the IEEE Computer Society since 2003, and a member of the IEEE Computer Society Technical Committee on Real-Time Systems since 2011. He has served as a program committee member and technical referee for professional conferences in the area of real-time and embedded systems.

Tingting Hu earned her master's degree in computer engineering in 2010 and PhD with the best dissertation award in computer and control engineering in 2015 from Politecnico di Torino, Turin, Italy. Between 2010 and 2016, she also worked as a research fellow with the National Research Council of Italy (CNR), Turin, Italy. From 2017–2018, she worked as a post-doc researcher at the University of Luxembourg with the Faculty of Science, Technology and Medicine. Since 2019, she works as a research scientist in the University of Luxembourg with the Faculty of Science, Technology and Medicine.

Her primary research interest concerns embedded systems design and implementation, spanning through topics such as real-time operating systems, industrial communication protocols, formal verification, and fault-tolerance for safety-critical systems. Currently, she is focusing on the research of model driven engineering for safety-critical, real-time embedded systems.

Her past research work led to the publication of research papers at the most prominent journals and leading conferences in the area of real-time and embedded systems. She is the co-author of *Embedded Software Development: The Open-Source Approach* published in 2015 with Taylor & Francis. In addition, she is one of the four main inventors of the European patent, "Limitation of Bit Stuffing in a Communication Frame of an Electronic Signal" (EP2908475B1, 2019).

Since 2017, she has been actively involved in teaching activities for two bachelor's courses: Computer Infrastructure and Network and Communication, as well as one master's course: Dependable Systems.

She serves as program committee member and technical referee for several primary conferences in her research area. She also works as industrial consultant for leading national industries in the provision of software design solutions for real-time embedded systems in the domain of industrial ovens, building automation, and motion control.

Ivan Cibrario Bertolotti received his Laurea degree (summa cum laude) in computer science from the University of Turin, Italy, in 1996. Since then, he has been a researcher with the National Research Council of Italy (CNR). Currently, he is with the Istituto di Elettronica e di Ingegneria dell'Informazione e delle Telecomunicazioni (IEIIT), Turin, Italy. His current research interests include real-time operating system design and implementation, industrial communication systems, and formal methods for software and protocol verification. Along the years, he published more than 100 peer-reviewed articles on these subjects.

In the last two decades, he has also been active as an industrial consultant for the software design of real-time embedded systems, as well as the evaluation and adoption of open-source components in these kinds of applications, working with leading international industries like STMicroelectronics and others. The application domains range from mobile multimedia terminals to system software for set-top boxes, building automation, distributed control of industrial ovens and medical refrigerators, and gravimetric dosing of plastics components.

From 2003–2011 and 2017–2019, Prof. Cibrario Bertolotti taught introductory and advanced courses on real-time operating systems at Politecnico di Torino, Turin, Italy. In the spring of 2009, he taught a real-time operating system course at the Graduate School in Information Engineering's PhD program, University of Padua, Italy. He coauthored two books and several book chapters on the same topics, and holds one Italian and three European/US patents.

Moreover, he has served as a program committee member and technical referee for the main international conferences and journals related to factory automation, factory communication systems, and industrial informatics. He has been a member of the IEEE Computer Society since 1997 and a member of the IEEE since 2006.

1 Introduction

Multicore processors are nowadays ubiquitous in desktop computing and are becoming more and more popular in many other application domains, ranging from mobile phones to hard real-time embedded systems. Yet, how to use them effectively with the help of an embedded real-time operating system is still little known to many practitioners. In the open-source arena, the matter is made even more complex due to the lack of comprehensive learning material in the scientific and technical literature. Thus, potential users easily run into the risk of misusing multicore processors, or not considering their merits and pitfalls in the right perspective.

The goal of the book is to provide readers with hands-on knowledge about the design and development cycle of a typical real-time application using the Real-Time Executive for Multiprocessor Systems (RTEMS) operating system, which is a representative and widely used Real-Time Operating System (RTOS) for embedded systems. The narrative starts from basic ideas (for instance, how to use an open-source toolchain) and then proceeds to discuss state-of-the-art concepts (like multicore scheduling and synchronization), which are in part still open to research.

Building on the extensive knowledge of leading RTEMS designers and developers, as well as academic researchers, the book aims at providing not only sound theoretical information but also valuable practical advice with a thorough description of the RTEMS Application Programming Interfaces. The topics covered in the book enable average readers to understand all aspects of the embedded software development process and readily apply the acquired knowledge in their next project. Moreover, fundamental theoretical concepts are introduced along the way, focusing on their consequences on the above-mentioned practical topics, which makes this book also good for graduate-level classroom use.

—

Part I of the book introduces the reader to embedded software development. First of all, it describes the fundamental tools used to compile and link application software and how the RTEMS operating system is configured for use. Then, the discussion continues with the basics of concurrent programming, real-time scheduling, and scheduling analysis. The chapters in this part are:

- Chapter 2, *Cross-Compilation Toolchain*. This chapter first describes the main components of a GNU-based[1] cross-compilation toolchain, focusing in particular on the linker command language and on GNU make, a tool commonly used to coordinate and automate the software build process. The second part of the chapter discusses the compile-time configuration of the RTEMS operating system.

---

[1]GNU stands for GNU's Not Unix!

1

- Chapter 3, *Concurrent Programming and Scheduling Algorithms*. The first main goal of this chapter is to lay out the theoretical foundations of concurrent programming. The discussion covers the all-important concept of process, or task, and how task state is represented within an operating system as it evolves over time. The second part of the chapter introduces the reader to real-time scheduling methods and techniques on single-processor systems, while the discussion of scheduling algorithms suitable for multicore systems is left to Part V.

- Chapter 4, *Scheduling Analysis and Interrupt Handling*. The main topic of this chapter is scheduling analysis, a set of mathematical tools to predict the worst-case timing behavior of a real-time system. The discussion starts from a high-level view of the system, abstracted as a set of tasks, and then shows how interrupt handling fits in the scheduling analysis framework. A set of practical considerations on interrupt handling, using a popular microprocessor architecture as a reference, concludes the chapter and helps illustrate how RTEMS implements some key activities, like context switch.

Part II discusses the concepts and mechanisms of task management and timekeeping, along with the two Application Programming Interfaces (APIs) that give users access to them. In particular:

- Chapter 5, *Task Management and Timekeeping, Classic API*. This chapter is devoted to the RTEMS scheduling algorithms for single-core systems and the facilities that RTEMS provides to manipulate tasks and account for the passage of time through its Classic API. It also contains a comparison with the POSIX standard API and a description of some lower-level aspects of interrupt handling on single-core systems made accessible by the RTEMS Interrupt Manager and often essential in embedded systems.

- Chapter 6, *Task Management and Timekeeping, POSIX API*. The chapter contains an extensive description of the POSIX API for task management and timekeeping. Special attention is given to the cancellation and signalling mechanisms, which do not have a direct counterpart in the Classic API.

Part III discusses the all-important topic of lock-based task synchronization and communication, as well as its interaction with scheduling and scheduling analysis. The three chapters in this part also introduce the reader to the main synchronization devices and message passing directives available in RTEMS through its Classic and POSIX APIs.

- Chapter 7, *Inter-Task Synchronization and Communication (IPC) Based on Shared Memory*. This chapter describes the fundamental concepts of race condition, critical region, and lock-based mutual exclusion. It introduces the reader to the classic inter-task synchronization and communication methods based on shared memory, namely, semaphores and monitors.

Two more specialized synchronization devices, barriers and events, are also included in the discussion due to their considerable practical interest.

- Chapter 8, *IPC, Task Execution, and Scheduling.* This chapter is devoted to priority inversion and deadlock, two very important issues related to lock-based synchronization and communication, which may impair the timings of any real-time system if not appropriately solved. To this aim, the chapter discusses several suitable design-time and runtime methods and techniques.
- Chapter 9, *IPC Based on Message Passing.* This chapter introduces readers to message passing, an IPC mechanism that does not rely on shared memory for data transfer, thus paving the way to a unified IPC technique that is also suitable for distributed systems in which multiple independent nodes are connected by a communication network. As in the rest of the book, theoretical concepts are presented together with their RTEMS implementation.

Part IV describes how RTEMS provides full-fledged TCP/IP network communication, which is becoming an ubiquitous requirement in modern embedded systems. It is composed of two chapters:

- Chapter 10, *Network Communication in RTEMS.* This chapter describes the internal structure of the RTEMS networking code and highlights the most important aspects of operating system / protocol stack integration, such as synchronization and the device driver interface.
- Chapter 11, *POSIX Sockets API.* This chapter complements the previous one and discusses in detail how users can access the RTEMS networking code in an operating system and protocol-independent way, by means of the standard POSIX Sockets API.

Part V concludes the book. Its two chapters describe the issues brought by multi-core embedded processors and how RTEMS supports them:

- Chapter 12, *Multicores in Embedded Systems.* The chapter outlines the motivation behind the widespread diffusion of multicore processors for embedded systems and provides an overview of their architecture. Then, it summarizes the challenges introduced by multicores in software development for embedded systems, focusing on the areas of task scheduling, schedulability algorithms and analysis, and proper inter-task communication and synchronization.
- Chapter 13, *Multicore Concurrency: Issues and Solutions.* The chapter summarizes the most common scheduling algorithms and synchronization devices for real-time multicore systems and illustrates how they are supported in RTEMS. The second part of the chapter provides information about lock-free and wait-free synchronization, which is often a valid alternative to lock-based synchronization described in Part III, as well as the use of spinlocks for task/interrupt handler synchronization.

Part I

Operating System Basics

2 Cross-Compilation Toolchain

CONTENTS

This chapter explains the general compilation approach adopted in embedded systems and then introduces the main components of a GNU-based toolchain. Among them, special attention is given to the linker command language used to write linker scripts and to GNU make, which is one of the most widespread tools for coordinating and automating the software build process. A short discussion of the RTEMS configuration system concludes the chapter.

2.1 FROM SOURCE CODE TO THE EXECUTABLE IMAGE

Unlike general-purpose systems, where code is compiled, built, and executed on the same machine, development for embedded systems usually requires the availability of a cross-compilation toolchain. It compiles and builds the source code on a *host* machine where the cross-compilation toolchain runs. Instead, the compiled code will execute on a *target* machine, in this case an embedded device. This is due to the

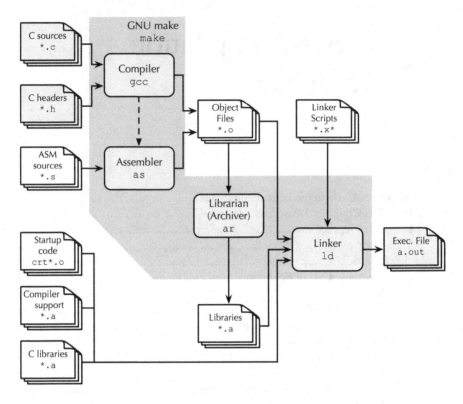

FIGURE 2.1   Simplified view of the C-language toolchain workflow.

resource limits on embedded devices, for instance, memory capacity and processor speed. Even when embedded boards are shipped with a pre-installed general-purpose operating system such as Linux, which consequently enables support for native compilation, these setups are generally not meant for real-time execution purpose.

Figure 2.1 demonstrates the compilation process, which translates source code into an executable image, using the GCC-based cross-compilation toolchain. As shown in the figure, a cross-compilation toolchain usually consists of the following components:

- The GCC *Compiler*. It translates a C/C++ source file, which in turn may include other headers or source files, and produces an object module in binary format. This generally involves a multi-step code generation process, which is better detailed in Section 2.1.1. Although it is generally called "the compiler", the gcc program is actually a *compiler driver*, able to perform different actions by invoking other toolchain components appropriately, depending on the input file type usually derived from its filename extension. The compiler driver behavior can be customized by means of command-line options. Depending on the target architecture, different options may be provided, to drive the compilation for that specific architecture. For instance, the `-march` option can be used to indicate the specific architecture belonging to the ARM family, such as `armv8-a` for the ARMv8-A architecure [14], or `armv7-a` for the ARMv7-A. The actions to be performed by `gcc` are further configured by means of a *specs* string or file. Both the compiler driver itself and its specs string are discussed in more detail in Section 2.1.1.

- The *librarian*—whose command name is `ar` because it was called *archivier* in the past—collects multiple object modules into a library. The same tool can also performs several other maintenance operations on a library. For instance, it is able to extract or delete a module from it.

  Other tools, like nm and `objdump`, perform more specialized operations related to object module and executable image contents and symbols. These tools will not be discussed further in the following due to space limitations. Interested readers may refer to their documentation [96] for more information.

- The *linker* `ld`, presented in Section 2.1.3, links object modules together and against libraries guided by command-line options or, more commonly, by one or more *linker scripts*. It resolves cross references to eventually build an *executable image*. Especially in embedded systems, the linking phase usually brings the application plus a variety of system code together into the executable image.

- There are several categories of system code used at link time:
  - The *startup* object files—usually called `crt*.o`—contain code that is executed first, when the executable image starts up. In standalone executable images, they also include system and hardware initialization code, often called *bootstrap* code.
  - The *compiler support library* `libgcc.a` contains utility functions needed by the code generator, but too big/complex to be instantiated inline. For instance, integer multiply/divide or floating-point operations on processors without hardware support for them.
  - The *standard C libraries*, `libc.a` and `libm.a`.
  - Possibly, the *operating system* itself. This is the case of most real-time operating systems and also RTEMS belongs to this category.

Another important component typically present in a toolchain belongs to a category by itself because it does not directly operate on source or object files, or executable images, but is responsible for coordinating and automating the software build process as a whole. One of the most widespread tools of this kind is the open-source *GNU make* program. In Figure 2.1, it is shown as a gray background that encompasses the other toolchain components and will be the subject of Section 2.3.

Building a toolchain is a complex affair, first of all due to its sheer size and complexity, but also because the toolchain components are themselves written in a high-level language and distributed as *source code*. For instance, the GNU compiler driver for the C/C++ programming languages is itself written in C/C++, and hence, a working C/C++ compiler is required in order to build it. There are several different ways to solve this "chicken and egg" problem, often called *bootstrap problem*, depending on the kind of compiler to be built.

As said previously, the kind of toolchain most frequently used in embedded software development is the *cross-compilation* toolchain. This kind of toolchain generates code for a certain architecture (the *target*), but runs on a different architecture (the development system, or *host*). This is because, due to limitations concerning their memory capacity and processor speed, embedded systems often cannot compile their own code.

The bootstrap problem becomes somewhat simpler in this case, because it is possible to use a native toolchain on the host to build the cross-compilation toolchain. The availability of a native toolchain is usually not an issue because most open-source operating system distributions already provide one ready-to-use. When using RTEMS, the process is further streamlined by the availability of a comprehensive tool, called *RTEMS Source Builder* (RSB) [106], able to build a complete open-source cross-compilation toolchain from source code in a fully automatic way.

To distinguish cross-compilation toolchain components from their native toolchain counterparts, their names are prefixed with a string that summarizes their target architecture. For instance, the C compiler driver for the ARM architecture and RTEMS operating system version 4.11 could be called `arm-rtems4.11-gcc` instead of simply `gcc`. In the following, we will keep using the short names for brevity.

2.1.1 THE COMPILER DRIVER

Figure 2.2 depicts the general outline of the compiler driver workflow. According to the high-level view given in the previous section, when the compiler driver is used to compile a C source file, in theory it should perform the following steps, by invoking a toolchain component for each of them:

1. Preprocess the source code with the C preprocessor `cpp`.
2. Compile the source code into assembly code, by means of the C compiler `cc1`.
3. Assemble it with the assembler `as` to produce the output object file.

The internal structure of the `gcc` compiler deserves a whole book by itself due to its complexity. Interested reader may refer to [121] for a comprehensive guide and to the official documentation for an authoritative description of the user-visible features of the compiler [118], as well as its internals [117]. In the following, we will focus only on some peculiarities of the `gcc`-based toolchain that deviate from the abstract view just presented, as well as on how the compiler driver itself works.

First of all, as shown in the figure, the preprocessor is *integrated* in the C compiler implemented by `cc1`. The C++ compiler `cc1plus` also uses the same approach. A *standalone preprocessor* `cpp` does exist, but it is not used during normal compilation. In any case, the behavior of the standalone preprocessor and the one implemented in the compiler is consistent because both make use of the same

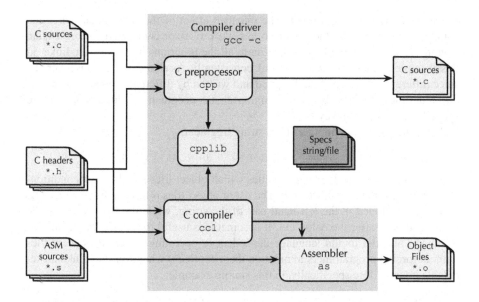

FIGURE 2.2 Simplified view of the compiler driver workflow.

preprocessing library, `cpplib`, which can also be used directly by application programs as a general-purpose macro expansion tool.

On the contrary, the assembler is implemented as a separate program, `as`, which is not part of the `gcc` distribution. Instead, it is distributed as part of the binary utilities package [96], which also includes the linker. The assembler will not be further discussed in this book due to space constraints.

One peculiar aspect of the `gcc`-based toolchain is that the compiler driver is programmable. Namely, it is driven by a set of rules contained in a "specs" string or file. The specs string can be used to customize the behavior of the compiler driver. It ensures that the compiler driver is as flexible as possible within its design envelope.

In the following, we will only provide an overview of the expressive power that specs strings have, and illustrate what can be accomplished with their help, by means of a couple of examples. A thorough documentation of specs string syntax and usage can be found in [117]. Basically, the rules contained in a specs string specify which sequence of programs the compiler driver should run, and their arguments, depending on the kind of file provided as input. A default specs string is built in the compiler driver itself and is used when no custom specs string is provided elsewhere.

The sequence of steps to be taken in order to compile a file can be specified depending on the *suffix* of the file itself. Other rules associated with some command-line options may change the arguments passed by the driver to the programs it invokes.

```
*link: %{mbig-endian:-EB}
```

For example, the specs string fragment listed above specifies that if the command-line option `-mbig-endian` is given to the compiler driver, then the linker must be invoked with the −EB option. The intended effect of the rule is that, when the compiler driver is given the option to target a processor configured for big-endian operations, it must pass this information to all the programs it invokes, the linker in this example, to make sure they all work consistently. As also shown by this example, since different programs were designed and written by different groups of people at different times, it is possible (and common) that conceptually analogous options are spelled out in different ways.

Let us now consider another specs string fragment:

```
*startfile: crti%O%s crtbegin%O%s new_crt0%O%s
```

In this case, the specs string specifies which object files should be unconditionally included at the start of the link. The list of object files is held in the startfile variable, mentioned in the left-hand part of the string, while the list itself is in the right-hand part, after the colon (:). It is sometimes useful to modify the default set of objects in order to add language-dependent or operating system-dependent files without forcing programmers to mention them explicitly whenever they link an executable image. More specifically, in this simple example:

- The *startfile: specification *overrides* the internal specs variable startfile and gives it a new value.
- crti%O%s and crtbegin%O%s are the standard initialization object files typical of C language programs. Within these strings %O represents the default object file suffix (by default, it is expanded to .o on Linux-based hosts) and %s specifies that the object file is a system file and shall be searched for in the system search path rather than in user-defined directories. The use of %O, %s, and other similar directives makes a specs string portable across host operating systems with different file naming conventions and allows the compiler to be installed in different places of the filesystem without affecting the user in any way.
- new_crt0%O%s replaces crt0%O%s (one of the standard initialization files of the C compiler) to provide, for instance, operating-system, language, or machine-specific initialization functions.

2.1.2   THE PREPROCESSOR

The preprocessor, called cpp in a GNU-based toolchain, performs three main activities that, at least conceptually, take place before the source code is passed to the compiler proper. They are:

1. File inclusion, invoked by the #include directive.
2. Macro definition, by means of the #define directive, and expansion.
3. Conditional inclusion/exclusion of part of the input file from the compilation process, depending on whether some macros are defined or not (for instance, when the #ifdef or #ifndef directives are used) and their value (#if directive).

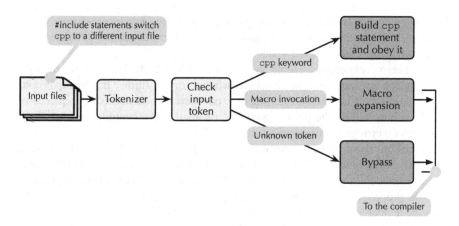

FIGURE 2.3 Simplified C-language preprocessor workflow.

According to the C language specification [71], the preprocessor works by *plaintext substitution*. However, since the preprocessor and the compiler grammar are the same at the *lexical (token) level*, in a GCC-based toolchain the preprocessor also performs tokenization of input files as an optimization. Hence, it provides a stream of tokens instead of plaintext to the compiler.

A token [3] is a data structure that, besides the token *text* (a sequence of characters), also contains information about its *nature* (for instance, whether the token represents a number, a keyword, or an identifier) and *debugging information* (the file and line number it was read from). Therefore, a token conveys additional information with respect to the portion of plaintext it corresponds to. This information is needed by the compiler anyway to further process its input, and hence, passing tokens instead of plaintext avoids duplicated processing.

Figure 2.3 contains a simplified view of the preprocessor workflow. Informally speaking, as it divides the input file into tokens, the preprocessor checks all of them and carries out one of three possible actions.

1. When the input token is a *preprocessor keyword*, like #define or #include, the preprocessor analyzes the tokens that follow it to build a complete statement and then obeys it.

   For example, after #define, it looks for a macro name followed by the (optional) macro body. When the macro definition statement is complete, the preprocessor records the association *name → body* in a table for future use.

   In this case, neither the preprocessor keyword nor the following tokens (the macro name and body in this example) are forwarded to the compiler. The macro body will become visible to the compiler only if the macro will be expanded later.

   The *name → body* table is initialized when preprocessing starts and discarded when it ends. As a consequence, macro definitions are not kept across multiple compilation units. However, the table is not empty at the very beginning of a compilation because the preprocessor itself pre-defines a number of macros.

2. When a macro is invoked, the preprocessor performs *macro expansion*. In the simplest case—that is, for *object-like* macros—macro expansion is triggered by encountering a macro name in the source code.

   The macro is expanded by replacing its name with its body. Then, the result of the expansion is examined again to check whether or not further macro expansions can be done. When no further macro expansions can be done, the sequence of tokens obtained by the preprocessor is forwarded to the compiler instead of the tokens that triggered macro expansion.

3. Tokens unknown to the preprocessor are simply passed to the compiler without modification. Since the preprocessor and compiler grammars are very different at the *syntactical level*, many kinds of token known to the compiler have no meaning to the preprocessor, even though the latter is perfectly able to build the token itself at the lexical level.

   For instance, it is obvious that type definitions are extremely important to the compiler, but they are completely transparent to the preprocessor.

The syntax of preprocessor statements is fairly simple. They always start with a *sharp character* (#) in column one. Spaces are allowed between # and the rest of the statement. The main categories of statement are:

- Macro definition: `#define`.
- File inclusion: `#include`.
- Conditional compilation: `#ifdef`, `#ifndef`, `#if`, `#else`, `#elif`, and `#endif`.
- Other, for instance: `#warning` and `#error`.

In the following, we will mainly focus on the macro definition and expansion process. Interested readers can refer to the full preprocessor documentation that comes with the `gcc` compiler [118] for further information about the other categories.

There are two kinds of macros: *object-like* and *function-like* macros. Object-like macros are the simplest and their handling by the preprocessor can be summarized in two main points:

- The name of the macro is replaced by its body when it is encountered in the source file. The result is reexamined after expansion to check whether or not other macros are involved. If this is the case, they are expanded as well. The process ends when no further macro expansions can be performed.
- Macro expansion does *not* take place when a macro is defined, but only when the macro is used. Hence, it is possible to have *forward references* to other macros within macro definitions.

For example, it is possible to define two macros, A and B, as follows:

```
#define B A+3
#define A 12
```

The definition of macro B does not produce any error although A has not been defined yet, because B's body is not expanded at the time of its definition. When B is encountered in the source file, after the previously listed definitions, it is expanded as: B → A+3 → 12+3. Since no other macro names are present, macro expansion ends at this point and the three tokens 12, +, and 3 are forwarded to the compiler.

Due to the way of communication between the preprocessor and the compiler, explained previously and outlined in Figure 2.3, the compiler does not know how tokens are obtained. Namely, it cannot distinguish between tokens coming from macro expansion and tokens taken directly from the source file. As a consequence, if B is used within a more complex expression, the compiler might get confused and interpret the expression in a counter-intuitive way.

Continuing the previous example, the expression B*5 is expanded by the preprocessor as B*5 → A+3*5 → 12+3*5. When the compiler parses the result, the evaluation of 3*5 is performed *before* +, due to the precedence rules of arithmetic operators, although this is probably not the behavior the programmer expects.

To solve this problem, it is often useful to put additional parentheses around macro bodies, as is shown in the following fragment of code:

```
#define B (A+3)
#define A 12
```

In this way, when B is invoked, it is expanded as B → (A+3) → (12+3) and the expression B*5 is seen by the compiler as (12+3)*5. In other words, the additional pair of parentheses coming from the expansion of macro B explicitly establishes the boundaries of macro expansion and overrides the arithmetic operator precedence rules.

The second kind of macro is represented by function-like macros. The main differences with respect to object-like macros can be summarized as follows:

- Function-like macros have a list of *parameters*, enclosed between parentheses (), after the macro name in their definition.
- Accordingly, they must be invoked by using the macro name followed by a list of *arguments*, also enclosed between ().

To illustrate how function-like macro expansion takes place, let us consider the following fragment of code as an example:

```
#define F(x, y) x*y*K
#define K 7
#define Z 3
```

When a function-like macro is invoked, its arguments are completely macro-expanded first. Therefore, for instance, the first step in the expansion of F(Z, 6) is: F(Z, 6) → F(3, 6).

Then, the parameters in the macro body are replaced by the corresponding, expanded arguments. Continuing our example:

- parameter x is replaced by argument 3, and

- y is replaced by 6.

After the replacement, the body of the macro becomes $3*6*K$. At this point, the modified body replaces the function-like macro invocation. Therefore, $F(3, 6) \rightarrow 3*6*K$.

The final step in function-like macro expansion consists of re-examining the result and check whether or not other macros (either object-like or function-like) can be expanded. In our example, the result of macro expansion obtained so far still contains the object-like macro name K and the preprocessor expands it according to its definition: $3*6*K \rightarrow 3*6*7$.

To summarize, the complete process of macro expansion when the function-like macro $F(Z, 6)$ is invoked is

$$F(Z, 6) \rightarrow F(3, 6) \qquad \text{(argument expansion)}$$
$$\rightarrow 3*6*K \qquad \text{(parameter substitution in the macro body)}$$
$$\rightarrow 3*6*7 \qquad \text{(expansion of the result)}$$

As already remarked previously about object-like macros, parentheses may be useful around parameters in function-like macro bodies, too, for the same reason. For instance, the expansion of $F(Z, 6+9)$ proceeds as shown below and clearly produces a counter-intuitive result, if we would like to consider F to be akin to a mathematical function.

$$F(Z, 6+9) \rightarrow F(3, 6+9) \qquad \text{(argument expansion)}$$
$$\rightarrow 3*6+9*K \qquad \text{(parameter substitution in the macro body)}$$
$$\rightarrow 3*6+9*7 \qquad \text{(expansion of the result)}$$

It is possible to work around this problem by defining $F(x, y)$ as $(x)*(y)*(K)$. In this way, the final result of the expansion is:

$$F(Z, 6+9) \rightarrow \cdots \rightarrow (3)*(6+9)*(7)$$

as intended.

2.1.3   THE LINKER

The main purpose of the linker, fully described in [34], is to combine a number of object files and libraries to build an executable image. In order to do this, the linker *resolves* inter-module symbol references, also by pulling object modules from libraries as required, and *relocates* code and data.

As explained previously, in a GNU toolchain the linker may be invoked directly as ld when performing a native compilation or <cross>-ld, where <cross> is the prefix denoting the target architecture, when cross-compiling. However, as described in Section 2.1.1, it is more often the compiler driver <cross>-gcc that automatically calls the linker as required. In both cases, the linking process is driven by a *linker script* written in the Link Editor Command Language.

Before describing in a more detailed way how the linker works, let us briefly review a couple of general linker options that are often useful, especially for embedded software development.

- The option -Map=<file> writes a *link map* to <file>. When the additional --cref option is given, the map also includes a cross reference table. Even though no further information about it will be given here, due to lack of space, the link map contains a significant amount of information about the outcome of the linking process. Interested readers may refer to the linker documentation [34] for more details about it.
- The option --oformat=<format> sets the format of the output file, among those recognized by ld. Being able to precisely control the output format helps to upload the executable image into the target platform successfully because upload tools often work only with a limited set of file formats. Reference [34] contains the full list of supported output formats, depending on the target architecture and linker configuration.
- The options --strip and --strip-debug remove symbolic information from the output file, leaving only the executable code and data. This step is sometimes required for executable image upload tools to work correctly because they might not properly handle any extra information present in the image. Symbolic information is mainly used to store debugging data in the executable image, for instance, the mapping between source code line numbers and machine code. For this reason, uploading this information into the target memory is useless in most cases, unless a debugger runs on the target itself.

When ld is invoked through the compiler driver, linker options must be preceded by the escape sequence -Wl to distinguish them from options directed to the compiler driver itself. A comma is used to separate the escape sequence from the string to be forwarded to the linker and no intervening spaces are allowed. For instance, gcc -Wl,-Map=f.map -o f f.c compiles and links f.c, and gives the -Map=f.map option to the linker.

As a last introductory step, it is also important to informally recall the main differences between *object modules*, *libraries*, and *executable images* as far as the linker is concerned. These differences, outlined below, will be further explained and highlighted in the following sections.

- Object files are unconditionally included in the final executable image. Instead, the object modules found in libraries, often called library modules, are used only on demand. More specifically, library modules are included by the linker only if they are needed to resolve pending symbol references.
- A library is simply a collection of unmodified object modules put together into a single file by the archiver or librarian ar.
- An executable image is formed by binding together object modules, either standalone or from libraries. However, it is not simply a collection, like a

library is, because the linker performs a significant amount of work in the
process.

Most linker activities revolve around *symbol* manipulation. Informally speaking,
a symbol is a convenient way to refer to the address of an object in memory in an
abstract way (by means of a human-readable name instead of a number) and even
before its exact location in memory is known.

The use of symbols is especially useful to the compiler during code generation.
For example, when the compiler generates code for a backward jump at the end of a
loop, two cases are possible:

1. If the processor supports *relative* jumps—that is, a jump in which the target address is calculated as the sum of the current program counter plus an offset stored
   in the jump instruction—the compiler may be able to generate the code completely and automatically by itself because it knows the "distance" between the
   jump instruction and its target. The linker is not involved in this case.
2. If the processor only supports *absolute* jumps—that is, a jump in which the target
   address is directly specified in the jump instruction—the compiler must leave a
   "blank" in the generated code. At most, the compiler may know the relative target
   address with respect to the beginning of the object code it is generating, but it
   does not know the final, absolute address where the code will eventually end up
   in memory. As will be better explained in the following, this blank will be filled
   by the linker when it performs symbol resolution and relocation.

Another intuitive example, regarding *data* instead of *code* addresses, is represented by global variables accessed by means of an `extern` declaration. Also in
this case, the compiler needs to refer to the variable by name—that is, through a
symbol—when it generates code, because it does not know its memory address at
all. Like before, the code that the compiler generates will be incomplete because
it will include "blanks," in which symbols are present in place of actual memory
addresses.

On the one hand, when the linker collects object files in order to produce the
executable image, it becomes possible to associate symbol definitions with the corresponding references, by means of a name-matching process known as *symbol resolution* or (according to an older nomenclature) *snapping*. On the other hand, symbol
values (to continue our examples, addresses of variables, and the exact address of
machine instructions) become known when the linker *relocates* object contents in
order to lay them out into memory. Only at this point can the linker "fill the blanks"
left by the compiler.

As an example of how symbol resolution takes place for data, let us consider the
two extremely simple source files listed in Figure 2.4. In this case, symbol resolution
proceeds as follows:

- When the compiler generates code for `f()` it does not know where (and if)
  variable `i` is defined. Therefore, in `f.o` the address of `i` is left blank, to be
  filled by the linker.

File f.c	*File* g.c

`extern int i;`
 	
`void f(void) {`
`    i = 7;`
`}`
	`int i;`

FIGURE 2.4 A simple example of symbol resolution.

- This is because the compiler works on exactly one *compilation unit* at a time, defined as the set of source and header files directly or indirectly included by means of #include statements by each individual top-level source file passed to the compiler.
- Therefore, when the compiler is working on f.c it does not consider g.c in any way, even though both files appear together on the command line.
- During symbol resolution, the linker observes that i is defined in g.o and associates the definition with the reference made in f.o.
- After the linker relocates the contents of g.o, the address of i becomes known and can eventually be used to complete the code in f.o.

It is also useful to remark that initialized data need a special treatment when the initial values must be in non-volatile memory. In this case, the linker must cooperate with the startup code (by providing memory layout information) so that those data can be initialized correctly when the application starts up. Further information on this point will be given in Section 2.2.2.

2.2 LINKER SCRIPTS

As mentioned previously, the linking process is driven by a set of commands, specified in a linker script. A linker script can be divided into three main parts, to be described in the following sections. Together, these three parts fully determine the overall behavior of the linker, because:

1. The *input and output* part picks the input files (object files and libraries) that the linker must consider and directs the linker output where desired.
2. The *memory layout* part describes the position and size of all memory areas available on the target system (also called banks), that is, the space the linker can use to lay out the executable image.
3. The *section and memory mapping* part specifies how input files contents, divided and organized into *sections*, must be mapped and relocated into memory banks.

If necessary, the linker script can be split into multiple files that are then bound together by means of the INCLUDE <filename> directive. The directive takes a file name as argument and directs the linker to include that file "as if" its contents appeared in place of the directive itself. The linker supports nested inclusion, and hence, INCLUDE directives can appear both in the main linker script and in an included script.

This is especially useful when the linker script becomes complex or it is convenient to divide it into parts for other reasons, for instance, to distinguish between architecture or language-dependent parts and general parts.

2.2.1 INPUT AND OUTPUT SEQUENCES

Input and output linker script commands specify:

- Which *input* files the linker will operate on, either object files or libraries. This is done by means of one or more INPUT() commands, which take input file names as arguments.
- The *sequence* in which they will be scanned by the linker, to perform symbol resolution and relocation. The sequence is implicitly established by the order in which input commands appear in the script and by the left-to-right order of their arguments.
- The special ways in which a specific file or group of files will be handled. For instance, the STARTUP() command labels a file as being a startup file rather than a normal object file.
- Where to look for *libraries*, when just the library name is given. This is accomplished by specifying one or more search paths by means of the SEARCH_DIR() command.
- Where the *output*—namely, the file that contains the executable image— will go, through the OUTPUT() command.

Most of these commands have a command-line counterpart that, sometimes, is more commonly used. For instance, the -o command-line option acts the same as OUTPUT() and mentioning an object file name on the linker command line has the same effect as putting it in an INPUT() linker script command. In general, the order between files given on the command line and the ones specified in a linker script depends on where the linker script is mentioned on the command line, although special linker script commands exist to override the default.

The *entry point* of the executable image—that is, the instruction that shall be executed first—can be set by means of the ENTRY(<symbol>) command in the linker script, where <symbol> is a symbol. However, it is important to remark that the only direct effect of ENTRY is to keep a record of the desired entry point in the executable image itself. Then, it is the program that loads the executable image into memory, often called the *loader* in linker's terminology, which is responsible to obey the request.

When no loader is used—that is, the executable image is uploaded by means of an upload tool residing on the development host, and then runs on the target's "bare metal"—the entry point is usually defined by hardware. For example, most processors start execution from a location indicated by their *reset vector* upon powerup. Any entry point set in the executable image is ignored in this case.

All together, the input linker script commands eventually determine the linker *input sequence*. Let us now focus on a short fragment of a linker script that contains several input commands and describe how the input sequence is built from them.

```
INPUT(a.o, b.o, c.o)
INPUT(d.o, e.o)
INPUT(libf.a)
```

Normally, the linker scans input files once and in the order established by the input sequence, which is defined by:

- The left-to-right order in which files appear within the INPUT() command. In this case, b.o follows a.o in the input sequence and e.o follows d.o.
- If there are multiple INPUT() commands in the linker script, they are considered in the same sequence as they appear in the script.

Therefore, in our example the linker scans the files in the order: a.o, b.o, c.o, d.o, e.o, and libf.a. As mentioned previously, object files and libraries can also be specified on the linker command line. In this case:

- The command line may also include an option (-T) to refer to the linker script.
- The input files specified on the command line are combined with those mentioned in the linker script depending on where the linker script has been referenced.

For instance, if the command line is gcc ... a.o -Tscript b.o and the linker script script contains the command INPUT(c.o, d.o), then the input sequence is: a.o, c.o, d.o, and b.o.

As mentioned previously the *startup file* is a special object file because it often contains low-level hardware initialization code and sets up the execution environment for application code. As a consequence, its position in memory with respect to other object modules may be constrained by the hardware startup procedure.

The STARTUP(<file>) command forces <file> to be the very first object file in the input sequence, regardless of where the command appears. For example, the linker script fragment:

```
INPUT(a.o, b.o)
STARTUP(s.o)
```

leads to the input sequence s.o, a.o, and b.o, although s.o is mentioned last.

Let us now describe how the linker transforms the input sequence into the *output sequence* of object modules that will eventually be used to build the executable image. We will do this by means of an example, with the help of Figure 2.5. In our example, the input sequence is composed of an object file g.o followed by two libraries, liba.a and libb.a, in this order. They are listed at the top of the figure, from left to right. For clarity, libraries are depicted as lighter gray rectangles, while

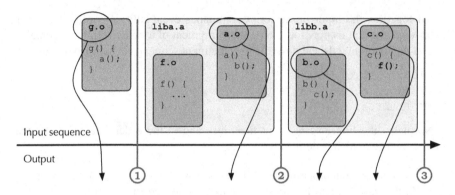

FIGURE 2.5 Linker's handling of object files and libraries.

object files correspond to darker gray rectangles. In turn, object files contain function definitions and references, as is also shown in the figure.

The construction of the output sequence proceeds as follows:

- Object module g.o is unconditionally placed in the output sequence. In the figure, this action is represented as a downward-pointing arrow. As a consequence the symbol a, which is referenced in the body of function g(), becomes undefined at point ①.
- When the linker scans liba.a, it finds a definition of a in module a.o and resolves it by placing a.o into the output. This makes b undefined at point ②, because the body of a contains a reference to b.
- Since only a is undefined at the moment, only module a.o is put in the output. More specifically, module f.o is not, because the linker is not aware of any undefined symbols related to it.
- When the linker scans libb.a, it finds a definition of b and places module b.o in the output. In turn, c becomes undefined. Since c is defined in c.o, that is, another module within the *same library*, the linker places this object module in the output, too.
- Module c.o contains a reference to f, and hence, f becomes undefined. Since the linker scans the input sequence only once, it is unable to refer back to liba.a at this point. Even though liba.a defines f, the linker cannot consider this definition. At point ③ f is still undefined.

In other words, the linker implicitly handles libraries as *sets*. Namely, the linker picks up object modules from a library on demand, and places them into the output. If this action introduces additional undefined symbols, the linker looks into the library again, until no more references can be resolved. At this time, the linker moves to the next object file or library.

As also shown in the example, this default way of scanning the input sequence is problematic when libraries contain circular cross references. More specifically, we say that a certain library *A* contains a circular cross-reference to library *B* when one of

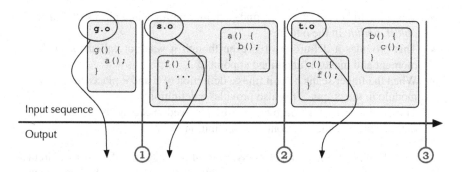

FIGURE 2.6 Object files versus libraries at link time.

A's object modules contains a reference to one of *B*'s modules and, symmetrically, one of *B*'s modules contains a reference back to one module of library *A*. More complex circular references are possible, too, involving more than two libraries.

When this occurs, regardless of the order in which libraries *A* and *B* appear in the input sequence, it is always possible that the linker is unable to resolve a reference to a symbol, even though one of the libraries indeed contains a definition for it. This is what happens in the example for symbol f.

In order to solve the problem, it is possible to group libraries together. This is done by means of the command GROUP(), which takes a list of libraries as argument. For example, the command GROUP(liba.a, libb.a) groups together libraries liba.a and libb.a and instructs the linker to handle both of them as a *single set*.

Going back to the example, the effect of GROUP(liba.a, libb.a) is that it directs the linker to look back into the set, find the definition of f, and place module f.o in the output.

It is possible to mix GROUP() and INPUT() within the input sequence to transform just *part* of it into a set. For example, given the following input sequence:

```
INPUT(a.o, b.o)
GROUP(liba.a, libb.a)
INPUT(libc.a)
```

the linker will first examine a.o, and then b.o. Afterwards, it will handle liba.a and libb.a as a single set. Last, it will handle libc.a on its own.

As will become clearer in the following, the use of GROUP() makes sense only for libraries, because object files are handled in a different way in the first place. Figure 2.6 further illustrates the differences. In particular, the input sequence shown in Figure 2.6 is identical to the one previously considered in Figure 2.5, with the only exception that libraries have been replaced by object modules. The input sequence of Figure 2.6 is processed as follows:

- Object module g.o is placed in the output and symbol a becomes undefined at point ①.

- When the linker scans s.o, it finds a definition for a and places *the whole object module* in the output.
- This provides a definition of f even though it was not called for at the moment and makes b undefined at point ②.
- When the linker scans t.o, it finds a definition of b and it places the whole module in the output. This also provides a definition of c.
- The reference to f made by c can be resolved successfully because the output sequence already contains a definition of f.

As a result, there are no unresolved symbols at point ③. In other words, circular references between object files are resolved automatically because the linker places them into the output as a whole.

2.2.2 MEMORY LAYOUT

The MEMORY command is used to describe the memory layout of the target system as a set of memory areas, often called *blocks* or *banks*. For clarity, command contents are usually written using one line of text for each block. Its general syntax is:

```
MEMORY
{
    <name> [(<attr>)] : ORIGIN = <origin>, LENGTH = <len>
    ...
}
```

where:

- ORIGIN and LENGTH are keywords of the linker script language.
- <name> is the human-readable name assigned to the block so that the other parts of the linker script can refer to that memory block by name.
- <attr> is optional. It gives information about the type of memory block and affects which kind of information the linker is allowed to store into it. For instance, R means read-only, W means read/write, and X means that the block may contain executable code.
- <origin> is the starting address of the memory block.
- <len> is the block length, in bytes.

For example, the following MEMORY command describes the Flash memory bank and the main RAM bank of the LPC1768 microcontroller, as defined in its user manual [90].

```
MEMORY
{
    rom (rx)  : ORIGIN = 0x00000000, LENGTH = 512K
    ram (rwx) : ORIGIN = 0x10000000, LENGTH =  32K
}
```

From the point of view of the linker, objects in memory may have two distinct memory addresses that often, but not always, coincide. To better describe them, let us now consider a definition of an initialized, global variable in the C language, for example:

```
int a = 3;
```

- After the linker has allocated variable a, it resides somewhere in RAM memory, for instance, at address 0x1000. RAM memory is needed because it must be possible to modify the value of a during program execution. However, since RAM memory contents are not preserved when the system is powered off, the initial value of a (3 in this case) must be stored in some non-volatile memory.
- On a general-purpose system, executable images are usually stored on a mass storage device as files within a filesystem. They are brought into memory when needed by an operating system component known as the loader. In this case, the initial value of the variable does not have a memory address of its own.
- Instead, in an embedded system a simpler approach is often taken, and initial values are stored directly within a Flash memory bank. To continue our example, the initial value may be stored at address 0x0020.
- In order to initialize a, the value 3 must be copied from Flash to RAM memory. In a standalone system, the copy must be performed by either the bootloader or the startup code, but in any case *before* the main C program starts. This is because the code generated by the compiler assumes that initialized variables contain their initial value. The whole process is summarized in Figure 2.7.
- To setup the initialized global variable correctly, the linker must therefore associate two memory addresses to initialized global data. Address 0x0020 is the *Load Memory Address* (LMA) of a because this is the memory address where its *initial contents* are stored.
- The second address, 0x1000 in our example, is the *Virtual Memory Address* (VMA) of a because this is the address used by the processor to refer to a at *runtime*.

Often, the VMA and LMA of an object are the same. For example, the address where a function is stored in memory is the same address used by the CPU to call it. When they are not, a copy is necessary, as illustrated previously.

This kind of copy can sometimes be avoided by using the const keyword of the C language, so that read-only data are allocated only in ROM. However, this is not strictly guaranteed by the language specification because const only determines the data *property* at the language level but does not necessarily affect their *allocation* at the memory layout level. In other words, data properties express how they can be manipulated in the program, which is not directly related to where they are in

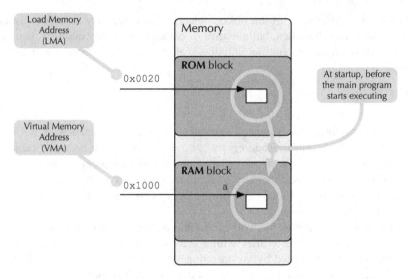

FIGURE 2.7 Load Memory Address (LMA) and Virtual Memory Address (VMA) of an initialized variable.

memory. As a consequence, the relationship between these two concepts may or may not be kept by the toolchain during object code generation.

From the practical point of view, it is important to remark that the linker follows the same order when it allocates memory for initialized variables in RAM and when it stores their initial value in ROM. Moreover, the linker does not interleave any additional memory object in either case. As a consequence, the layout of the ROM area that stores initial values and of the corresponding RAM area is the same. Only their starting addresses are different.

In turn, this implies that the *relative position* of variables and their corresponding initialization values within their areas is the same. Hence, instead of copying variable by variable, the startup code just copies the whole area in one single sweep. The base addresses and size of the RAM and ROM areas used for initialized variables are provided to the startup code, by means of symbols defined in the linker script as described in Section 2.2.3.

It is also worth remarking that there is an unfortunate clash of terminology between virtual memory addresses as they are defined in the linker's nomenclature and virtual memory addresses in the context of virtual memory systems.

2.2.3 LINKER SYMBOLS

As described previously, the concept of *symbol* plays a central role in linker's operations. Symbols are mainly defined and referenced in object files but they can also be defined and referenced in a linker script. Symbols all belong to the same category, regardless of where they are defined. Namely:

- A symbol defined in an object module can be referenced in the linker script. By defining symbols appropriately, the object module can modify the inner workings of the linker script and affect section mapping and memory layout. For example, it is possible to set the stack size of the executable image from one of the object modules.
- Symmetrically, a symbol defined in the linker script can be referenced by an object module, and hence, the linker script can determine some aspects of the object module's behavior. For example, as mentioned in Section 2.2.2, the linker script can communicate the base addresses and size of the RAM and ROM areas used for initialized variables to the startup code.

In a linker script, an *assignment*, denoted by means of the usual = (equal sign) operator, gives a value to a symbol. The value is calculated as the result of an expression written on the right-hand side of the assignment. The expression may contain most C-language arithmetic and Boolean operators. It may involve both constants and symbols. As explained in more details in the following, the result of an expression may be *absolute* or *relative* to the beginning of an output section depending on the contents of the expression itself (mainly, the use of the ABSOLUTE() function) and also where the expression is in the linker script.

The special (and widely used) symbol . (dot) is the *location counter*. It represents the absolute or relative output location (depending on the context) that the linker is about to fill while it is scanning the linker script and its input sequence to lay out objects into memory. With some exceptions, the location counter may generally appear wherever a normal symbol is allowed. For example, it appears on the right-hand side of an assignment in the following example.

```
__stack = .
```

This assignment sets the symbol __stack to the value of the location counter. Assigning a value to . moves the location counter. For example, the following assignment:

```
. += 0x4000
```

allocates 0x4000 bytes starting from where the location counter currently points and moves the location counter after the reserved area.

An assignment may appear in three different positions in a linker script and its position partly affects how the linker interprets it.

1. *By itself.* In this case, the assigned value is absolute and, contrary to the general rule outlined previously, the location counter . cannot be used.
2. As a statement *within a* SECTIONS *command.* The assigned value is *absolute* but, unlike in the previous case, the use of . is allowed. It represents an *absolute* location counter.
3. Within an *output section description*, nested in a SECTIONS command. The assigned value is *relative* and . represents the *relative* value of the location counter with respect to the beginning of the output section.

As an example, let us consider the following linker script fragment, which summarizes the concepts just introduced. More thorough and formal information about output sections is given in Section 2.2.4.

```
SECTIONS
{
    .  = ALIGN(0x4000);
    . += 0x4000;
    __stack = .;
}
```

In this example:

- The first assignment aligns the location counter to a multiple of 16 kbyte (`0x4000`).
- The second assignment moves the location counter forward by 16 kbyte. That is, it allocates 16 kbyte of memory for the stack.
- The third assignment sets the symbol `__stack` to the top of the stack. The startup code will refer to this symbol to set the initial stack pointer.

2.2.4   SECTION AND MEMORY MAPPING

The contents of each input object file are divided by the compiler (or the assembler) into several categories according to their characteristics, like:

- code (`.text`),
- initialized data (`.data`),
- uninitialized data (`.bss`).

Each category corresponds to its own *input section* of the object file, whose name has also been listed above. For example, the object code generated by the C compiler is placed in the `.text` section of the input object files. Libraries follow the same rules because they are just collections of object files.

The part of linker script devoted to *section mapping* tells the linker how to fill the memory image with *output sections*, which are generated by collecting input sections. It has the following syntax:

```
SECTIONS
{
    <sub-command>
    . . .
}
```

where:

- The `SECTIONS` command encloses a sequence of sub-commands, delimited by braces.

- A sub-command may be:
  - an ENTRY command, used to set the initial entry point of the executable image as described in Section 2.2.1,
  - a symbol assignment,
  - an overlay specification (seldom used in modern programs),
  - a section mapping command.

A section mapping command has a relatively complex syntax, illustrated in the following.

```
<section> [<address>] [(<type>)] :
  [<attribute> ...]
  [<constraint>]
  {
    <output-section-command>
    ...
  }
  [> <region>] [AT> <lma_region>]
  [: <phdr> ...] [= <fillexp>]
```

Most components of a section mapping command, namely, the ones shown within brackets ([ ]), are optional. Within a section mapping command, an *output section command* may be:

- a symbol assignment, outlined in Section 2.2.3,
- *data values* to be included directly in the output section, mainly used for padding,
- a special *output section keyword*, which will not be further discussed in this book,
- an *input section description*, which identifies the input sections that will become part of the output section.

An input section description must be written according to the syntax indicated below and indicates which input sections must be mapped into the output section.

```
<filename> ( <section_name> ... )
```

It consists of:

- A <filename> specification that identifies one or more object files in the input sequence. Some wildcards are allowed, the most common one is *, which matches all files in the input sequence. It is also possible to exclude some input files, by means of EXCLUDE_FILE (...), where ... is the list of files to be excluded. This is often useful in combination with wildcards to refine the result produced by the wildcards themselves.
- One or more <section_name> specifications that identify which input sections, within the files indicated by <filename>, we want to refer to.

The order in which input section descriptions appear is important because it defines the order in which input sections are placed in the output sections. For example, the following input section description:

```
* ( .text .rodata )
```

places the .text and .rodata sections of all files in the input sequence in the output section. The sections appear in the output in the same order as they appear in the input.

Instead, this slightly different description:

```
* ( .text )
* ( .rodata )
```

first places all the .text sections, and then all the .rodata sections.

Let us now examine the other main components of the section mapping command one by one. The very first part of a section mapping command specifies the output section *name*, *address*, and *type*. In particular:

- <section> is the name of the output section and is mandatory.
- <address>, if specified, sets the *VMA* of the output section. When it is not specified, the linker sets it automatically, based on the output memory block <region>, if specified, or the current location counter. Moreover, it takes into account the strictest alignment constraint required by the input sections that are placed in the output sections and the output sections alignment itself, which is be specified with an optional [<attribute>] and will be explained later.
- The most commonly used special output section <type> is NOLOAD. It indicates that the section shall not be loaded into memory when the program is run. When omitted, the linker creates a normal output section specified with the section name, for instance, .text.

Immediately thereafter, it is possible to specify a set of output section *attributes*, according to the following syntax:

```
[AT( <lma> )]
[ALIGN( <section_align> )]
[SUBALIGN( <subsection_align> )]
[<constraint>]
```

- The AT attribute sets the *LMA* of the output section to address <lma>.
- The ALIGN attribute specifies the alignment of the output section.
- The SUBALIGN attribute specifies the alignment of the input sections placed in the output section. It overrides the "natural" alignment specified in the input sections themselves.
- <constraint> is normally empty. It may specify under which constraints the output sections must be created. For example, it is possible to specify

that the output section must be created only if all input sections are read-only [34].

The *memory block mapping* specification is the very last part of a section mapping command and comes after the list of output section commands. It specifies in which memory block (also called *region*) the output section must be placed. Its syntax is:

```
[> <region>] [AT> <lma_region>]
[: <phdr> ...] [= <fillexp>]
```

where:

- `> <region>` specifies the memory block for the output section *VMA*, that is, where it will be referred to by the processor.
- `AT> <lma_region>` specifies the memory block for the output section *LMA*, that is, where its initial contents reside.
- `<phdr>` and `<fillexp>` are used to assign the output section to an *output segment* and to set the *fill pattern* to be used in the output section, respectively.

Segments are a concept introduced by some executable image formats, for example, the executable and linkable format (ELF) [33] format. In a nutshell, they can be seen as groups of sections that are considered as a single unit and handled all together by the loader.

The fill pattern is used to fill the parts of the output section whose contents are not explicitly specified by the linker script. This happens, for instance, when the location counter is moved or the linker introduces a gap in the section to satisfy an alignment constraint.

2.3   GNU MAKE AND MAKEFILES

The GNU make tool [47], fully described in [116], manages the *build process* of a software component, that is, the execution of the correct sequence of commands to transform its source code modules into a library or an executable program as efficiently as possible. Since, especially for large components, it rapidly becomes unfeasible to rebuild the whole component every time, GNU make implements an extensive inference system that allows it to:

1. decide which parts of a component shall be rebuilt after some source modules have been updated, based on their *dependencies*, and then
2. automatically execute the appropriate sequence of *commands* to carry out the rebuild.

Both dependencies and command sequences are specified by means of a set of rules, according to the syntax that will be better described in the following. These rules can be defined explicitly in a GNU make input file, often called `Makefile` by convention. Moreover, `make` contains a rather extensive set of predefined built-in rules, which are implicitly applied unless overridden in a `Makefile`. GNU make can also retrieve explicit user-defined rules from other sources. The main ones are:

- One of the files `GNUmakefile`, `makefile`, or `Makefile` if they are present in the current directory. The first file found takes precedence on the others, which are then silently ignored.
- The file specified by means of the `-f` or `--file` options on the command line.

It is possible to include a `Makefile` into another by means of the `include` directive. In order to locate the file to be included, GNU make looks in the current directory and any other directories mentioned on the command line, using the `-I` option. As will be better explained in the following, the `include` directive accepts any file name as argument and even names computed on the fly by GNU make itself. Hence, it allows programmers to use any arbitrary file as (part of) a `Makefile`.

Besides options, the command line may also contain additional arguments, which specify the *targets* that GNU make must try to update. If no targets are given on the command line, GNU make pursues the first target defined in the `Makefile`.

2.3.1   EXPLICIT RULES

The general format of an explicit rule in a `Makefile` is:

```
<target> ... : <prerequisites> ...
        <command line>
        ...
```

In an explicit rule:

- The `<target>` is usually a file that will be (re)generated when the rule is applied.
- The `<prerequisites>` are the files on which `<target>` depends and that, when modified, trigger the regeneration of the target.
- The sequence of `<command line>`s are the actions that GNU make must perform in order to regenerate the target, in shell syntax.
- Every command line must be preceded by a tab character and is executed in *its own* shell.

It is extremely important to pay attention to the last aspect of command line execution, which is sometimes neglected, because it may have very important consequences on the effects commands have. For instance, the following rule does *not* list the contents of directory `somewhere`.

```
all:
        cd somewhere
        ls
```

TABLE 2.1

GNU Make Command Line Execution Options

Option	Description
@	Suppress the automatic echo of the command line that GNU make normally performs immediately before execution.
–	When this option is present, GNU make ignores any error that occurs during the execution of the command line and continues anyway.

This is because, even though the `cd` command indeed changes the current directory to `somewhere`, it does so only within the shell it is executed by. Since the `ls` command execution takes place in a new shell, the previous notion of current directory is lost when the new shell is created.

As mentioned previously, the prerequisites list specifies the target dependencies. GNU make looks at the prerequisites list to deduce whether or not a target must be regenerated by applying the rule. More specifically, GNU make applies the rule when one or more prerequisites are *more recent* than the target. For example, the rule:

```
kbd.o : kbd.c defs.h command.h
        cc -c kbd.c
```

specifies that the object file `kbd.o` (target) must be regenerated when at least one file among `kbd.c`, `defs.h`, and `command.h` (prerequisites) has been modified. In order to regenerate `kbd.o`, GNU make invokes `cc -c kbd.c` (command line) within a shell.

The shell, that is, the command line interpreter used for command line execution is by default `/bin/sh` on unix-like systems, unless the `Makefile` specifies otherwise by setting the `SHELL` variable. Notably, it does not depend on the user login shell to make it easier to port the `Makefile` from one user environment to another.

Unless otherwise specified, by means of one of the command line execution options listed in Table 2.1, commands are echoed before execution. Moreover, when an error occurs in a command line, GNU make abandons the execution of the current rule and (depending on other command-line options) may stop completely. Command line execution options must appear at the very beginning of the command line, before the text of the command to be executed, and are not passed to the shell. In order to do the same things in a systematic way GNU make supports options like `--silent` and `--ignore`, which apply to all command lines or, in other words, change the default behavior of GNU make.

2.3.2 VARIABLES

A variable is a name defined in a `Makefile`, which represents a text string. The string is the value of the variable. The value of a certain variable VAR—by convention, GNU make variable names are often capitalized—is usually retrieved and used (that is, *expanded*) by means of the construct $(VAR) or ${VAR}. In order to introduce a dollar character somewhere in a Makefile without calling for variable expansion, it is possible to use the escape sequence $$, which represents one dollar character, $.

In a Makefile, variables are expanded "on the fly," while the file is being read, except when they appear within a command line or in the right-hand part of a variable assignment made by means of the assignment operator "=". This aspect of variable expansion is especially important and we will further elaborate on it in the following, because the behavior of GNU make departs significantly from what is done by most other language processors, for instance, the C compiler.

Another difference with respect to other programming languages is that the $() operators can be nested. For instance, it is legal, and often useful, to state $($(VAR)). In this way, the *value* of a variable (like VAR) can be used as a variable *name*. For example, let us consider the following fragment of a Makefile:

```
MFLAGS = $(MFLAGS_$(ARCH))
MFLAGS_Linux  = -Wall -Wno-attributes -Wno-address
MFLAGS_Darwin = -Wall
```

- The variable MFLAGS is set to different values depending on the contents of the ARCH variable. In the example, this variable is assumed to be set elsewhere to the host operating system name, either Linux or Darwin in the example.
- This is a compact way to put different operating system-dependent compiler flags in the variable MFLAGS without using conditional directives or writing several separate Makefiles, one for each operating system.

A variable can get a value in several different ways, listed here in order of decreasing priority.

1. As specified when GNU make is *invoked*, by means of an assignment statement put directly on its command line. For instance, the command make VAR=v invokes GNU make with VAR set to v.
2. By means of an *assignment* in a Makefile, as will be further explained in the following.
3. Through a shell *environment* variable definition.
4. Some variables are set *automatically* to useful values during rule application.
5. Finally, some variables have an *initial* value, too.

When there are multiple assignments to the same variable, the highest-priority one silently prevails over the others.

GNU make supports two kinds, or *flavors*, of variables. It is important to further elaborate on this difference because they are defined and expanded in different ways:

TABLE 2.2

GNU Make Assignment Operators

Operator	Description
VAR = ...	Define a recursively-expanded variable
VAR := ...	Define a simply-expanded variable
VAR ?= ...	Define the recursively-expanded variable VAR only if it is still undefined
VAR += ...	Append ... to variable VAR (see text)

1. *Recursively expanded* variables are defined by means of the operator =, informally mentioned previously. The evaluation of the right-hand side of the assignment, as well as the expansion of any references to other variables it may contain, are delayed until the variable being defined is itself expanded. Evaluation and variable expansion then proceed recursively.
2. *Simply expanded* variables are defined by means of the operator :=. The value of the variable is determined once and for all when the assignment is executed. The expression on the right-hand side of the assignment is evaluated immediately, expanding any references to other variables.

Table 2.2 lists all the main assignment operators that GNU make supports. It is worth mentioning that the "append" variant of the assignment preserves (when possible) the kind of variable it operates upon. In particular:

- If VAR is undefined it is the same as =, and hence, it defines a recursively expanded variable.
- If VAR is already defined as a simply expanded variable, it immediately expands the right-hand side of the assignment and appends the result to the previous definition.
- If VAR is already defined as a recursively expanded variable, it appends the right-hand side of the assignment to the previous definition without performing any expansion.

In order to better grasp the effect of delayed variable expansion, let us consider the following two examples.

```
X = 3
Y = $(X)
X = 8
```

In this first example, the final value of Y is 8 because the right-hand side of its assignment is expanded only when Y is used. Let us now consider a simply expanded variable.

```
X = 3
```

```
Y := $(X)
X = 8
```

In this case, the final value of Y is 3 because the right-hand side of its assignment is expanded immediately, when the assignment is performed. As can be seen from the previous examples, delayed expansion of recursively expanded variables has unusual, but often useful, side effects. Let us just briefly consider the two main benefits of delayed expansion:

- Forward variable references in assignments, even to variables that are still undefined, are not an issue.
- When a variable is eventually expanded, it makes use of the "latest" value of the variables it depends upon.

2.3.3 PATTERN RULES AND AUTOMATIC VARIABLES

Often, all files belonging to the same group or category (for example, object files) follow the same generation rules. In this case, rather than providing an explicit rule for each of them and make the Makefile hard to read and maintain, it is more appropriate and convenient to define a *pattern rule*.

As shown in the code example that follows, and as its name says, a pattern rule applies to all files that match a certain pattern, which is specified in the rule in place of the target.

```
%.o :   %.c
    cc -c $<


kbd.o :   defs.h command.h
```

In particular:

- Informally speaking, in the pattern the character % represents any non-empty character string.
- The same character can be used in the prerequisites, too, to specify how they are related to the target.
- The command lines associated with a pattern rule can be customized, based on the specific target the rule is being applied to, by means of *automatic variables* like $< in the example.
- It is possible to augment the prerequisites of a pattern rule on a target-by-target basis, by means of explicit rules without command lines, as shown at the end of the example.

More precisely, a target pattern is composed of three parts: a *prefix*, a % character, and a *suffix*. The prefix and/or suffix may be empty. A target name (which often is a file name) matches the pattern if it starts with the pattern prefix and ends with the pattern suffix. The non-empty sequence of characters between the prefix and the suffix is called the *stem*.

TABLE 2.3
Main GNU Make Automatic Variables

Var.	Description	Example value
$@	Target of the rule	kbd.o
$<	First prerequisite of the rule	kbd.c
$^	List of all prerequisites of the rule, delimited by blanks	kbd.c defs.h command.h
$?	List of prerequisites that are more recent than the target	defs.h
$*	Stem of the rule (only for pattern rules)	kbd

Since, as said previously, rule targets are often file names, directory specifications in a pattern are handled specially, to make it easier to write compact and general rules that apply to target files residing in different directories. In particular:

- If a target pattern does not contain any slash—which is the character that separates directory names in a file path specification—all directory names are removed from target file names before comparing them with the pattern.
- Upon a successful match, directory names are restored at the beginning of the stem. This operation is carried out before generating prerequisites.
- Prerequisites are generated by substituting the stem of the rule in the right-hand part of the rule, that is, the part that follows the colon (:).
- For example, file src/p.o satisfies the pattern rule %.o : %.c. In this case, the prefix is *empty*, the stem is src/p and the prerequisite is src/p.c because the src/ directory is removed from the file name before comparing it with the pattern and then restored.

When it applies a rule GNU make automatically defines several *automatic variables*, which become available in the corresponding command lines. Table 2.3 contains a short list of these variables and describes their contents. As an example, the rightmost column of the table also shows the value that automatic variables would get if the rules above were applied to regenerate kbd.o, mentioned in the previous example, because defs.h has been modified.

To continue the example, let us assume that the Makefile we are considering contains the following additional rule. The rule updates library lib.a, by means of the ar tool, whenever any of the object files it contains (main.o, kbd.o, and disk.o) is updated.

```
lib.a : main.o kbd.o disk.o
    ar rs $@ $?
```

After applying the previous rule, kbd.o becomes more recent than lib.a, because it has just been updated. In turn, this triggers the application of the second rule shown above. While the second rule is being applied, the automatic variable corresponding to the target of the rule ($@) is set to `lib.a` and the list of prerequisites more recent than the target ($?) is set to `kbd.o`.

To further illustrate the use of automatic variables, we can also remark that we could use $^ instead of $? in order to completely rebuild the library rather than update it. This is because, as mentioned in Table 2.3, $^ contains the list of all prerequisites of the rule.

It is also worth noting that GNU make comes with a large set of predefined implicit *built-in* rules. Most of them are pattern rules, and hence, they generally apply to a wide range of targets and it is important to be aware of their existence. They can be printed by means of the command-line option `--print-data-base`, which can also be abbreviated as `-p`.

For instance, there is a built-in pattern rule to generate an object file given the corresponding C source file:

```
%.o: %.c
        $(COMPILE.c) $(OUTPUT_OPTION) $<
```

The variables cited in the command line have got a built-in definition as well, that is:

```
COMPILE.c = $(CC) $(CFLAGS) $(CPPFLAGS) $(TARGET_ARCH) -c
OUTPUT_OPTION = -o $@
CC = cc
```

As a consequence, just by defining some additional variables, for example `CFLAGS`, it is often possible to customize the behavior of a built-in rule instead of defining a new one. When an explicit rule in the `Makefile` overlaps with a built-in rule because it has the same target and prerequisites, *the former* takes precedence. This priority scheme has been designed to avoid undue interference of implicit built-in rules the programmer may be unaware of, with any explicit rule written in the `Makefile`.

2.3.4 DIRECTIVES AND FUNCTIONS

GNU make provides an extensive set of *directives* and built-in *functions*. In general, directives control how the input information needed by GNU make is built, by taking it from various input files, and which parts of those input files are considered. Provided here is a glance at two commonly-used directives, namely:

- The `include <file> ...` directive instructs GNU make to temporarily stop reading from the current `Makefile` at the point where the directive appears, read the additional `file`(s) mentioned in the directive, and then continue.
  The file specification may contain a single file name or a list of names, separated by spaces. In addition, it may also contain variable and function expansions, as well as any file name wildcards known to the shell.

- The ifeq (<exp1>, <exp2>) directive evaluates the two expressions exp1 and exp2. If they are textually identical then GNU make uses the Makefile section between ifeq and the next else directive; otherwise it uses the section between else and endif.
  In other words, this directive is similar to conditional statements in other programming languages. Directives ifneq, ifdef, and ifndef also exist and have the expected intuitive meaning.

Concerning functions, the general syntax of a function call is

$(<function> <arguments>)

where:

- <function> represents the function name and <arguments> is a list of one or more arguments. At least one blank space is required to separate the function name from the first argument. Arguments are separated by commas.
- By convention, variable names are written in all capitals, whereas function names are in lowercase, to help readers distinguish between the two.

Arguments may contain references to:

- Variables, for instance: $(subst a,b,$(X)). This statement calls the function subst with 3 arguments: a, b, and the result of the expansion of variable X.
- Nested function calls, like: $(subst a,b,$(subst c,d,$(X))). Here, the third argument of the outer subst is the result of the inner, nested subst.

As for directives, in the following we are about to informally discuss only a few GNU make functions that are commonly found in Makefiles. Interested readers should refer to the full documentation of GNU make [47], for in-depth information.

- The function $(subst <from>,<to>,<text>) replaces <from> with <to> in <text>. Both <from> and <to> must be simple text strings. For example:

$(subst .c,.o,p.c q.c) $\longrightarrow$ p.o q.o

- The function $(patsubst <from>,<to>,<text>) is similar to subst, but it is more powerful because <from> and <to> are patterns instead of text strings. The meaning of the % character is the same as in pattern rules.

$(patsubst %.c,%.o,p.c q.c) $\longrightarrow$ p.o q.o

- The function `$(wildcard <pattern> ...)` returns a list of names of existing files that match one of the given patterns. For example, the expression `$(wildcard *.c)` evaluates to the list of all C source files in the current directory.
  `wildcard` is commonly used to set a variable to a list of file names with common characteristics, like C source files. Then, it is possible to further work on the list with the help of other functions and use the results as targets, as shown in the following example.

```
SRC = $(wildcard *.c)
ELF = $(patsubst %.c,%.elf,$(SRC))

all: $(ELF)

%.elf: %.c
        $(CC) -o $@ $<
```

- The function `$(shell <command>)` executes a shell `<command>` and captures its output as return value. For example, when executed on a Linux system:

$$\$(shell\ uname)\ \longrightarrow\ Linux$$

In this way, it is possible to set a variable to an operating system-dependent value and have GNU make do different things depending on the operating system it is running on, as shown in Section 2.3.2, without providing separate `Makefiles` for all of them, which would be harder to maintain.

2.4 BASIC DESCRIPTION OF RTEMS AND ITS CONFIGURATION SYSTEM

In this section we describe the high-level concepts related to configuring RTEMS for a specific application. This description is relevant for RTEMS versions 4-5, but is anticipated to change drastically starting with RTEMS version 6. The complete details of RTEMS configuration can be found in the RTEMS Classic API Guide's chapter called Configuring a System.

A challenging aspect for any OS design and implementation is how to configure the resources it manages. GPOSs specify the configuration during the OS build process, for example, the Linux KConfig files and related utilities are run to generate a special `.config` file that is used during the make command to select and configure kernel subsystems. The configuration is done by expert kernel developers and distribution packagers that decide on the base images and the set of loadable modules that users may need over a wide variety of applications. RTOSs defer configuration to the end-user whom is expected to tailor the resource configuration toward their application and better customize the OS support.

RTOS configuration is accomplished by direct or indirect configuration. The direct approach exposes the RTOS resources to the application during the compile or link phase. Indirect approaches provide an API and data structures for an application developer to create their configuration, for example, specifying the resources they need in a structured document such as an XML file. An indirect approach is advantageous because it offers additional structure and support to the configuration phase, whereas the direct approach is simple to implement.

The configuration of RTEMS is split in two phases. The first phase is done prior to compilation of RTEMS itself, and accomplishes the goal of configuring the compiler with options for building the RTEMS base image. The second phase is done during compilation of an application prior to linking it with the RTEMS base image and populates several data structures that are used by RTEMS to manage resources especially as they relate to allocation of internal objects. We describe each of these phases in the following.

2.4.1   RTEMS COMPILE-TIME CONFIGURATION

The build system of RTEMS relies on the autotools framework to configure build files for invoking make. The automake and autoconf programs are both relied upon to customize the compilation for different options that an end-user may need. Autoconf input files (with `.ac` extension) are used to generate a *configure* script that sets the compilation options. The most important options for RTEMS include:

- `--target=`
- `--enable-rtemsbsp=`
- `--enable-smp`
- `--enable-tests`

Several other options exist for special purposes. The `target` option sets the cross-compiler to use, which is primarily important for selecting a compiler toolchain for correct ISA and version of RTEMS being built. The option to `enable-rtemsbsp` allows selection of the BSP that the user needs; most users will have a single BSP of interest to them, and this option will avoid building all the available BSPs for the ISA chosen in the `target`. Use the `enable-smp` option to compile RTEMS with support for multicore BSPs. Although optional, the `enable-tests` is recommended to build the expansive suite of tests included with RTEMS; this option can be parameterized with `=samples` to build the sample applications, or if the option is not used then only a base image of RTEMS is compiled against which users can later link their applications to create loadable binary files. Interested readers and users should refer to the RTEMS documentation for additional information about these and other build-time configure options.

2.4.2   APPLICATION COMPILE-TIME CONFIGURATION

The resources that an application may require from RTEMS are configured statically during compilation through a set of C preprocessor macros that are used in a single monolithic header file named `confdefs.h`. This approach is colloquially called the *confdefs configuration*, and examples can be found in many of the RTEMS test programs. Most of these tests consolidate the configuration macros in a local header named `system.h`.

As an example, consider the configuration of the `hello` sample, found in `testsuites/samples/hello/init.c`:

```
#define CONFIGURE_APPLICATION_DOES_NOT_NEED_CLOCK_DRIVER
#define CONFIGURE_APPLICATION_NEEDS_SIMPLE_CONSOLE_DRIVER

#define CONFIGURE_MAXIMUM_TASKS                 1

#define CONFIGURE_RTEMS_INIT_TASKS_TABLE

#define CONFIGURE_INIT_TASK_ATTRIBUTES RTEMS_FLOATING_POINT

#define CONFIGURE_INITIAL_EXTENSIONS RTEMS_TEST_INITIAL
_EXTENSION

#define CONFIGURE_INIT
#include <rtems/confdefs.h>
```

The hello world test is very simple and does not require a functional clock, which is disabled by the first line `#define CONFIGURE_APPLICATION_DOES_NOT _NEED_CLOCK_DRIVER`. Hello just needs a console driver and a single task, which are configured in the next two lines. That one task is the initalization task, and the macro `CONFIGURE_RTEMS_INIT_TASKS_TABLE` instructs RTEMS to set up an initialization task structure, which by default only contains a single init task. The init task needs to be made a floating point task because some architectures use floating point registers in their implementation of `printf`. The `CONFIGURE_INITIAL_EXTENSIONS` is used to install a set of extensions that are specialized for the RTEMS testsuite. As a high-level switch, the `CONFIGURE_INIT` macro will cause confdefs to interpret all the macros and generate the configuration tables. This macro should only be defined once by an application prior to including `confdefs.h` or else the configuration tables will be generated multiple times and the linker will generate errors for name reuse. All application configurations must end with `#include <rtems/confdefs.h>`. Once this file is included (with `CONFIGURE_INIT` defined) then RTEMS is configured.

More complicated applications that use other resources managed by RTEMS will need to make use of other configuration macros. For example, an application with three tasks would increase the number 1 to 3 in the definition of `CONFIGURE_MAXIMUM_TASKS`, and if that application uses a barrier for synchronization it would need to configure that resource by the `CONFIGURE_MAXIMUM_BARRIERS` macro. To ease porting and application development there is also a `CONFIGURE_UNLIMITED_OBJECTS` macro that removes limits on many of the configured resources and relies on dynamic allocation to satisfy requests to create resources. This macro is not recommended

for use in deployed applications, and should normally be combined with the
CONFIGURE_UNIFIED_WORK_AREAS macro that causes RTEMS to allocate system resources from the same dynamic memory pool as the C program heap. The option for unlimited objects does allow for a very simple and reliable configuration to use for prototyping:

```
#define CONFIGURE_APPLICATION_NEEDS_CLOCK_DRIVER
#define CONFIGURE_APPLICATION_NEEDS_CONSOLE_DRIVER
#define CONFIGURE_UNIFIED_WORK_AREAS
#define CONFIGURE_UNLIMITED_OBJECTS
#define CONFIGURE_RTEMS_INIT_TASKS_TABLE
#define CONFIGURE_INIT
#include <rtems/confdefs.h>
```

2.5 SUMMARY

After outlining how a toolchain generates an executable image starting from the application source code in Section 2.1, this chapter focused on two aspects of a GNU-based toolchain that are very important from the practical point of view, but often neglected: the linker command language in Section 2.2 and GNU make in Section 2.3.

The last part of the chapter, Section 2.4, went from general toolchain behavior to more specific aspects related to the RTEMS operating system. It provided a summary of the two main operating system configuration opportunities, which take place when RTEMS itself is compiled for a certain target platform, and when an application is compiled and linked against the operating system.

3 Concurrent Programming and Scheduling Algorithms

CONTENTS

This chapter lays out the theoretical foundations of concurrent programming, starting from the all-important concept of *process*, or *task*. Then, it introduces the fundamental tools used to represent the state of a process and express how it evolves over time, that is, the task control block and the task state diagram.

The second part of the chapter presents the main concepts and techniques of task-based, real-time scheduling on single processor systems. The discussion of scheduling algorithms suitable for multicore systems is left to Part V of the book.

3.1 FOUNDATIONS OF CONCURRENT PROGRAMMING

3.1.1 FROM INTERRUPT HANDLING TO MULTIPROGRAMMING

Historically, the very first computers were completely sequential machines. Instruction execution normally advanced along increasing memory addresses, and diversions from a sequential control flow could only be performed synchronously, by means of branch or call instructions embedded in the instruction stream. On those machines the only way to become aware of external asynchronous events, for instance, the user pressing a key on the keyboard, was to periodically and repeatedly *poll* input–output (I/O) devices, like the keyboard controller in our example.

In essence, each individual polling operation consisted of querying a suitable device register containing status information, analyzing the status to understand whether an event of interest occurred, and conditionally branching to the corresponding event-handling function. In turn, the event-handling function further interacted with the device, to retrieve the information associated with the event and process it.

However, despite the apparent simplicity of such an approach, it soon became evident that, with the ever-increasing number and complexity of I/O devices attached to a computer system, it was becoming more and more difficult to properly interlace polling operations and ordinary application programs, in order to ensure that all I/O devices were serviced with acceptable timings. The connection of I/O devices with quite diverse timing requirements to the same computer further exacerbated the issue.

On one hand, querying all devices within a single polling function would require the application program to invoke the function frequently enough to satisfy the device with the strictest timing requirements and would likely introduce excessive overhead. On the other hand, dividing devices into classes depending on their timing requirements and introducing multiple polling function, one for each category, would lead to the proliferation of these "polling points" in the application code.

From the programming point of view, another important shortcoming of this approach was an inherent lack of modularity because programmers were forced to deal with tightly intertwined fragments of code serving very different purposes. In turn, this made application code harder to understand and maintain, especially as its complexity grew.

In order to alleviate the issue, computer designers aimed at giving devices the ability to draw the attention of the processor when needed, through a mechanism that has become known as *interrupt*. While not interrupted, the processor still executes instructions as described previously, without performing any polling. When an interrupt request arrives, and provided certain conditions are fulfilled, the processor temporarily abandons sequential execution, usually at the boundary between two consecutive instructions, and diverts to a piece of code called interrupt handler. In turn, the interrupt handler is in charge of inspecting the interrupting device and properly handling the event it signaled.

When the interrupt handler terminates, often by executing a special "return from interrupt" instruction, the processor resumes from where it left when the interrupt request was accepted. Even though technical details may vary from one computer to another, an important characteristic of interrupt handling is that it is *transparent* from the point of view of the interrupted code. In other words, the acceptance of one or more interrupt requests while a certain piece of code is being executed does not change its functional semantics in any way and, when looking at the system at a higher level of abstraction, it gives the illusion that ordinary program execution and interrupt handling proceed *concurrently*.

This is accomplished by saving the relevant part of the processor context (like its program counter and general purpose registers' content) when the interrupt is accepted, often on a dedicated portion of memory organized as a stack, and restoring it when the interrupt handler eventually returns. On the contrary, and this is an aspect of extreme interest for real-time systems, interrupts do affect the *timings* of the interrupted code. This is because, quite obviously, the execution of an interrupt handler necessarily makes the underlying, interrupted code run slower.

There is some controversy on when, and by whom, this interrupt-based device handling strategy was introduced for the very first time, but certainly one of the earliest documented examples can be found in [120]. There, the authors describe how they modified a Remington Rand Univac 1103 computer to support interrupts, using an approach that would then become commercially available on model 1103A. Despite dating back to 1958, their proposal already contained most of the elements typical of present-day computer architectures, which will be discussed in more detail in Section 4.2:

1. The ability to selectively enable and disable interrupt requests originating from a specific device, by means of appropriate instructions.
2. Support of device-specific interrupt handlers, so that the processor can conveniently execute different code depending on the interrupt source.
3. A priority hierarchy among devices, which comes into effect when multiple devices submit an interrupt request at the same time.
4. An automatic mechanism to prevent a new interrupt request from being accepted while an interrupt handler is being executed.

Regarding the last point, more recent architectures often follow a more complex approach, called interrupt *nesting*, which allows an incoming interrupt to be accepted even during the execution of an interrupt handler, subject to some conditions on the incoming interrupt's priority. When this happens, interrupt handlers are nested into each other and executed in a last-in, first-out fashion. In exchange for its higher complexity, this approach reduces the handling latency of high-priority interrupts because interrupt handlers are no longer constrained to be executed in a strictly sequential, first-in, first-out way.

At the same time, the work also highlights two possible pitfalls of concurrent code execution, still very relevant nowadays:

- When the application code and the interrupt handler share data, like I/O buffers, appropriate interlocks must be put in place to avoid data corruption. By intuition, data corruption may occur, for instance, if the application code reads from a buffer before the I/O device has filled it completely with fresh input data. In this case, the application code may make use of a mix of new and old data left in the buffer by the previous I/O operation.
- Any issue with interlocks may easily lead to *time-dependent* errors. Namely, an incorrect program may run flawlessly multiple times, and fail only when interrupts are accepted at specific, unfortunate locations within the application program. For instance, accessing an I/O buffer without proper interlocks is bound to cause issues only when an interrupt handler that modifies the same buffer is executed while an access is in progress.

As a side note, it must also be said that polling-based device management is not without merits, especially in hard real-time systems where overall efficiency is of secondary importance with respect to meeting the strict and demanding timing requirements sometimes imposed by external devices.

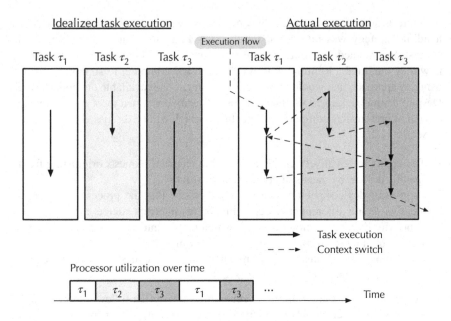

FIGURE 3.1 Idealized and actual task execution flow in multiprogramming.

With time, the observation that interrupts enable a computer to perform two distinct and independent activities apparently at the same time—normal instruction execution and interrupt handling—led to the desire of extending this concept to the user level, by means of techniques known as *multiprogramming*.

These techniques are very widely used nowadays and their benefits are particularly evident, for instance, with personal computers, in which users ordinarily interact with multiple applications at the same time and they all seemingly proceed in parallel, or *concurrently*, as the need arises. For instance, gone are the days in which users had to wait for the printer to finish printing before they could continue with their word processing program, as all personal computers are now able to print in background.

As illustrated in Figure 3.1, in a multiprogrammed execution environment, the operating system repeatedly switches the processor back and forth from one activity, or *task*, to another. As a result, users perceive that the execution of tasks $\tau_1, \ldots, \tau_3$ is proceeding concurrently, as depicted on the left of the figure, whereas the processor actually executes instructions as shown on the right. If properly implemented, this *context switch* is completely transparent to, and independent from, the activities themselves, which may even be unaware of it.

Operating systems often use *thread* as an equivalent term for task. Depending on the context, the two terms will be used interchangeably in the following. Historically the term *process*, to be introduced in Section 3.1.2, was first used to denote an activity within a concurrent system.

However, as explained in Section 6.3 regarding the POSIX standard, nowadays the two words *process* and *thread* (or *task*) are commonly defined and understood in a hierarchical way. With rare exceptions [2], a process is a container for one or more threads and different processes not only have their own independent control flows, one per thread, but also distinct memory address spaces. On the contrary, all threads (or tasks) living within the same process implicitly share memory.

Even more importantly, designing and organizing software around the concurrent execution of multiple tasks turned out to be a very useful and general abstraction. In particular, it is readily applicable also when it corresponds to a true execution parallelism at the hardware level. This is the case of multiprocessor and multicore systems, which will be the subject of Part V of this book. In those systems, either multiple processors or a single processor with multiple cores share a common memory.

Each processor or core is able to carry out its own sequential flow of instructions independently from the others, as well as handling interrupts and performing context switches. Especially when contrasting single-core with multicore systems, the term *pseudo-parallelism* is sometimes used to remark that, despite multiprogramming, a single-core computer is still executing exactly one activity at any given instant of time, in a strictly sequential way.

3.1.2   COOPERATING SEQUENTIAL PROCESSES

The idea that any concurrent system, regardless of its nature and complexity, can be designed as a set of cooperating sequential activities executed concurrently was first introduced in a seminal work by Dijkstra [42]. These concurrent activities are usually called *processes*, to adhere to Dijkstra's original nomenclature. As outlined previously, especially within the context of real-time operating systems, they are also commonly referred to as *tasks*. For the sake of consistency, we will also call them tasks in this book.

Each task is autonomous for what concerns execution and holds all the information needed to represent the execution of a sequential program that evolves with time. By intuition, this information must necessarily include not only the program instructions but also the state of the processor (program counter, registers) and memory (variables).

In other words, each task can be regarded as the execution of a sequential program by "its own" processor although, as shown in Figure 3.1, depending on the number of physical processors and cores available on the system with respect to the number of tasks to be executed, cores may actually be switched from one task to another by the operating system. Even if we restrict our attention to a single-processor, single-core system, there are in principle many different possible strategies the operating system can use to execute tasks, for instance, by deciding where context switch points should be.

At one end of the spectrum, strategies like the ones used in *cyclic executives* [18] confine scheduling decisions to when a task activates or voluntarily ceases execution. To improve system performance from the timing point of view, tasks may be split into multiple parts but, also in this case, task splitting points are fixed and pre-determined. As a consequence, under appropriate assumptions concerning the tasks' structure, like their periodicity, all scheduling decisions can be taken once and for all in advance, and merely carried out at runtime, giving rise to the *offline* scheduling strategies.

Instead, *online* strategies—like the ones to be discussed in this chapter—lie at the opposite end of the spectrum because scheduling decisions are taken at runtime, based on suitable task attributes. Sometimes, these attributes may be fixed and as simple as task *priorities* that, informally speaking, express the relative importance of the tasks in the system. In other, more complex strategies, they may instead vary with time to indicate, for instance, how close a task is to violating one of its timing constraints. As a consequence, in both cases a scheduling decision may lead to a context switch anywhere within tasks.

Referring back to Figure 3.1, different strategies lead to different sequences of operations performed by the system (one of them is depicted on the right of the figure) to realize the idealized task execution shown on the left. As a consequence, tasks execution may *interleave* in different ways. The multiprogramming mechanism ensures that, in the long run, all tasks make progress even though, as shown in the timeline of processor activity over time at the bottom of the figure, the processor indeed executes only one task at a time.

Comparing the left and right sides of the figure also explains why the adoption of a task-based model simplifies the design and implementation of a concurrent system. By means of this model, system design is carried out at the task level, a clean and easy to understand abstraction, without worrying about the low-level mechanisms behind its implementation, which stay hidden within the underlying operating systems. In principle, it is not even necessary to know whether the hardware really supports true execution concurrency, or the degree of such a parallelism. For this reason, properly design task-based systems can easily be ported to multiprocessor and multicore systems.

As outlined previously, the responsibility of choosing which tasks will be executed at any given time by the available processors, and for how long, falls on the operating system and, in particular, on an operating system component known as *scheduler*. In a real-time operating system, the scheduler works according to algorithms to be discussed in Section 3.2. Since tasks are first-class entities in any modern operating system, RTEMS provides an extensive API for task management, which will be described in Chapter 5.

Of course, if a set of tasks must cooperate to solve a certain problem, not all possible scheduling decisions will produce meaningful results. For example, if a certain task τ_2 relies upon some data computed by another task τ_1, the execution of τ_2 must not start before the conclusion of τ_1. Therefore, one of the main goals of the branch of computer science known as *concurrent programming* is to define a set of task communication and synchronization primitives. When used appropriately, these primitives ensure that the *results* of the concurrent program will be correct by introducing and enforcing appropriate constraints on scheduling decisions. They will be discussed in Chapters 7 and 9.

Co

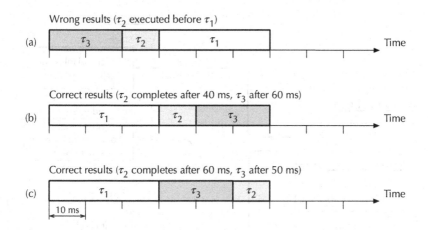

FIGURE 3.2 Role of interleaving in functional correctness and timings.

3.2 SCHEDULING POLICIES, MECHANISMS, AND ALGORITHMS

3.2.1 TASK INTERLEAVING AND TIMINGS

An aspect of paramount importance in any real-time system is that, even if the proper application of concurrent programming techniques guarantees that the concurrent program will be functionally correct—that is, its results will be correct—the scheduling decisions taken by the operating system may still affect the behavior of the system in other important ways. This is due to the fact that, even when all constraints set forth by the interprocess communication and synchronization primitives are met, there are still many acceptable interleavings. For instance, choosing one interleaving or another does not affect the functional aspects of the computation, but may significantly change the timings of the tasks involved.

To further illustrate this concept, Figure 3.2 shows three different interleavings of tasks τ_1, τ_2, and τ_3 when they are executed on a single-core processor. We suppose all tasks are ready for execution at $t = 0$ and their execution requires $C_1 = 10$ ms, $C_2 = 30$ ms, and $C_3 = 20$ ms of processor time, respectively. To make the example simpler, we also neglect for the time being that most operating systems are able to switch from one task to another during their execution, as described in Section 3.1, and assume that individual tasks are scheduled as indivisible units instead.

Since we further assume that τ_1 produces some data used by τ_2, it turns out that schedule (a), shown at the top of the figure, is unsuitable from the functional point of view because it does not satisfy the precedence constraint between τ_1 and τ_2 we just stated and would lead τ_2 to produce incorrect results. On the other hand, schedules (b) and (c) are both satisfactory from this point of view. However, they are very dissimilar from the timing perspective. Namely, as also shown in the figure, the completion time of τ_2 and τ_3 are very different in the two cases. If we were dealing with a real-time system in which, for example, τ_2 and τ_3 must conclude within a relative

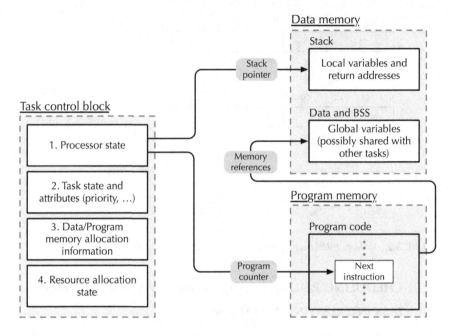

FIGURE 3.3 Task state and task control block (TCB).

deadline of $D_2 = 60$ ms and $D_3 = 55$ ms, interleaving (c) would satisfy this requirement, whereas interleaving (b) would not.

In order to address this issue, real-time systems use specially devised scheduling algorithms. Those algorithms, complemented by appropriate analysis techniques to be summarized in Section 4.1, guarantee that a concurrent program will not only be functionally correct, but it will also satisfy its timing constraints for all permitted interleavings.

3.2.2 TASK CONTROL BLOCK AND TASK STATE DIAGRAM

Before delving into more details on how operating systems handle and manage tasks, it is important to give a more precise definition of what a task really is, and what information it characterizes. In order to represent a task at runtime, operating systems store all the relevant information about it in a data structure, known as *task control block* (TCB). According to the general definition of task, it must contain all the information needed to represent the execution of a sequential program as it evolves over time.

As shown in Figure 3.3, there are four main components directly or indirectly linked to a TCB:

1. The TCB contains a full copy of the *processor state*. The operating system makes use of this piece of information to implement its context switch method, that is, to

switch the processor from one task to another. In fact, a context switch consists for the most part of saving the processor state of the previous task into its TCB, and then restoring the processor state from the TCB of the next task. A key point for the portability of an operating system is the clean separation between the context switch method, whose implementation determines *how* a context switch is carried out and often includes architecture-dependent aspects, and the scheduling strategy mentioned earlier, which is architecture-independent and dictates *when* a context switch takes place.

Within the processor state, two important entities are the *program counter*, which points to the next instruction that the processor is going to execute within the task's program code, and the *stack pointer*, which defines the boundary between full and empty elements in the task stack. Both are depicted as arrows in the figure. As can be inferred from the above description, the processor state is an essential part of the TCB and is always present, regardless of the operating system. Operating systems may instead differ on the details of *where* the processor state is stored. Conceptually, as shown in Figure 3.3, the processor state is part of the TCB and should be held within it. Some operating systems follow this approach literally, whereas others store part or all of the processor state elsewhere, and then make it accessible from the TCB through a pointer.

The second choice is especially convenient and efficient when the underlying processor provides hardware assistance to save and restore part of the processor state to/from a pre-defined, architecture-dependent location, which usually cannot be changed at will in software. For instance, ARM Cortex-M processors [9] autonomously save part of their state onto the current task stack when they start handling an interrupt or, more generally, an exception. Operating systems that base their context switch implementation upon the underlying exception-handling mechanism—thus performing context switches within exception handlers—may efficiently save the rest of the processor state in the same place. However, RTEMS does not follow this approach. Even though hardware exceptions may still result in a context switch, RTEMS performs all context switches within a task, rather than exception handling, context, as described in Section 4.2.

2. The *task state and attributes* are used by the operating system itself to schedule tasks and support inter-task synchronization and communication in an orderly way. A more detailed description of the task state and the way it is used by the scheduler will be given in the following, while information on inter-task synchronization can be found in Chapters 7 and 9.

3. The *data and program memory allocation* information held in the TCB keep a record of the memory areas currently assigned to the task. The extent and complexity of this information heavily depends on the purpose and sophistication of the operating system.

On one hand, very simple operating systems may only support a fixed number of statically created tasks and may not need to keep this information at all. This is because data and program memory areas are assigned to tasks at link time and the assignment never changes over time. On the other hand, when the operating

system supports the dynamic creation of tasks at runtime, even though their code
and data have been pre-loaded into memory, it must typically allocate the TCB,
the task stack, and possibly other data structures from a memory pool and keep
a record of the allocation. Moreover, if the operating system is also capable of
loading executable images from a mass storage device, it should also keep a record
of where the program code and data have been placed in memory.

4. When the operating system is in charge of resource allocation and release, the
task control block also contains the *resource allocation state* of the corresponding
task. The word resource is used here in a very broad sense. It certainly includes
all hardware devices connected to the system, but it may also refer to *software*
resources. Having the operating system work as a mediator between tasks and
resources regarding allocation and release is a universal and well-known feature
of virtually all general-purpose operating systems because the goal of this kind
of operating system is to support the coexistence of multiple application tasks,
developed by a multitude of programmers. For this reason, resource sharing and
allocation must be kept under tight control.

In a real-time embedded system, especially small ones, the scenario is very different because the task set to be executed is often well known in advance. The
relationship between tasks and resources may also be different, leading to a reduced amount of *contention* for resource use among tasks. For instance, in a
general-purpose operating system it is very common for application tasks to compete among each other to use a graphics coprocessor, and sharing this resource in
an appropriate way is essential.

On the contrary, in a real-time system devices are often dedicated to a single
purpose and can be used directly only by a single task. For example, an analog to
digital converter is usually managed by a cyclic data acquisition task. Accessing
the device itself is of no interest to any other task in the system although, of
course, those may make use of the acquired data. Therefore, in real-time operating
systems resources are often permanently and implicitly allocated to a single task
or group of tasks, and the operating system itself is only marginally involved in
their management.

A proper definition of the information included in a TCB is important not only to
thoroughly understand what a task *is*, but also how tasks are *managed* by the operating system. In fact, it is easy to notice that TCB contents also represent the information that the operating system must save and restore to implement a context switch,
in order to steer the processor from executing one task to another in a transparent
way.

Another essential part of understanding how operating systems manage task execution is to have a precise idea of how the *task state* evolves over time. By itself,
the TCB holds the task state but, being a data structure, it only gives a static depiction of it. A commonly used way to describe in a formal way all the possible states
a task may be in during its lifespan is to define a directed graph, called *task state
diagram* (sometimes abbreviated as TSD). The details of how the task state diagram
is organized and laid out vary from one operating system to another, but the most

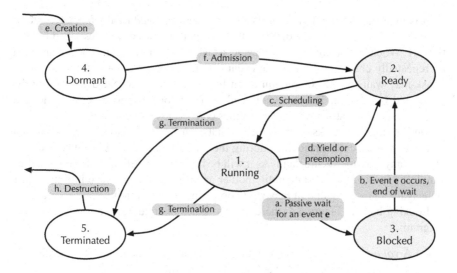

FIGURE 3.4 Abstract view of a task state diagram (TSD).

important concepts are common to all of them. Figure 3.4 depicts a simplified task state diagram to be used as an example. First of all, in a task state diagram:

- *nodes* represent possible task states, and
- *arcs* represent transitions from one state to another.

The three most important states for what concerns task scheduling and synchronization are the ones shown in gray. More specifically:

1. A task is *running* when it is actively being executed by a core, and hence, it is making progress. The number of tasks in the *running* state is limited by the total number of cores available in the system.
2. When the number of tasks eligible for execution in the system exceeds the number of available cores at a given instant, quite a common occurrence, only some of them can be brought into the *running* state and actually executed. The others stay in the *ready* state without making any progress for the time being.
3. Tasks often have to wait for an external event to occur. For example:
   - A periodic task, after completing its activity in the current period, has to wait until the next period begins by performing a *time-related* wait.
   - A task that issues an I/O request to a device must often perform an *I/O* wait until the operation is complete. The completion event is usually signaled by means of an interrupt request.
   - More in general, waits are also required to satisfy *precedence* constraints and ensure an orderly *communication* among tasks.
   In all these cases, tasks move to the *blocked* state and stay there, without competing with other tasks for execution, until the event they are waiting for occurs.

It is important to highlight that the form of wait realized by the *blocked* state of the task state diagram is very different than the polling-based wait described in Section 3.1. When performing a polling cycle, a task waits for the occurrence of an event by repeatedly checking whether the event of interest occurred or not. For instance, a task may wait for the completion of an I/O operation by repeatedly querying the associated I/O device. This is usually called *active* (or *busy*) *wait* because the task actively executes instructions and consumes core cycles while waiting.

On the contrary, tasks in the blocked state perform a *passive wait* and do not consume any execution resources during their wait because the scheduler does not allocate any core to execute them. As it will be better described in the following, they simply lie in the *blocked* state until another agent makes them move into the *ready* state again. Only at that point they start competing for execution resources again.

Looking back at Figure 3.4, there are two kinds of state transition in a task state diagram:

- A *voluntary* transition is taken under the control of the task that undergoes it, as a result of an explicit action it has performed.
- An *involuntary* transition is not under the control of the task affected by it. Instead, it is the consequence of an action taken by another task, the operating system, or the occurrence of an external event.

If we restrict ourselves for the time being to transitions involving the main task state diagram states discussed so far:

a. The transition from the *running* to the *blocked* state is an example of voluntary transition because it is always under the control of the affected task. In particular, it is performed when the task invokes one of the synchronization primitives to be discussed in Chapters 5, 7, and 9, in order to wait for a certain event **e**.
b. Instead, the transition from the *blocked* to the *ready* state is involuntary because it takes place when event **e** eventually occurs. Depending on the nature of **e**, the agent responsible for waking up the waiting task may be another task (when the wait is due to inter-task synchronization), the operating system timing facility (when the task is waiting for a time-related event), or an interrupt handler (when the task is waiting for an external event, such as an I/O operation), but it is never under the control of the task affected by the transition.

   Somewhat contrary to intuition, the waiting task is returned to the *ready* state and starts competing again for execution against the other tasks, but it does not go directly to the *running* state. However, this is in accordance to the general concept of modularity and separation of duties among operating system components. In this case, the synchronization mechanism is responsible for deciding whether or not a task is eligible for execution (by placing it in the *ready* or *blocked* state), whereas the scheduler determines which tasks should actually be executed at any given time (by moving them back and forth between the *ready* and *running* states).
c. As just mentioned, the operating system scheduler is responsible for picking up tasks in the *ready* state for execution and moving them into the *running* state, according to the outcome of its scheduling algorithm, whenever a core is available

for use. This is usually called task *scheduling* and is another example of involuntary transition.

d. The transition from the *running* to the *ready* state is more complex because it may be either voluntary or involuntary. A running task may voluntarily signal its willingness to relinquish the core it is being executed on by means of an operating system request known as *yield*, which brings the task back to the *ready* state. In turn, this leads the operating system to run its scheduling algorithm and choose a task to run among the ones in the *ready* state. Depending on the scheduling algorithm and the characteristics of the other tasks in the *ready* state, the choice may or may not fall on the task that just yielded. In other words, the effect of a yield is just to ask the operating system to reconsider the scheduling decision it previously made.

Another possibility is that the operating system itself decides to run the scheduling algorithm. Depending on the operating system, this may occur periodically or whenever a task transitions into the *ready* state from some other states for any reason. The second kind of behavior is more common with real-time operating systems because, by intuition, when a task becomes ready for execution, it may be "more important" than one of the running tasks from the point of view of the scheduling algorithm.

When this is the case, the operating system forcibly moves one of the tasks in the *running* state back into the *ready* state, with an action called *preemption*. Then, it will choose one of the tasks in the *ready* state and move it into the *running* state. On a multicore system, as better described in Part V of this book, the choice is also affected by the *affinity* of a task, that is, the set of cores on which it is allowed to run.

From the practical standpoint, operating systems strive to make decisions about moving tasks from one state to another as sparingly as possible to improve efficiency. Accordingly, RTEMS schedules only when an event that may change the outcome of a previous decision occurs, and even in that case it performs the minimum amount of work necessary to implement the desired scheduling algorithm correctly. The most complex event from this point of view takes place when a task blocks, and is way more complex than a yield.

Besides the essential states and transitions presented so far, most real-world operating systems implement additional ones. The most common additional states are summarized in the following.

4. General-purpose operating systems usually move new tasks directly into the *ready* state upon creation. This is the case, for instance, if task creation is accomplished through the POSIX API [68]. A more common choice in real-time operating systems, also pursued by RTEMS, is to split the creation of a task from the start of its execution. In this case, newly created tasks are put in a special *dormant* state.

5. When a task deletes itself or another task deletes it, it immediately ceases execution but some operating systems may not delete its TCB immediately. In these cases, the deleted task goes into the *terminated* state. It stays in that state until

the operating system completes the cleanup operations associated with task termination. For instance, deleted Linux tasks by default go to the special "zombie" state until their parent explicitly invokes a dedicated system call to wait for their termination and retrieve their final exit status. More in general, this happens in virtually all Unix-like operating systems [86].

When there are additional states in the task state diagram, further transitions are also needed to connect them to the main task state diagram states, namely:

e. The *creation* transition instantiates a new TCB, which describes the task being created. As discussed previously, the transition may lead the new task into the *dormant* or *ready* state depending on the operating system at hand.
f. The *admission* transition starts the execution of a dormant task, usually under the initiative of a task that is already executing in the system. Since a task cannot perform any action when it does not exist or it has not been started, both this transition and the previous one are necessarily involuntary.
g. Tasks cease execution by means of a *termination* transition. Since most operating systems allow a task to terminate itself or another task, the transition can be either voluntary or involuntary. In the second case, the transition may originate from the *running* or from the *ready* state.
h. Finally, the *destruction* transition permanently removes a task from the system and destroys its TCB.

Even though the presence of additional states between the *termination* and the *destruction* transitions may seem unimportant, it must be taken into due account from the practical point of view. This is because some or all of the resources allocated to the task—for instance, the TCB and possibly others—cannot be freed and later reused unless the task is properly disposed of. Following an incomplete or incorrect task termination and destruction procedure may easily lead to hard-to-spot memory leaks or corruption.

3.2.3   REAL-TIME SCHEDULING ALGORITHMS

In the previous sections, we introduced the notion of a scheduling algorithm rather informally. At the same time, we also hinted at the importance of a task—and its *priority*—as one of the main criteria used by real-time scheduling algorithms to select which tasks should be brought into the *running* state at any given time.

It is now time to have a deeper and more formal look at how two of the most widespread algorithms work. The discussion will start from the simplest algorithms, which define priorities as fixed values assigned to tasks upon creation, because this is what most real-time operating systems provide and what most applications use. More complex algorithms undoubtedly have advantages with respect to simple ones, but they may easily go against the all-important requirement that real-time operating systems implementers want efficient, practically implementable algorithms.

Due to lack of space, the discussion will be kept at an introductory level, with the goal of providing interested readers with enough background information to further investigate the matter by means of more advanced books, like [29, 31, 36, 85]. Further details about the actual implementation of the RTEMS scheduling algorithms can be found in its documentation [105]. Moreover, Reference [108] is an authoritative survey of the history and evolution of real-time scheduling algorithms.

Last but not least, in the following we are going to present only scheduling algorithms for single-core processors, postponing the discussion of multicore scheduling to Chapter 13.

First of all, let us observe that, in any application comprising multiple concurrently-executed tasks, the exact order in which tasks execute is not completely specified and constrained by the application itself. As described in Section 3.2.1, some constraints on task execution order are necessary to ensure that the results produced by the application are correct in all cases. In fact, despite these correctness-related constraints, the application will still exhibit a significant amount of nondeterminism.

Namely, the execution of its tasks may still interleave in different ways without violating any of those constraints. Going back to the example of Figure 3.2, interleavings (b) and (c) are equivalent from the functional correctness point of view. However, they are not at all equivalent with respect to timings. Therefore, if some tasks have a deadline on how much time it takes to complete them, a constraint also known as *response time deadline*, only *some* of the interleavings that are acceptable from the point of view of correctness will also be adequate to satisfy those additional constraints.

As a result, in a real-time system it is necessary to *further restrict* the nondeterminism, beyond what is necessary to guarantee functional correctness, to ensure that the task execution sequence will not only produce correct results in all cases but will also lead tasks to meet their deadlines. This is exactly what is done by real-time scheduling algorithms.

When using one of these algorithms, and under appropriate hypotheses, the scheduling analysis techniques to be briefly presented in Section 4.1 are able to establish whether or not all tasks in the system will be able to meet their deadlines and, using more complex techniques, calculate the worst-case response time of each task, too.

It turns out that assessing the timing behavior of an arbitrarily complex concurrent application is very difficult. For this reason, it is first of all necessary to introduce a simplified *task model*, which imposes some restrictions on the structure of the application to be considered for analysis and its tasks. The simplest model, also known as *basic* task model, has the following characteristics:

1. The application consists of a fixed number of tasks, and that number is known in advance. All tasks are created and started at the same time, when the application as a whole starts executing.
2. Tasks are *periodic*, with fixed and known periods, so that each task can be seen as an infinite sequence of *instances* or *jobs*. Each task instance becomes ready for execution at regular time intervals, that is, at the beginning of each task period.

3. Tasks are completely *independent* of each other. They neither synchronize nor communicate in any way, and they do not wait for external events.
4. As outlined above, timing constraints are expressed by means of *deadlines*. For a given task, a deadline represents an upper bound on the response time of its instances that must always be satisfied. In the basic task model the deadline of each task is equal to its period. In other words, the previous instance of a task must always be completed before the next one becomes ready for execution. Deadlines defined in this way are often called *implicit deadlines* in literature.
5. The worst-case execution time of each task—that is, the maximum amount of processor time it may possibly need to complete any of its instances when the task is executed in isolation—is fixed and can be computed offline.
6. All system's overheads, for example, context switch times, are negligible.

Although the basic task model is very intuitive and simple, it still leads to very important results concerning theoretical scheduling analysis. Moreover, it is the foundation and starting point of Rate Monotonic Analysis (RMA), which is probably the most widespread analysis method for real-time systems [109]. At the same time, it also has some shortcomings that hinder its application to real-world scenarios and must be relaxed to make scheduling analysis useful in practice. More specifically:

- The requirement about task independence rules out time-related waits and inter-task communication as described in Chapters 5, 7, and 9. This is unacceptable in practice because it goes against the way concurrent systems are usually designed—as a set of tasks that cooperate, and hence, necessarily interact with one another.
- The deadline of a task is not always the same as its period. For instance, a deadline shorter than the period—often called *constrained deadline*—is of particular interest to model tasks that are executed infrequently but, when they are, must be completed with tight timing constraints. This is typical, for instance, of tasks that must react to and handle abnormal conditions in a system.
- Some tasks are *aperiodic*. This may happen, for instance, when the execution of a task is triggered by an event external to the system. Again, this is typical of alarms and many forms of network communication, in which the arrival of an incoming frame is all but periodic in nature.
- It may be difficult to determine an upper bound on a task execution time which is at the same time *accurate* and *tight*. For instance, many iterative algorithms may take a different number of iterations depending on their input data. For complex algorithms, the worst case may be difficult to identify theoretically, and in any case, there may be a significant difference between the average and the worst-case number of iterations.
- Modern hardware architectures include hardware components (like caches, for example), in which the average time needed to complete an operation may differ from the worst-case time by several orders of magnitude. As a consequence, they bring even more uncertainty to worst-case task execution time calculations.

TABLE 3.1

Notation for Real-Time Scheduling Algorithms and Analysis

Symbol	Meaning
N	Number of tasks in the system
τ_i	The i-th task, $1 \le i \le N$
T_i	The period of task τ_i
D_i	The relative deadline of task τ_i
C_i	The worst-case execution time of task τ_i
R_i	The worst-case response time of task τ_i
φ_i	Initial phase of task τ_i
$\tau_{i,j}$	The j-th instance of the i-th task
$r_{i,j}$	The release time of $\tau_{i,j}$
$d_{i,j}$	The absolute deadline of $\tau_{i,j}$
$c_{i,j}$	The execution time of $\tau_{i,j}$
$f_{i,j}$	The response time of $\tau_{i,j}$
$hp(i)$	Indexes of tasks with priority higher than τ_i
B_i	Worst-case blocking time endured by τ_i
K	Number of semaphores in the system
S_k	The k-th semaphore, $1 \le k \le K$
$usage(k,i)$	Function that is 1 if τ_i makes use of S_k, and 0 otherwise
Q_i	Number of self-suspension points in task τ_i
P_i	The worst-case self-suspension time of task τ_i

Before proceeding further, it is also necessary to introduce some notation to be used throughout the book. Even though it is not completely standardized, the notation summarized in Table 3.1 and illustrated in Figure 3.5 is the one adopted by most textbooks and publications on the subject. In particular:

- The symbol τ_i has already introduced previously and represents the i-th task in the system. If N represents the number of tasks in the system, it is $1 \le i \le N$. Unless otherwise specified, tasks are enumerated by decreasing priority. In other words, if $i < j$, the priority of τ_i is greater than the priority of τ_j.
- A periodic tasks τ_i consist of an infinite number of repetitions, or *instances*. When it is necessary to distinguish an instance of τ_i from another, we use the notation $\tau_{i,j}$ to indicate the j-th instance of τ_i. Instances are enumerated according to their temporal order, so that $\tau_{i,j}$ precedes $\tau_{i,k}$ in time if and only if $j < k$. The first instance of τ_i is $\tau_{i,0}$.
- By definition, individual instances of a periodic task τ_i are *released*, that is, they become ready for execution, at regular time intervals. The distance between two adjacent releases is the *period* of the task, denoted by T_i.
- The symbol D_i represents the *relative deadline* of τ_i, that is, the deadline of an instance expressed with respect to the release time of the instance itself. According to the model, the relative deadline is therefore the same

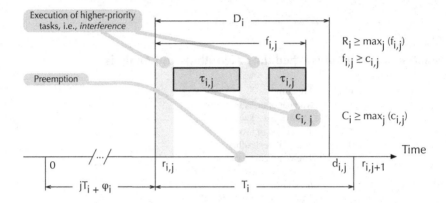

FIGURE 3.5 Notation for real-time scheduling algorithms and analysis.

for all instances of a given task. In the following it will be assumed that $\forall i \ D_i \leq T_i$.

- The worst-case execution time of τ_i is denoted as C_i, whereas $c_{i,j}$ represents the execution time of its j-th instance $\tau_{i,j}$. Taking into account that C_i can sometimes be a conservative estimate, we can write $C_i \geq \max_j(c_{i,j})$. As outlined above, the worst-case execution time of a task is an upper bound on the amount of processor time needed to complete any of its instances when the task is executed *in isolation*, that is, without the presence of any other tasks in the system. It is worth remarking that the task *execution* time shall not be confused with its *response* time, to be described next.

- The worst-case response time of τ_i, denoted as R_i, is an upper bound on the amount of time needed to complete any of its instances when the task is executed *together with* all the other tasks in the system. It is therefore $\forall i \ R_i \geq C_i$ because the presence of other tasks can only worsen the completion time of τ_i. For instance, the presence of a higher-priority task τ_j may lead the scheduler to preempt τ_i in favor of τ_j when the latter becomes ready for execution. This phenomenon is known as *interference*.

- Since the actual response time $f_{i,j}$ of task instance $\tau_{i,j}$ depends on the amount of interference that particular instance was subject to, it will vary from one instance to another. It still is $\forall j \ f_{i,j} \geq c_{i,j}$ and, according to the definition of worst-case response time, we can also write $R_i \geq \max_j(f_{i,j})$.

- The symbol $r_{i,j}$ is used to denote the release time of task instance $\tau_{i,j}$, that is:

$$r_{i,j} = \varphi_i + jT_i, \quad j = 0, 1, \ldots \tag{3.1}$$

where φ_i represents the initial *phase* of τ_i, namely, the absolute release time of its first instance $\tau_{i,0}$.

- Similarly, $d_{i,j}$ represents the *absolute* deadline of task instance $\tau_{i,j}$, which is given by:

$$d_{i,j} = r_{i,j} + D_i. \tag{3.2}$$

An important difference between D_i and $d_{i,j}$ is that the former is a relative quantity—measured with respect to the release time of each instance of τ_i—and is the same for all instances. On the contrary, the latter is an absolute quantity that represents the instant in time at which task instance $\tau_{i,j}$ must necessarily already be completed in order to satisfy its timing constraints. As a consequence, $d_{i,j}$ is different for each instance of τ_i.

Looking back at Figure 3.5 further highlights the difference between $c_{i,j}$ and $f_{i,j}$ when another, higher-priority task executes concurrently with $\tau_{i,j}$ on the same core, thus causing interference, which is represented by light gray, vertical bars in the figure. In this case:

- If a higher-priority task is executing at $r_{i,j}$, when $\tau_{i,j}$ is released, $\tau_{i,j}$ does not immediately transition to the running state. Instead, it stays in the ready state until the higher-priority task has been completed or leaves the running state for some other reasons. This corresponds to the leftmost interference bar in Figure 3.5.
- If a higher-priority task is released while $\tau_{i,j}$ is running, the operating system may temporarily stop its execution in favor of the higher-priority task, with an action known as *preemption*, and resume it at a later time. This gives rise to further interference, indicated by the rightmost bar in the figure.

As a result, the response time $f_{i,j}$ of $\tau_{i,j}$ may become significantly longer than its execution time $c_{i,j}$. A very important goal of defining a satisfactory real-time scheduling algorithm, along with an appropriate way of analyzing its behavior, is to ensure that $f_{i,j}$ is bounded for any instance $\tau_{i,j}$ of τ_i.

The *Rate Monotonic* (RM) scheduling algorithm, introduced by Liu and Leyland [84] assigns to each task τ_i in the system a *fixed* priority, which is inversely proportional to its period T_i. Tasks are then selected for execution according to their priority, that is, at each instant the operating system scheduler chooses for execution the ready task with the highest priority. Preemption of lower-priority tasks in favor of higher-priority ones is performed as soon as a higher-priority task becomes ready for execution.

The Rate Monotonic priority assignment takes into account only task periods T_i, and not their worst-case execution times C_i, thus favoring tasks with a shorter period. Intuitively, this makes sense because in the basic task model we are assuming $D_i = T_i$, and hence, tasks with a shorter period have less time available to complete their work before they miss their deadline. On the contrary, tasks with a longer period can afford giving precedence to more urgent tasks and still be able to finish their execution in time.

This informal reasoning can be confirmed with a mathematical proof of optimality. It has been proven that Rate Monotonic is the best scheduling policy among all the *fixed* priority scheduling policies on a single-core processor when the *basic task model* is considered. In particular, under the following assumptions:

1. We consider a fixed set of independent tasks.
2. Each τ_i is periodic with period T_i and has a known worst-case execution time C_i.
3. The relative deadline of each task is equal to its period, that is, $\forall i \ \ D_i = T_i$.
4. All tasks are released together for the first time at $t = 0$, that is, $\forall i \ \ \varphi_i = 0$.
5. Tasks are scheduled preemptively according to a *fixed* priority assignment.
6. There is only one execution core.

It has been proven [84] that, if a given set of periodic tasks with fixed priorities can be scheduled so that all tasks meet their deadlines by means of some other scheduling algorithm A, then the Rate Monotonic algorithm is also able to do the same.

Although the Rate Monotonic algorithm has been proven to be optimal among all fixed-priority scheduling algorithms under the hypotheses just discussed, it is still interesting to investigate whether it is possible to "do better" by relaxing some constraints on the structure and complexity of their scheduler. In particular, it is interesting to consider the scenario in which task priorities are no longer constrained to be fixed, but may change over time instead. The answer to this question was given by Liu and Layland in [84], by defining a dynamic-priority scheduling algorithm called Earliest Deadline First (EDF) and proving it is optimal among all possible scheduling algorithms, again under some hypotheses.

The EDF algorithm selects tasks according to their absolute deadlines. That is, at each instant, tasks with earlier deadlines receive higher priorities. According to (3.1) and (3.2), the absolute deadline $d_{i,j}$ of $\tau_{i,j}$ is:

$$d_{i,j} = \varphi_i + jT_i + D_i. \tag{3.3}$$

From this equation, it is clear that the priority of a given task τ_i changes dynamically from one instance to the next because it depends on the deadline of its currently active instance. On the other hand, the priority of a given task instance $\tau_{i,j}$ is still fixed because its deadline is computed once and for all by means of (3.3).

This property also gives a significant clue on how to simplify the practical implementation of EDF. In fact, EDF implementation does not require that the scheduler continuously monitors the current situation and rearranges task priorities when needed. This would very likely be too onerous. Instead, task priorities shall be updated only when a new task instance is released. Afterwards, when time passes, the priority order among active task instances does not change because their absolute deadlines do not move.

As happened for RM, the EDF algorithm works well according to intuition because it makes sense to give a higher priority to more "urgent" task instances, that is, instances that are getting closer to their deadlines without being completed yet. The reasoning has been confirmed in [84] by a mathematical proof. In particular, it has been proven that EDF is optimal under the following assumptions:

1. We consider a fixed set of independent tasks.
2. Each τ_i is periodic with period T_i and has a known worst-case execution time C_i.
3. The relative deadline of each task is equal to its period, that is, $\forall i \ D_i = T_i$.
4. All tasks are released together for the first time at $t = 0$, that is, $\forall i \ \varphi_i = 0$.
5. Tasks are scheduled preemptively according to their *dynamic* priority.
6. There is only one execution core.

The definition of optimality used in the proof is the same one adopted for Rate Monotonic. Namely, the proof shows that, if *any* task set is schedulable by *any* scheduling algorithm under the hypotheses of the theorem, then it is also schedulable by EDF. In spite of its proven optimality, EDF is rarely implemented in common real-time operating systems. RTEMS is a notable exception in this respect because it does have EDF as an option, which is used by default on multicore systems.

Even if we stay with a fixed priority assignment, to take advantage of its low implementation complexity, there are other ways to relax some constraints on task characteristics and devise scheduling algorithms more appropriate for use in real-world scenarios. For instance, it has already been mentioned that the assumption $D_i = T_i$, that is, assuming that all tasks have a relative deadline equal to their period, may be sometimes unrealistic.

If we extend the basic task model to cover the more general case $D_i \leq T_i$, Leung and Whitehead [82] were able to prove that the Deadline Monotonic Priority Order (DMPO) priority assignment is optimum under the following hypotheses:

1. We consider a fixed set of independent tasks.
2. Each τ_i is periodic with period T_i and has a known worst-case execution time C_i.
3. The relative deadline of each task does not exceed its period, that is, $\forall i \ D_i \leq T_i$.
4. All tasks are released together for the first time at $t = 0$, that is, $\forall i \ \varphi_i = 0$.
5. Tasks are scheduled preemptively according to their *fixed* priority.
6. There is only one execution core.

The deadline monotonic priority assignment is very similar in concept to RM. It assigns to each task a fixed priority inversely proportional to its relative deadline instead of its period, as RM would do. Once more, the meaning of the term *optimum* must be understood in the same way as for RM and EDF.

This extension is especially convenient to deal with tasks that are not periodic in nature, and hence, called *aperiodic* tasks. Tasks of this kind still consist of an infinite sequence of identical instances. However, their release does not take place at a regular rate. For instance, aperiodic tasks may be triggered by:

- User commands that require a response from the system.
- External events, such as alarms, generated at unpredictable times.

In many settings of practical interest, it is possible to determine the minimum interarrival time interval of an aperiodic task. In this case, we call it a *sporadic* task. For example, a minimum interarrival time can safely be assumed for user-generated events, due to the inherent speed limits of human beings. Mechanical devices, like keys and relays, also give rise to signals with a guaranteed minimum interarrival time, when de-bounced appropriately.

Then, the occurrence of triggering events can be rate-limited to ensure that, once a sporadic task has been triggered, it will not be triggered again until at least the minimum interarrival time has elapsed.

One simple way of expanding the basic process model to include sporadic tasks is to interpret the period T_i as the *minimum* interarrival time interval of τ_i. This is an obviously conservative choice, because a sporadic task τ_i can actually be released much less frequently than T_i, but it nevertheless guarantees that any scheduling analysis technique applied to the task set, if successful, ensures that the system can sustain the maximum release rate of τ_i. More sophisticated methods of handling sporadic tasks do exist, but their description is beyond the scope of this book. A more comprehensive treatment of this topic can be found in References [29, 31, 85].

For sporadic tasks, the assumption $D_i = T_i$ usually becomes unreasonable because, for instance, they may encapsulate an alarm handler. In many systems, alarms occur infrequently, leading to a relatively high minimum interarrival time T_i. However, when they do occur, they must be handled within a deadline that is much shorter than their period, that is, $D_i \ll T_i$. This is exactly the scenario that the deadline monotonic priority order has been designed to handle.

Similarly, it can also be observed that the hypothesis $\forall i \; \varphi_i = 0$, which states that all tasks are released simultaneously at $t = 0$ and defines a *synchronous* periodic system, is not always satisfied and, practically speaking, can be a challenge. Specialized synchronization devices, like the *barriers* to be discussed in Section 7.5, can often be used to this purpose.

Nevertheless, *asynchronous* periodic systems, characterized by having $\varphi_i \neq 0$ for some i, are also of practical interest. In these systems, tasks are never all released simultaneously and it has been proven that the deadline monotonic priority assignment is no longer optimal for them [82]. However, an optimal fixed-priority assignment does exist, it has been proposed by Audsley [15], and is known as Optimal Priority Assignment (OPA).

Although OPA is considerably more complex than the deadline monotonic priority order or RM, all its complexity is confined to the priority assignment algorithm. Once priorities have been assigned to tasks, they are fixed and can be implemented by the same lower-level scheduling mechanism used by the deadline monotonic priority order and RM. In other word—unlike what happens to some extent with EDF—the complexity of OPA does not impact the efficiency of the performance-critical operating system component that moves tasks between the *ready* and *running* states of the task state diagram.

3.3 SUMMARY

This chapter contains the basics of concurrent programming. It first defined key concepts like the cooperating sequential processes model, the task control block, and the task state diagram. Then, it described the most widespread real-time scheduling algorithms for single-core systems, known as rate monotonic (RM) and earliest deadline first (EDF).

The discussion of how the extension of these algorithms to multicore processors affects their optimality and performance is left to Part V of the book, along with a description of several scheduling algorithms specifically tailored to multicore processors.

4 Scheduling Analysis and Interrupt Handling

CONTENTS

The first part of this chapter contains an introduction to schedulability analysis, a set of mathematical tools to predict the worst-case timing behavior of a task-based system. Although only the most basic techniques can be discussed here due to lack of space, additional references are provided to interested readers.

In the second part of the chapter, we move to more practical considerations on interrupt handling, using a popular microprocessor architecture as a reference. Further information is also given about how the RTEMS operating system implements some key activities in this area, most notably task context switch. This information is essential to fully understand how operating systems implement the theoretical concepts and algorithms presented in Chapter 3.

The chapter ends with a discussion of how interrupt handling fits in the schedulability analysis framework, a topic of significant practical importance since most real-time systems comprise multiple interrupt sources whose impact cannot be neglected.

4.1 BASICS OF REAL-TIME SCHEDULING ANALYSIS

In the previous section, it has been stated that several scheduling algorithms are optimal within the scope of the basic task model, possibly extended in various ways. Those important theoretical results are valid, subject to certain assumptions about the general characteristics of the scheduling algorithm, task properties, and the underlying system.

However, they still do not answer a rather important practical question that arises during the design of a real-time software application. Namely, software designers want to know whether or not a certain task set they are working with can successfully be scheduled by means of one of the scheduling algorithms made available by the operating system of their choice.

A useful consequence of having introduced a formal task model is that it becomes possible to define precisely the meaning of "successful scheduling" or, in more formal terms, *schedulability* of a task set. More specifically, we say that a task set is schedulable by a given scheduling algorithm if all tasks in the set meet their deadline—and hence, they all satisfy the timing constraints set forth in the system specification. This happens if and only if $\forall i, j \ \ f_{i,j} \leq d_{i,j}$ or, if we resort to the concept of worst-case response time, $\forall i \ \ R_i \leq D_i$.

Although, in some simple cases, a straightforward "yes or no" answer to the schedulability question may be enough, more often than not designers also need to be confident about how much timing margin their systems have. In other words, designers may not be content to just know that all tasks in the set will meet their deadline and may also want to know the actual value of R_i. This information enables them to judge, for instance, how far or how close their tasks are from missing their deadlines, in case there are some unexpected extra delays in their execution.

In this chapter, the scope of the analysis will be limited to fixed-priority scheduling algorithms for synchronous systems, such as RM and the deadline monotonic priority order, running on a single-core processor. Similar, but considerably more complex analysis methods also exist for EDF, OPA, and other scheduling algorithms. Readers are referred to other publications, for instance [15, 29, 31, 36, 85], for a complete description of those methods.

4.1.1 UTILIZATION-BASED SCHEDULABILITY TESTS

The simplest family of scheduling analysis methods can be applied to single-core systems and is based on a quantity called *utilization factor*, usually denoted as U and defined as:

$$U = \sum_{i=1}^{N} \frac{C_i}{T_i} \tag{4.1}$$

where, according to the notation presented in Table 3.1, C_i represents the worst-case execution time of task τ_i and T_i is its period.

By intuition, the fraction C_i/T_i represents the fraction of processor time spent executing task τ_i in the worst case. The utilization factor is therefore a measure of the computational load imposed by a given task set on the core that executes it. Accordingly, the computational load associated with a task increases when its worst-case execution time C_i increases and/or its period T_i decreases.

Although U is derived from a very simple calculation and its value may be imprecise due to uncertainties in the C_i described earlier, it still provides useful insights about the schedulability of the task set it refers to. Even more importantly, the methods to be discussed in the following can be readily applied during system design. Thus, they may raise important early warnings about the soundness of the design itself and save valuable time, because it makes little sense to try and implement a system that is broken by design.

First of all, an important theoretical result identifies task sets that are certainly *not schedulable*. Namely, if $U > 1$ for a given task set, then the task set is not schedulable by a single-core processor, regardless of the scheduling algorithm in use. Unlike the other results to be discussed here, this one can also be extended to multicore systems to state that, on an M-core system, a task set is certainly not schedulable if $U > M$.

Besides the formal proof—which can be found in [84]—this result is quite intuitive. Basically, it states that it is impossible to allocate to the tasks a fraction of processor time U that exceeds the total processor time available for use, that is, 1 on a single-core processor. It should also be noted that this result merely represents a *necessary* condition and, by itself, it does not provide any useful information about the schedulability of a task set when $U \leq 1$.

For the RM priority assignment, a *sufficient* test provides more insights, assuming the task set is synchronous and conforms to the basic task model introduced in Section 3.2.3. More specifically, it is possible to determine a threshold value for U so that, if U is below the threshold, the task set can certainly be scheduled, independently of all its other characteristics, on a single-core processor. Namely, in [40, 84] it has been proven that if

$$U = \sum_{i=1}^{N} \frac{C_i}{T_i} \leq N(2^{1/N} - 1), \tag{4.2}$$

then the task set is certainly schedulable when using the RM priority assignment. The necessary and sufficient schedulability tests are summarized in the lower and mid part of Figure 4.1. At the same time the figure highlights that when the processor utilization factor U falls in the range $(N(2^{1/N} - 1), 1)$, this utilization-based test provides no definitive answers.

The hyperbolic bound for RM described in [20] gives rise to a schedulability test that has the same computational complexity as the one given in (4.2) and makes use of the same underlying hypotheses, but is less pessimistic. Namely, if:

$$\prod_{i=1}^{N} \left(\frac{C_i}{T_i} + 1 \right) \leq 2, \tag{4.3}$$

then the task set is certainly schedulable when using the RM priority assignment. Nevertheless, this is still a *sufficient* test, that is, a task set might still be schedulable although it does not pass the test. In this case, more complex analysis techniques, like the one described in Section 4.1.2, are necessary to exactly assess schedulability. However, it must also be considered that the extra computational complexity may render these techniques impractical for large task sets.

A schedulability test similar to (4.2), under the same hypotheses, also exists for EDF [84]. In this case, the threshold value for U is exactly 1, which makes the test

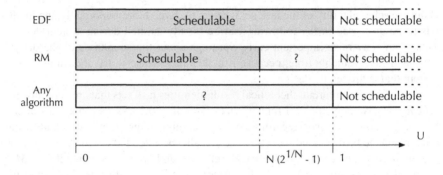

FIGURE 4.1 Utilization-based schedulability tests for a single-core processor.

both necessary and sufficient. That is, a task set of N periodic, synchronous tasks conforming to the basic process model is schedulable with the EDF algorithm if and only if:

$$U = \sum_{i=1}^{N} \frac{C_i}{T_i} \leq 1. \tag{4.4}$$

As also shown in the upper part of Figure 4.1, with respect to the corresponding tests for RM, the utilization-based schedulability tests for EDF leaves no area of uncertainty.

4.1.2 RESPONSE TIME ANALYSIS

In the previous section we have observed that, for certain values of U, utilization-based tests do not provide enough information about the schedulability of a task set with the RM priority assignment. Hence, researchers developed more complex tests, able to provide a definitive answers, without any areas of uncertainty. Among them, we will focus on a method known as response time analysis (RTA) [16, 17], an *exact* (both necessary and sufficient) schedulability test that can be applied to any fixed-priority assignment scheme on single-core processors.

This test not only gives a "yes or no" answer to the schedulability question but also calculates the worst-case response times R_i individually for each task. Therefore, it becomes possible to compare them with the corresponding deadlines D_i not only to assess whether all tasks meet their deadlines or not, but also to judge how far or how close they are from missing their deadlines.

According to response time analysis, the worst-case response time R_i of task τ_i in a synchronous system can be calculated by considering the following recurrence relationship:

$$w_i^{(k+1)} = C_i + \sum_{j \in \text{hp}(i)} \left\lceil \frac{w_i^{(k)}}{T_j} \right\rceil C_j, \tag{4.5}$$

in which $\lceil \cdot \rceil$ is the ceiling function, which gives the least integer greater than or equal to its argument, and $hp(i)$ denotes the set of indices of the tasks with a priority higher than τ_i. For RM, the set contains the indices j of all tasks τ_j with a period $T_j < T_i$. If we adhere to the convention that tasks are enumerated in order of decreasing priority, it is also $hp(i) = \{1, \dots, i-1\}$.

Informally speaking, $w_i^{(k+1)}$ and $w_i^{(k)}$ are the $(k+1)$-th and the k-th estimate of R_i, respectively, and Equation (4.5) provides a way to calculate the next estimate of R_i starting from the previous one. The first approximation $w_i^{(0)}$ of R_i is set to $w_i^{(0)} = C_i$, which is the smallest possible value of R_i. It has been proven that the succession $w_i^{(0)}, w_i^{(1)}, \dots, w_i^{(k)}, \dots$ defined by (4.5) is monotonic and nondecreasing. Two cases are then possible:

1. If the succession does not converge, there exists at least one scheduling scenario in which τ_i does not meet its deadline D_i, regardless of the specific value of D_i.
2. If the succession converges, it converges to R_i, and hence, it will be $w_i^{(k+1)} = w_i^{(k)} = R_i$ for some k. In this case, τ_i meets its deadline in every possible scheduling scenario if and only if the worst-case response time provided by response time analysis is $R_i \leq D_i$.

Unlike the U-based scheduling tests discussed in Section 4.1.1, this method no longer assumes that the relative deadline D_i is equal to the task period T_i and is also able to handle the more general case in which $D_i \leq T_i$. In addition, the method works with any fixed-priority ordering, and not just with the Rate Monotonic priority assignment, as long as $hp(i)$ is defined appropriately, a preemptive scheduler is in use, we are considering a synchronous task set that conforms to the basic task model and it is executed on a single-core processor. As a consequence, it is also readily applicable to the deadline monotonic priority order.

4.1.3   TASK INTERACTIONS AND SELF-SUSPENSION

Another interesting property of response time analysis is that it is more flexible than U-based tests and is easily amenable to further extensions, for instance, to consider the effect of task interactions. These extensions aim at removing one important limitation of the basic task model used so far and bring it closer to how real-world tasks behave.

For simplicity, in this book, the discussion will only address the following two main kinds of interaction:

1. Task interactions due to *mutual exclusion*, a ubiquitous necessity when dealing with shared data, as shown in Chapter 7.
2. Task *self suspension*, which takes place when a task passively waits for any kind of external event.

Readers are referred, for instance, to [31, 36, 85] for more detailed and comprehensive information about the topic.

The second kind of interaction is important because it takes place in many input–output operations. In those cases, a task within the device driver typically starts an operation, by programming the device control registers appropriately, and then passively waits for the results. In turn, the device signals that the requested operation has been completed by raising an interrupt request, whose handler wakes up the waiting task.

Other examples, involving only tasks rather than devices, include semaphore-based task synchronization, outlined in Chapter 7, and synchronous message passing operations, which are discussed in Chapter 9.

Regardless of the nature of the interaction, tasks are bound to experience a certain amount of *blocking* from that interaction, whenever such an interaction is based upon one of the wait- and lock-based synchronization methods to be described in this part of the book. What a proper design of these synchronization methods and of the way tasks use them can do is to guarantee that the worst-case blocking time endured by each individual task τ_i, denoted as B_i, is bounded. The worst-case blocking time can then be calculated and used to refine the response time analysis method discussed in Section 4.1.2, in order to determine worst-case response times.

Wait- and lock-free synchronization methods also exists and are especially useful in the context of multicore concurrency, although they may imply a significant drawback in design and implementation complexity. They will be discussed in more detail in Chapter 13.

When staying with more traditional synchronization methods, the value of B_i can then be used to extend response time analysis and consider the blocking time in worst-case response time calculations. Namely, the basic recurrence relationship (4.5) can be rewritten as:

$$w_i^{(k+1)} = C_i + B_i + \sum_{j \in hp(i)} \left\lceil \frac{w_i^{(k)}}{T_j} \right\rceil C_j. \tag{4.6}$$

It has been proven that the new recurrence relationship still has the same properties as the original one. In particular, if the succession $w_i^{(0)}, w_i^{(1)}, \ldots, w_i^{(k)}, \ldots$ converges, it still provides the worst-case response time R_i for an appropriate choice of $w_i^{(0)}$. On the other hand, if the succession does not converge, τ_i is surely not schedulable. As before, setting $w_i^{(0)} = C_i$ provides a sensible initial value for the succession.

The main difference is that the new formulation may be *pessimistic*, instead of necessary and sufficient, because the bound B_i on the worst-case blocking time might not be tight. Therefore it might be practically impossible for a task to ever incur in a blocking time equal to B_i, and hence, actually experience the worst-case response time calculated by means of (4.6).

Mutual exclusion blocking

As better illustrated in Chapter 8, using a plain *semaphore* as a mutual exclusion device is not enough to ensure that the B_i are bounded, due to a well-known phenomenon called *unbounded priority inversion*. However, appropriate extensions to semaphore's semantics, also called *protocols* and also presented in Chapter 8, have been devised to solve this issue.

If we consider the *priority inheritance* protocol—proposed by Sha, Rajkumar, and Lehoczky [110]—it has been proven that, if there are a total of K semaphores $S_1, \ldots, S_K$ in the system and critical regions are *not nested*, the worst-case blocking time experienced by each instance of task τ_i due to task interaction is bounded, and a bound is given by:

$$B_i^{\text{PI}} = \sum_{k=1}^{K} \text{usage}(k, i) C(k). \tag{4.7}$$

In the equation above,

- usage(k, i) is a function that returns 1 if semaphore S_k is used by (at least) one task with a priority less than the priority of τ_i, and also by (at least) one task with a priority higher than or equal to the priority of τ_i, *including τ_i itself*. Otherwise, usage(k, i) returns 0.
- $C(k)$ is the worst-case execution time among all critical regions associated with, or guarded by, semaphore S_k.

It should be noted that the bound provided by (4.7) is often "pessimistic" when applied to real-world scenarios, because

- It assumes that if a certain semaphore *can possibly block* a task, it *will indeed block* it.
- For each semaphore, the blocking time suffered by τ_i is always assumed to be equal to the worst-case execution time of the longest critical region guarded by that semaphore, even though the blocking tasks actually do not stay in the region for that long.

However, even being pessimistic, it is an acceptable compromise between the tightness of the bound it calculates and its computational complexity. Better algorithms exist and are able to provide a tighter bound of the worst-case blocking time, but the complexity of the analysis also becomes higher, which may make it problematic for large task sets.

Another possibility foreseen in Chapter 8 is to use the *priority ceiling* protocol, proposed by the same authors [110]. In this case, it can be proved that the worst-case blocking time experienced by each activation of task τ_i due to task interactions is bounded by:

$$B_i^{\text{PC}} = \max_{k=1}^{K} \left\{ \text{usage}(k, i) C(k) \right\}, \tag{4.8}$$

where usage(k, i) and $C(k)$ have the same meaning as in (4.7). This formula is also valid for a variant of the priority ceiling protocol, called *immediate* priority ceiling or priority ceiling *emulation* protocol. This variant is of great practical interest because, together with priority inheritance, is one of the protocols specified by the POSIX standard [68].

By comparing (4.7) and (4.8), we can easily see that it is $B_i^{PC} \leq B_i^{PI}$ for any given task set. Nevertheless, we must also recall that the priority ceiling protocol is less flexible than priority inheritance, because it requires a priori knowledge of all tasks that make use of a certain semaphore and their priority.

Task self-suspension

The analysis of task self-suspension presented here is based on [102], which addresses schedulability analysis in the broader context of real-time synchronization for multiprocessor systems. The reference also contains further, more detailed information, as well as the formal proof of all the statements to be discussed next.

Somewhat contrary to intuition, the effects of self-suspension are not necessarily *local* to the worst-case response time of the task that is experiencing it. On the contrary, the self-suspension of a high-priority task may also increase the worst-case response time of lower-priority tasks and, possibly, make them no longer schedulable.

This is because, after the self-suspension of a high-priority task ends, the task becomes ready for execution again and will preempt any lower-priority task. It can be proven that this new preemption opportunity may imply a greater impact on the worst-case response time of lower-priority tasks with respect to the case in which the high-priority task runs until completion without self-suspending.

Nevertheless, an upper bound B_i^{SS} on the worst-case blocking endured by task τ_i due to its own self-suspension, as well as the self-suspension of higher-priority tasks, still exists and can be calculated efficiently as

$$B_i^{SS} = P_i + \sum_{j \in \mathrm{hp}(i)} \min(C_j, P_j). \qquad (4.9)$$

In the above formula:

- P_i is the worst-case self-suspension time of task τ_i.
- $\mathrm{hp}(i)$ denotes the set of task indexes with a priority higher than τ_i.
- C_j is the execution time of task τ_j.

According to (4.9), the worst-case blocking time B_i^{SS} due to self-suspension endured by task τ_i is given by the sum of its own worst-case self-suspension time P_i plus a contribution from each of the higher-priority tasks, that is, the tasks whose index belongs to $\mathrm{hp}(i)$. The individual contribution of task $\tau_j, j \in \mathrm{hp}(i)$, to B_i^{SS} is given by its worst-case self-suspension time P_j, but it can never exceed its worst-case execution time C_j.

When considering the effects of mutual exclusion and self-suspension together, it turns out that these two sources of blocking are not independent from each other, because the self-suspension of a task has an impact on how it interacts with other tasks. There are several different ways to consider the combined effect of self-suspension and mutual exclusion on worst-case blocking time calculation. Perhaps the most intuitive one, presented in References [85, 102], makes use of the notion of task *segments*—that is, portions of task execution delimited by a self-suspension point. Accordingly, if task τ_i performs Q_i self-suspensions during its execution, it contains $Q_i + 1$ segments.

Task segments are considered to be completely independent from each other for what concerns blocking due to mutual exclusion. The analysis then proceeds in a conservative way, by assuming that each task goes back to the worst possible mutual-exclusion blocking scenario after each self-suspension. Following this approach, the worst-case blocking time B_i^{PI} or B_i^{PC} of τ_i due to mutual exclusion, calculated as specified in (4.7) or (4.8), becomes the worst-case blocking time endured by *each individual task segment*.

Hence, the worst-case blocking time of task τ_i due to mutual exclusion, B_i^{TI}, is given by:

$$B_i^{TI} = (Q_i + 1)B_i^{PI} \quad \text{(priority inheritance protocol), or} \quad (4.10)$$
$$B_i^{TI} = (Q_i + 1)B_i^{PC} \quad \text{(priority ceiling protocol).} \quad (4.11)$$

By combining (4.9) and (4.10)–(4.11), the total worst-case blocking time B_i of τ_i, considering both self-suspension directly and its effect on mutual exclusion blocking, can be written as:

$$
\begin{aligned}
B_i &= B_i^{SS} + B_i^{TI} \\
&= P_i + \sum_{j \in \text{hp}(i)} \min(C_j, P_j) + (Q_i + 1) \sum_{k=1}^{K} \text{usage}(k, i)C(k), \quad (4.12)
\end{aligned}
$$

for the priority inheritance protocol, or:

$$
\begin{aligned}
B_i &= B_i^{SS} + B_i^{TI} \\
&= P_i + \sum_{j \in \text{hp}(i)} \min(C_j, P_j) + (Q_i + 1) \max_{k=1}^{K} \{\text{usage}(k, i)C(k)\}, \quad (4.13)
\end{aligned}
$$

for the priority ceiling or the immediate priority ceiling protocols. In the formulas above, C_j and $C(k)$ have got two different meanings that should not be confused despite the likeness in notation, namely:

- C_j is the worst-case execution time of a *specific task*, τ_j in this case, while
- $C(k)$ is the worst-case execution time of *any task* within any of the critical regions guarded by S_k.

A distinct advantage of the approach just described is that it is fairly simple and requires very limited knowledge about the internal structure of the tasks. For instance, it is necessary to know *how many* self-suspension points there are in each task, but it is not essential to know exactly *where* they are. This kind of information is simple to collect and maintain as software evolves with time. However, the disadvantage of using a very limited amount of information is that it makes the method extremely conservative. Thus, the bound B_i calculated according to (4.12)–(4.13) is definitely not tight and may widely overestimate the actual worst-case blocking time in some cases.

More sophisticated and precise methods do exist, such as the one described in Reference [77]. However, as we have seen in several other cases, the price to be paid for a tighter upper bound is that much more information needs to be collected and, perhaps even more importantly, maintained as the software evolves. For instance, in the case of [77], we need to know not only how many self suspension points there are in each task but also their exact location within the task. Namely, we need to know the worst-case execution time of each individual task segment, instead of the worst-case execution time of the task as a whole.

4.2   PRACTICAL CONSIDERATIONS ON INTERRUPT HANDLING

While the previous sections have given a more theoretical view of interrupt handling, here we will focus on three aspects of more practical interest, namely, how contemporary processors handle interrupt requests at the hardware level and how the RTEMS operating system manages interrupts and context switches. Finally, we will also provide some hints on how to take into account the interrupt load of a system in schedulability analysis.

4.2.1   EXCEPTION HANDLING IN THE CORTEX-M PROCESSOR

Similar to many other contemporary processors, the Cortex-M also handles other kinds of events, such as *faults*, in the same way as interrupts, that is, in a unified way. All these events are collectively referred to as *exceptions*. For this reason, in this section we will generally talk about exception handling, rather than interrupt handling.

A property that all exceptions have in common is that their occurrence *may* disrupt the normal instruction execution flow and direct the processor to execute a fragment of code associated with them, called *exception handler*, as discussed in Section 3.1. In the statement above we say "may" because the mere occurrence of an exception is necessary, but not sufficient, to ensure that the processor will start handling it immediately, if at all.

In fact, as it will become clearer in the following, a rather complex prioritization mechanism internal to the processor lies in between the occurrence of an exception—which is often related to a *hardware*-generated event—and the corresponding *software* action, that is, the execution of its handler. This mechanism, which plays a central role in Cortex-M exception handling, is mainly driven by an *exception priority* value associated with each source of exception.

For the time being, it is enough to say that a higher priority improves the exception handling *latency*—that is, the time elapsed between the occurrence of an exception and the execution of its handler—because the processor handles this exception in preference of others. The main categories of exceptions are described in the following.

Interrupt requests

This kind of exception is raised by a peripheral device, in order to signal to the processor, and eventually to the software, the occurrence of an event of interest concerning the device itself, often the completion of an I/O operation. For instance, an Ethernet controller may use an interrupt request to indicate that one or more network frames have been received, or that a frame previously enqueued for transmission has been sent onto the network. These interrupts are naturally raised asynchronously with respect to the code the processor is currently executing.

On Cortex-M processors, interrupt requests can also be triggered by means of a software action. In this case, they are often referred to as *software interrupts*, which are raised synchronously with respect to code execution. Regardless of their origin (hardware or software) interrupt requests are all handled in the same way without any further distinction. Like ordinary interrupt requests, also *Non-Maskable Interrupt* (NMI) requests can be issued by either hardware or software. The main difference is that their priority is among the highest in the system, immediately below the priority of the reset exception.

Many operating systems, including RTEMS, disable regular interrupts internally, around very short critical sections that must be executed as an indivisible unit. Due to their high priority, non-maskable interrupts are left enabled, though. As a consequence, the corresponding interrupt handlers may not use any operating system services because they could violate those critical sections.

Also belonging to this category are two more exceptions that are generated within the processor itself, rather than coming from external peripheral devices. They are:

- The *SysTick* exception, generated periodically by the 24-bit count-down system timer, which is very important because operating systems often derive all their timing information from it. If needed, the same exception can also be issued by software.
- The *PendSV* exception, which can only be triggered by software and whose exception handler is used by some operating systems to perform scheduling.

As described in Section 4.2.3, RTEMS takes a different approach and makes only limited use of the PendSV exception within its architecture-dependent layer, to trigger a scheduling and context switch operation that is then completely carried out in software within a task context. This is because RTEMS avoids using architecture-specific scheduling and context switch methods—another notable example being the Intel x86 hardware context switching facility—in order to enhance portability and not tie the operating system to any specific processor architecture.

Faults

Generally speaking, faults are a consequence of an abnormal event detected by the processor, either internally or while communicating with memory and other devices. These events are of great interest (and concern) to software because most often they indicate serious hardware or software issues that may prevent the system from continuing with normal activities. More specifically, the following kinds of fault are foreseen in Cortex-M processors.

- *UsageFault*. This fault occurs when an instruction cannot be executed for various reasons. For instance, the instruction may be undefined or may contain a misaligned address that prevents it from accessing memory correctly. For divide instructions, another reason for raising a *UsageFault* is an attempt to divide by zero.
  Some of the above-mentioned fault sources (like dividing by zero) can be *masked* in software, that is, the processor can be instructed to ignore them without generating any fault, whereas others (such as encountering an undefined instruction) cannot.

- *BusFault*. This fault is generated when an error occurs on the data or instruction bus while accessing memory. It can be generated as a consequence of an explicit memory access performed by an instruction during its execution, and also by fetching an instruction from memory. This fault does not report errors generated by the memory protection mechanism, which instead trigger a *MemManage* fault.
  It should also be noted that the Cortex-M is a memory-mapped input-output (I/O) architecture and whenever we refer to a "memory" address, we actually mean an address within the processor's address space, which may refer to either a memory location or an I/O register.

- *MemManage*. This fault occurs when a memory access is blocked by the memory protection mechanism. An optional Memory Protection Unit (MPU) provides a programmable way of protecting memory regions against data read and write operations, as well as instruction fetches, also depending on the current privilege mode of the processor.
  Even processors not equipped with a MPU may still set forth some predefined, non-programmable constraints on memory accesses and generate faults when they are violated. For instance, LPC17xx processors [90] forbid data access and instruction fetch in unimplemented regions of the address space, as well as instruction fetch from addresses assigned to I/O peripherals. They generate a *BusFault* fault in both cases.

A special kind of fault is *HardFault*. It can be generated for two different reasons. First, the processor generates a *HardFault* when it detects an error during exception processing, thus making normal exception handling impossible. The second reason why a *HardFault* can be generated is a mechanism known as *fault escalation*. A full description of this mechanism is beyond the scope of this book but, summarily speaking, under certain conditions fault escalation may transform, or escalate, some other exceptions into a *HardFault*.

This may happen when a new exception occurs while the processor is already handling another exception. Since the underlying processor architecture supports exception nesting, as discussed in Section 4.2.2, in this case the processor may either accept and handle the incoming exception immediately—thus nesting it into the previous one by means of a mechanism that resembles a function call—or postpone it, depending on the relative priority of the two exceptions.

However, there are exceptions that must necessarily be handled synchronously with respect to code execution, namely, before the processor proceeds to the next instruction, even though another higher-priority exception is already being handled. A typical example is the *UsageFault* exception because, as described previously, it indicates that the current instruction cannot be executed.

In those cases, if the priority of the incoming exception would be insufficient for immediately handling it, the processor automatically escalates the priority of the incoming exception to the priority of *HardFault*, one of the highest in the system. If priority escalation is not yet enough to make the incoming exception active because the processor is already running at *HardFault* priority or higher, the general assumption of the architecture is that the occurrence of the new exception is unrecoverable and fatal.

In the majority of cases, the processor reacts by suspending normal instruction execution completely and entering a lockup state. Although there are some conditions that may lead the processor to abandon this loop and resume normal instruction execution, in most cases the only way out of this state is to perform a reset.

Supervisor call (SVC)

This exception is raised by the execution of a SVC (supervisor call) assembly instruction. It is a typical example of synchronous exception because it is generated at the exact point when the processor encounters this assembly instruction in the instruction stream. The SVC exception is in a class by itself because it is often used as a way to enter the operating system kernel and request it to perform a function.

However, as discussed in Section 4.2.3, in the case of RTEMS, the kernel is entered by means of an ordinary function call and the SVC exception is only used to restore the portion of processor context the hardware saved onto a task stack because an exception had been accepted while that task was executing.

Reset

This exception is raised by the processor power-up sequence or a warm reset. It is handled like other exceptions but, as it will better explained in the following, several extra operations are carried out to ensure the processor starts executing code in a consistent way. For instance, the reset exception initializes the Master Stack Pointer (MSP), which is one of the two stack pointers in the processor and the one it will actually use immediately after reset, and sets the initial program counter from which code execution will start. The processor execution mode is also different. Moreover, if the reset exception occurs while the processor is already running (warm reset), instruction execution can stop at an arbitrary point.

TABLE 4.1

Exception and Execution Priorities in the Cortex-M3 Processor

Exception	Priority p
Reset	-3
Non-Maskable Interrupt (NMI)	-2
HardFault	-1
UsageFault, BusFault, MemManage, SVC	$0 \leq p \leq 255$, or -1^*
Interrupts (including SysTick and PendSV)	$0 \leq p \leq 255$

Execution	Priority q
Base level	256
Active exceptions	Minimum value a among active exceptions, if any
BASEPRI	Value $1 \leq b \leq 255$ of the register if $b \neq 0$
PRIMASK	0 if the mask bit is set
FAULTMASK	-1 if the mask bit is set

* Subject to priority escalation

4.2.2   EXCEPTION PRIORITIES AND ENTRY/EXIT SEQUENCE

As recalled in Section 3.1, in the past accepting and handling an interrupt from a certain source mainly depended on whether that interrupt source was enabled and whether the processor was servicing another interrupt already. Interrupt prioritization was used only to disambiguate multiple interrupt requests issued at the same time. Although this is still partly true today, in the Cortex-M the decision of whether or not an incoming exception request should become *active*, that is, whether or not the processor should immediately start handling it, also depends on other factors.

- As shown on the top half of Table 4.1, each kind of exception has its own *exception priority*, an integer value. In the priority hierarchy, lower values correspond to higher priorities. Priority values can be either fixed or programmable.

  For instance, the priority of a NMI request is always -2, whereas for interrupt requests the priority can be individually set to any non-negative value p between 0 and 255 included. Depending on the processor implementation, only some high-order bits of the priority value may be significant, and hence, numerically different priority values may correspond to the same actual priority if they differ only in some of their low-order bits.

  When there are multiple, simultaneous incoming exceptions, the one with the highest priority (lowest priority value) prevails on the others and determines the *incoming exception priority i*. If there are two or more simultaneously incoming exceptions with the same priority, the one with the numerically lowest *exception number*, a unique and fixed number defined by the processor architecture and associated to each exception, takes precedence.

- At the same time, the processor also keeps track of its current *execution priority e*. The execution priority is calculated as the minimum of several values, listed in the bottom half of Table 4.1:
  - The base level of execution priority, which is one more than the highest priority value supported by the processor, 256 in this case.
  - The minimum priority value a among all exceptions that became active in the past and are still being handled, if any. There may be more than one such exceptions because, as it will be better described in the following, the Cortex-M architecture supports nested exception handlers.
  - The value $1 \leq b \leq 255$ of the unsigned 8-bit register BASEPRI, if it is not zero. If BASEPRI is zero, its value is not taken into account in the calculation.
  - The values 0 and/or -1, depending on whether or not the 1-bit registers PRIMASK and/or FAULTMASK are set, respectively.

The values of i and e are then compared to determine if the incoming exception must become active or it must stay pending. In particular:

- If $i < e$, the incoming exception becomes active and the processor starts handling it immediately. As a consequence, if the processor was already handling another exception, the handling of the new exception is nested into the old one. In other words, the processor temporarily stops handling the old exception in favor of the new one, and will go back to it at a later time, with a mechanism similar to an ordinary function call.
- Otherwise, it is $i \geq e$ and the incoming exception stays pending. The processor will re-evaluate the possibility of making it active whenever e changes and becomes numerically higher than it was. The re-evaluation is still carried out as described above, incorporating into i the priority of any further exception requests that arrived in the meantime. For instance, re-evaluation takes place when the handling of an active exception terminates, because this increases the value of a and, in turn, may increase the value of e.

Therefore, programmers can set BASEPRI, PRIMASK and/or FAULTMASK to mask off and postpone the handling of some exceptions by lowering the value of e. For instance, setting BASEPRI to a non-zero, positive value t prevents the processor from accepting any incoming exception whose priority value i is $i \geq t$, unless priority escalation takes place.

The relative priority between two exceptions determines whether the arrival of one exception can preempt the handling of the other or not. Concerning this aspect, it is important to mention that the comparison between i and e is possibly affected by a processor feature known as *priority grouping*.

Even though a thorough discussion is beyond the scope of this book, by means of priority grouping it is possible to instruct the processor to ignore some low-order bits of the priority value when comparing i and e, so that priority levels that are distinct in principle "look the same" in the comparison. As a result, exceptions whose priorities only differ in these low-order bits, but are equal in the others, will not be nested into each other.

Before discussing in detail what happens when the processor accepts an exception request it is necessary to briefly introduce the concept of processor *execution mode*. From this point of view the Cortex-M approach is considerably simpler than others and only has two execution modes, called *thread* and *handler* mode.

Thread mode

Thread mode is the normal task execution mode and is also the mode the processor goes into when it accepts and handles a reset exception. The SPSEL bit of the CONTROL register determines which stack pointer the processor uses when in thread mode. The two possible choices are the Main Stack Pointer (MSP) and the Process Stack Pointer (PSP), which usually refer to distinct stacks in memory. Knowing exactly which stack pointer is in use at any given time is extremely important to fully understand how the underlying operating system mechanisms for exception handling and multitasking work.

Handler mode

As its name says, this mode is used by the processor to execute all exception handlers except the reset handler. Code executed in handler mode makes use of the MSP, regardless of the settings of the SPSEL bit. When the processor is executing in thread mode, accepts an exception request, and makes it active, it automatically enters handler mode. The opposite transition, from handler mode back to thread mode, occurs when an exception handler returns, there are no other active exceptions, and the processor started handling the current exception while it was executing in thread mode.

A peculiar case of mode transition happens when code running in thread mode executes a SVC instruction that, as described in Section 4.2.1, unconditionally issues an exception request. The exception is accepted synchronously with respect to the current instruction flow and grants controlled access to handler mode through a trusted software routine—the SVC exception handler—implemented by the operating system.

The presence of two distinct execution modes for task and exception handler execution, respectively, is extremely common across modern processor architectures, although the names given to these modes may differ from one architecture to another. Indeed, RTEMS assumes these two modes are available on all architectures it supports. As it will be better explained in Section 4.2.3, it always performs context switches in thread mode to enhance portability.

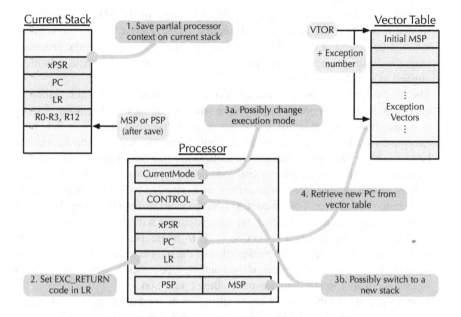

FIGURE 4.2   Exception entry sequence in the ARM Cortex-M processors.

Exception entry sequence

The main phases that make up the exception entry sequence, that is, the sequence of actions the processor performs when it makes an exception active and starts handling it, are depicted in Figure 4.2.

1. The first action is to save part of the current execution context on the current stack, the one that the processor is using when the exception request is accepted. The minimum amount of information that is saved into a *basic* exception frame consists of registers R0 through R3, R12, the link register LR (also called R14), the program counter PC (R15), and the program status register xPSR, for a total of 32 bytes.

   When the processor implements the optional floating-point extension, part of the floating-point context is saved as well, into an *extended* exception frame. However, the part of floating-point context to be saved on the stack is rather large and requires 68 additional bytes. Hence, RTEMS tries to avoid saving and restoring the floating point context to the extent possible.

   The architecture also supports a *lazy* context switch strategy [10], that is, a mechanism to automatically push the floating-point context only if and when the context is about to be modified by the execution of a floating-point instruction. However, this approach can only be used on single-core systems, as it does not work properly on multicores. For simplicity, the discussion that follows will not explore the management of the floating-point context further.

In both cases, the processor updates either the PSP or the MSP appropriately, so that it points at the base of the exception frame. Depending on the value of the STKALIGN bit of the Configuration and Control Register CCR, the processor may also further adjust the stack pointer to make sure that the exception frame is aligned to a multiple of 8 bytes.

The reason behind saving the execution context is that accepting and handling an exception shall not prevent the processor from going back to its current activity at a later time. On the contrary, for the reasons described in Section 3.1, most exceptions shall be handled transparently with respect to any code that happens to be executing when they arrive. This is particularly true for interrupts and, more in general, any other exception requests that occur asynchronously with respect to current processor activities, because they are very often totally unrelated to them. The choice of which part of the context is saved is instead motivated by the crucial goal of making the resulting stack layout compatible with the ARM Architecture Procedure Calling Standard (AAPCS) [12]. In particular, upon exception entry the processor saves the caller-saved portion of the integer context. If the exception handler makes use of other parts of the context, it becomes its own responsibility to save them appropriately to prevent context corruption.

In this way, any AAPCS-compliant function can be used as an exception handler, an especially important feature when exception handlers are written in a high-level language because compilers generate AAPCS-compliant code by default. Hence, they can also generate exception handling code without treating it as a special case.

To sum up, the processor hardware saves the context on the stack exactly like an AAPCS-complaint software procedure does when it is about to call another. As a result, an exception handler call performed by hardware is indistinguishable from a regular software-initiated procedure call, from the point of view of the procedure being called (often named *callee*).

This is a generic pattern to be realized on all architectures, regardless of how much assistance is provided by hardware. In the case of the Cortex-M, all caller-saved registers are automatically pushed on the stack by the processor itself, but on other architectures some assembly instructions may be needed to complement what hardware has initiated. In general, RTEMS saves enough context to call code written in C and proceeds with exception handling from there.

2. Set the link register LR to an appropriate *exception return* (EXC_RETURN) code. When an exception return code is loaded into the program counter PC, as part of a function epilogue, it directs the processor to perform an exception handler return sequence instead of an ordinary return from a procedure call.

Once again, this aspect of the exception entry sequence has been designed to permit any AAPCS-compliant function to be used directly as an exception handler. Indeed, the AAPCS stipulates that a procedure call must store the return address into the link register LR before setting the program counter PC to the procedure entry point. This is typically accomplished by executing a branch and link instruction BL with a PC-relative target address. Symmetrically, the callee returns by storing back into PC the value saved into LR at the time of the call. This can be done, for instance, by means of a branch and exchange instruction BX, using LR as argument.

The information provided by the EXC_RETURN code allows the processor to locate the exception frame to be restored, interpret it in the right way, and bring back the processor to the execution mode in effect when the exception was accepted. Namely, the 5 low-order bits of the EXC_RETURN code indicate:

- whether the processor was using MSP or PSP as stack pointer when the exception frame was created,
- the kind of exception frame to be restored, basic or extended,
- the execution mode the processor must go back to.

The 4 higher-order bits of EXC_RETURN are always set to 0xF to indicate the value being loaded into PC is indeed an EXC_RETURN code that the processor must handle specially, rather than a regular memory address. The remaining bits are currently unused.

It should be noted that the processor interprets the value being loaded into PC as a possible EXC_RETURN value only in specific cases, better detailed in [8, 9]. In other cases, for instance, when the PC is loaded while the processor is in thread mode (and hence, no exception handler can possibly be active), the value is taken literally, as a memory address. To avoid improper behavior if an EXC_RETURN value is mistakenly loaded into PC in these cases, the hardware protects the address range 0xF0000000–0xFFFFFFFF against instruction execution.

3. Switch to handler mode if the processor was in thread mode when the exception was accepted. If the processor was already handling another exception, it stays in handler mode. As explained previously, as a consequence of the mode switch, the processor may also start using a new stack. A noteworthy exception to this rule is the reset exception, which is handled in thread mode with the processor automatically configured to use the MSP.

Additional operations performed by the processor, not shown in Figure 4.2 for simplicity, include storing the exception number of the exception just accepted in the IPSR sub-register—which is part of the xPSR register—and updating several System Control Space (SCS) registers to reflect exception acceptance.

Another side effect of accepting an exception is that is clears the per-core state of any pending synchronization instructions, namely, LDREX and STREX. Therefore, any synchronization procedure using those instruction that was pending upon exception entry will need to be retried after execution resumes. This topic is extremely important for inter-core synchronization in multicore systems, to be presented in Chapter 13.

4. The very last action performed by the processor upon exception entry is to retrieve the target PC—that is, the entry point of the exception handler—from the exception vector table and jump to it. The exception vector table is a memory-resident array of 32-bit integers, holding memory addresses called exception vectors. More specifically, the i-th entry of the table holds the entry point of the handler for exception number i. No ambiguity can arise because exception numbers are fixed and, unlike priorities, are unique to each exception.

Only the first 16 exception numbers are explicitly defined by the architecture spec-ification. The total number of vectors is not fixed and depends on the number of exceptions supported by specific members of the Cortex-M family, as well as con-figuration and implementation options. The very first entry (at index 0) is used in a special way because no exception is ever assigned exception number zero. In-stead, this entry contains the initial value loaded into MSP upon reset.

The starting address of the vector table is held in a register called Vector Table Offset Register (VTOR). The 7 low-order bits of VTOR are reserved and are al-ways interpreted as zero, therefore the minimum alignment of the vector table in memory is 128 bytes. Further alignment constraints may come into effect in some cases, depending on the total number of entries in the table.

The VTOR register is reset to zero when the processor accepts a reset exception, before handling it. As a consequence, the initial values of PC and MSP upon re-set are not retrieved from the exception table in effect when the reset exception was accepted, but from the one at address zero. In all cases, the VTOR register determines the address the processor emits to access the exception vector table. Depending on the specific device, this may or may not be the physical address of the vector table in memory, which may be further changed by address remapping, external to the processor. In those cases, it is necessary to refer to the device— rather than the processor—documentation to ascertain which registers control the mapping and how.

For instance, in the NXP LPC17xx microcontroller family [90] it is possible to remap at address 0x00000000 (where the vector table begins by default) an image of the bootstrap ROM (which is normally accessible at physical address 0x1FFF0000) instead of the on-chip flash memory (which is normally mapped at address 0x00000000). Remapping is controlled by bit 0 of the device-specific MEMMAP register.

The case of nested exceptions follows the same general rules, namely, the pro-cessor pushes the exception frame that contains the execution context of the current exception handler on the current stack using the active stack pointer, which will nec-essarily be the MSP. This course of action enables the last-in, first-out saving and restoration of exception frames, in the same way as ordinary stack frames are han-dled in regular function calls. The mechanism guarantees that exception handlers are nested properly, correctly preserving their execution context in the process.

Return from an exception

As stated previously, the processor starts an exception return sequence when an EXC_RETURN code is loaded into the PC at the end of an exception handler, with the ultimate goal of transparently resuming the activity it was performing when the exception became active. In order to do this, the processor must basically revert all the steps depicted in Figure 4.2.

1. First of all, the processor examines and interprets the EXC_RETURN code to deter-mine which stack pointer (MSP or PSP) it shall use to locate the exception frame

to be restored, assess the structure and contents of the exception frame (basic or extended), and decide the processor mode (handler or thread) to be entered after restoration.

2. Then, the processor performs several integrity checks, fully illustrated in [8, 9], to ensure that returning from an exception is legal considering the current execution context. For instance, the exception currently being handled and whose number had been recorded in IPSR upon exception entry, must be active in order to legitimately return from it. Furthermore, the processor must currently be executing in handler mode and, if it is about to return to thread execution mode, the value to be restored into IPSR must be zero, thus indicating that no exceptions are active any more. Any failed check raises a *UsageFault* exception, which is then handled as usual.

3. Finally, the processor restores the contents of the exception frame located as described in the previous steps. Among other things, the context includes the exception number being handled when the current exception was accepted, in the IPSR sub-register of xPSR, and the PC at which the exception being concluded was accepted.

A direct consequence of context restoration is that, in the case of nested exceptions, the processor resumes execution from where it was previously interrupted and the IPSR contains the exception number of the exception whose handling is being resumed.

Instead, if the exception handler from which the processor is returning is the last of a chain of nested exceptions (or the only one, in case exceptions were not nested at all), the IPSR that was formerly stored in the exception stack being restored is zero, and this is also the value that must be loaded into the register upon exception return to signal that the processor is no longer handling any exceptions.

4.2.3   RTEMS CONTEXT SWITCH AND EXCEPTION HANDLING

In Section 3.2.2 we introduced the general concept of task control block (TCB), saying it plays a key role in multitasking operating systems and, more specifically, in the context switching mechanism. Accordingly, Figure 3.3 portrays TCB contents in an abstract way. In this section, we will discuss in more detail how RTEMS implements the portion of TCB related to context switching on the Cortex-M architecture.

In order to show how context switching and exception handling interact, we will also illustrate how RTEMS makes use of software-triggered exceptions to reschedule a core after an interrupt. However, although RTEMS makes use of these exceptions to facilitate and streamline context switching, it still performs all context switches from within a task context. This approach makes the design more consistent and uniform because there is no longer any difference between a context switch triggered by an explicit, voluntary action performed by a task (for instance, when it blocks), and an involuntary context switch (caused by an interrupt handler that readied a higher-priority task).

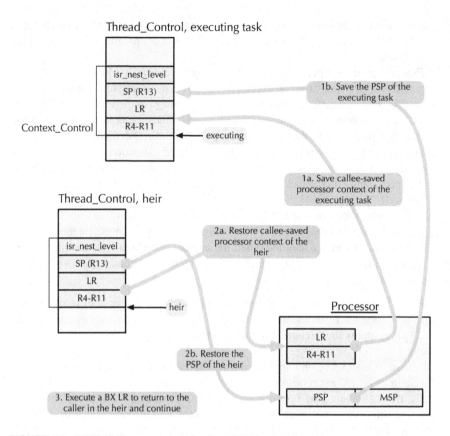

FIGURE 4.3   RTEMS Context switch for the ARM Cortex-M processors.

As shown at the top of Figure 4.3, an RTEMS TCB is represented by a `Thread_Control` data structure, which embeds an architecture-dependent sub-structure called `Context_Control`. The sub-structure contains all the callee-saved registers specified in the Cortex-M AAPCS [12]. Caller-saved registers are not stored within it because the RTEMS function in charge of context switching, `_CPU_Context_switch`, is an AAPCS-compliant C function. Hence, when any higher-level function calls it to perform a context switch, it is the compiler's responsibility to properly save these registers beforehand, typically on the task stack, and restore them after `_CPU_Context_switch` returns. In summary, during a context switch:

- If the calling function needs to preserve the content of some caller-saved registers of the executing task, it is its responsibility to save them on the task stack, by means of compiler-generated code, before calling the context switch function `_CPU_Context_switch`.
- This function is given two pointers as arguments, `executing` and `heir`.

The first points to the `Context_Control` of the currently executing task, the second to the `Context_Control` of the next task to be executed. This is called the *heir* in RTEMS documentation, in a reference to who is next task to sit on the processor "throne".

- The context switch function preserves all callee-saved registers plus the PSP, by saving them in the `Context_Control` of the executing task.
- In order to restore the context of the heir, `_CPU_Context_switch` loads all callee-saved registers plus the PSP from the `Context_Control` within the heir's TCB and returns to the caller.
- The compiler-generated function call epilogue code will then restore caller-saved registers, if they were live at the time of the call. Since `_CPU_Context_switch` sets the PSP to point to the heir's stack, the restoration will take place from there, as it should.

Figure 4.3 summarizes the main steps performed by `_CPU_Context_switch`. One aspect shown in the figure and not yet discussed for simplicity is the fact that `Context_Control` includes a field called `isr_nest_level`. This field is not part of the processor context and corresponds to a field with the same name that RTEMS maintains in the per-core data structure `Per_CPU_Control`. It is saved from the data structure into the TCB of the executing thread and restored into the data structure from the TCB of the heir thread alongside the other parts of the context discussed previously. It is an integer that represents the per-core current interrupt service routine (ISR) nesting level.

So far we described how a context switch is implemented when it is triggered by the executing task in a synchronous way by explicitly calling `_CPU_Context_switch`, either directly or, more ofter than not, by means of higher-level functions. This typically happens when the executing task voluntarily yields the processor or when it executes a blocking synchronization primitive, thus moving into the *blocked* state of the task state diagram, as explained in Section 3.2.2.

Another important reason to perform a context switch is to preempt a lower-priority task when a formerly blocked higher-priority task becomes ready for execution. This happens as a result of an event, for instance, a device interrupt. Timed wait operations belong to this category, too, because also in that case the waiting task is unblocked as a result of a timer interrupt. In RTEMS, like in other operating systems, this goal is accomplished by making the hardware-assisted exception entry/exit mechanism depicted in Figure 4.2 and the software-controlled context switch shown in Figure 4.3 work together.

```
void _ARMV7M_Interrupt_service_leave( void )
{
  Per_CPU_Control *cpu_self = _Per_CPU_Get();

  --cpu_self->thread_dispatch_disable_level;
  --cpu_self->isr_nest_level;

  /*
   * Optimistically activate a pendable service call if a thread dispatch is
   * necessary.  The _ARMV7M_Pendable_service_call() will check that a thread
   * dispatch is allowed.
```

```
  */
  if ( cpu_self->dispatch_necessary ) {
    _ARMV7M_SCB->icsr = ARMV7M_SCB_ICSR_PENDSVSET;
  }
}
```

Firstly, the function _ARMV7M_Interrupt_service_leave, to be called while leaving an exception handler, checks whether a task *dispatch* is necessary, by checking the per-core flag dispatch_necessary. This flag is set by higher-level synchronization primitives when they detect that a task with a priority higher than the currently executing task has been woken up. In RTEMS, task dispatching is the sequence of two separate activities, organized in a manager versus worker fashion:

- The execution of the scheduling algorithm makes the decision of what the next task to be executed will be and plays the management role. The implementation of the scheduling algorithm is also portable and architecture-independent.
- After the manager has designated the heir, the context switch code performs the context switch from the currently executing task to the heir. This piece of code is the worker, and also embeds all architecture-dependencies of task dispatching as a whole.

The same division of duties and sequence of operations also take place when a task voluntarily blocks and the operating system must necessarily choose its heir. If a dispatch is needed, the function _ARMV7M_Interrupt_service_leave triggers a *PendSV* exception, described in Section 4.2.1. RTEMS configures this exception to have the lowest exception priority in the whole system. Therefore, it will stay pending and will be serviced only when no other higher-priority exceptions are being handled.

As shown on the top left part of Figure 4.4, when the *PendSV* exception becomes active the processor saves on the task stack an exception frame according to the general exception entry mechanism illustrated in Figure 4.2. The PC in this exception frame points to the task instruction that has been interrupted by *PendSV* exception handling. This is also the PC from which the execution of the current task must eventually resume.

```
void _ARMV7M_Pendable_service_call( void )
{
  [...]

  {
    ARMV7M_Exception_frame *ef;

    cpu_self->isr_nest_level = 1;

    _ARMV7M_SCB->icsr = ARMV7M_SCB_ICSR_PENDSVCLR;
    _ARMV7M_Trigger_lazy_floating_point_context_save();

    ef = (ARMV7M_Exception_frame *) _ARMV7M_Get_PSP();
    --ef;
    _ARMV7M_Set_PSP( (uint32_t) ef );

    /*
```

Task Stack

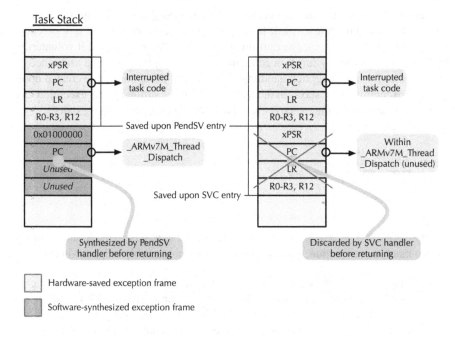

FIGURE 4.4   RTEMS task stack diagrams while dispatching after an IRQ.

```
 * According to "ARMv7-M Architecture Reference Manual" section B1.5.6
 * "Exception entry behavior" the return address is half-word aligned.
 */
ef->register_pc = (void *)
  ((uintptr_t) _ARMV7M_Thread_dispatch & ~((uintptr_t) 1));

ef->register_xpsr = 0x01000000U;
  }
}
```

The processor then executes _ARMV7M_Pendable_service_call, the C
function registered as *PendSV* exception handler, in handler mode. After ensuring
that task dispatch is allowed (by means of a fragment of code omitted in the previous listing), this function synthesizes a new exception frame, represented by the data
type ARMV7M_Exception_frame, pushes it on the stack by decrementing the PSP
(on this architecture, stacks grow towards lower addresses), and returns. This corresponds to the darker gray exception frame visible on the bottom left of Figure 4.4.

The exception frame's PC points to the _ARMV7M_Thread_dispatch function and contains a default xPSR, set to 0x01000000. The other fields of the exception frame are not set explicitly to save time, because they are not going to be
used in the following. When the *PendSV* exception handler returns, the LR register
still contains the EXC_RETURN code stored by hardware. As a result, the processor
restores the software-synthesized exception frame and resumes execution from the
_ARMV7M_Thread_dispatch function in thread mode.

This is a key point because the net result is that, immediately after one or more interrupts that triggered a reschedule have been serviced, the interrupted task executes the `_ARMV7M_Thread_dispatch` function exactly "as if" it called it voluntarily, by means of an ordinary function call, although it took place with the assistance of this peculiar form of exception handling. The original exception frame created upon *PendSV* exception entry stays on the task stack, so that ordinary task execution can be resumed at a later time, when it eventually goes back to the running state.

```
static void __attribute__((naked)) _ARMV7M_Thread_dispatch( void )
{
  __asm__ volatile (
    "bl _Thread_Dispatch\n"
    /* FIXME: SVC, binutils bug */
    ".short 0xdf00\n"
    "nop\n"
  );
}
```

The `_ARMV7M_Thread_dispatch` function is extremely simple and consists of two machine instruction, specified by means of an assembly language insert:

- First, it calls `_Thread_Dispatch`. This RTEMS function executes the scheduling algorithm in a completely architecture-independent way to calculate the heir of the currently executing task. If necessary, this management action is followed by an architecture-dependent context switch realized, as described previously, by calling the `_CPU_Context_switch` function.

  Since all these operations are performed in thread mode, the processor behavior is exactly the same as for the synchronous context switch discussed previously, and the higher-layer code can be kept unaware of the distinction.

  If a context switch takes place, the current task no longer continues its execution within `_Thread_Dispatch` until another context switch brings it back to the running state.

- Secondly, when `_Thread_Dispatch` returns, it executes a SVC 0 instruction that triggers a synchronous exception. As a result, the hardware pushes on the stack a new exception frame, depicted at the bottom right of Figure 4.4. The PC of this frame points to the instruction that follows the SVC within `_ARMV7M_Thread_dispatch`.

  As a side note, the assembly language insert encodes SVC 0 in hexadecimal because some versions of the assembler (part of the GNU binutils toolchain package) have a bug that prevents them from recognizing the SVC mnemonic correctly.

```
void _ARMV7M_Supervisor_call( void )
{
  Per_CPU_Control *cpu_self = _Per_CPU_Get();
  ARMV7M_Exception_frame *ef;

  _ARMV7M_Trigger_lazy_floating_point_context_save();
```

```
ef = (ARMV7M_Exception_frame *) _ARMV7M_Get_PSP();
++ef;
_ARMV7M_Set_PSP( (uint32_t) ef );

cpu_self->isr_nest_level = 0;

if ( cpu_self->dispatch_necessary ) {
  _ARMV7M_Pendable_service_call();
}
}
```

Then, the processor executes the SVC handler, _ARMV7M_Supervisor_call, in handler mode. This function discards the exception frame pushed by hardware, adjusts the PSP accordingly and, unless another dispatch has become necessary in the meantime, returns. Since the LR register contains an EXC_RETURN code also in this case, the one calculated while entering the SVC handler, the hardware determines it is returning from an exception and restores the exception frame previously saved during *PendSV* exception entry.

On the other hand, if a new dispatch has become necessary—because, for instance, other interrupts have been handled in the meantime and other, even higher-priority tasks have been woken up—the whole process is repeated by calling the *PendSV* exception handler again.

In two cases, the EXC_RETURN code calculated by hardware while pushing a certain exception frame is used to restore a different exception frame. More specifically:

1. The EXC_RETURN code calculated upon entering the *PendSV* exception handler is used to restore the software-synthesized exception frame.
2. The EXC_RETURN code calculated upon entering the SVC exception handler is used to restore the exception frame saved while entering the *PendSV* exception handler.

The whole mechanism still works because RTEMS ensures that exception frame formats are nevertheless the same, and hence, they are indistinguishable from each other.

As a final remark, we can confirm that RTEMS saves all processor registers (both caller-saved and callee-saved) upon preemption, thus making it completely transparent. This is because:

- Caller-saved registers are saved by hardware while entering the *PendSV* exception handler, within an exception frame, as shown in Figure 4.2.
- Callee-saved registers are saved by the context switch function, in the Context_Control structure of the task TCB, as depicted in Figure 4.3.

The Context_Control structure also contains the task PSP. When restored, it enables the processor to properly retrieve and restore also the exception frame content from the task stack.

4.2.4   INTERRUPTS IN SCHEDULABILITY ANALYSIS

If we compare a contemporary, fully prioritized exception handling mechanism, such as the one described in Section 4.2.1, with the assumptions of schedulability analysis set out in Section 4.1, we discern some important analogies.

Firstly, an interrupt handler can be seen as a sporadic pseudo-task, whose activation is triggered by the corresponding interrupt source. Then, the arrival of an interrupt request is an external event that moves the corresponding pseudo-task to the *ready* state of the task state diagram. The priority of this pseudo-task is dictated by the priority of the interrupt source and is fixed, but is always implicitly higher than any other regular task in the system, because the arrival of an interrupt request immediately preempts any regular task that had been executing.

Secondly, the relative priority of different interrupt sources determines whether or not the arrival of a fresh interrupt request would preempt a currently executing interrupt handler, giving rise to interrupt nesting. If no nesting takes place, the interrupt request stays pending until the processor execution priority level allows it to be accepted and become active. This mechanism is equivalent to an ordinary task staying in the *ready* state of the task state diagram while a higher-priority task is executing, until the scheduler moves it to *running* state, the only difference being that the scheduling decision is taken by hardware instead of software.

Therefore, in principle, some of the schedulability analysis techniques described in Section 4.1, most notably response time analysis, can be used to assess the impact of interrupt handling on tasks response times. This is true provided it is possible to calculate, measure, or estimate the worst-case execution time of each interrupt handler (which then becomes the C_i of the corresponding pseudo-task) and several other practical requirements are met.

Interrupt arrival rate

An important hypothesis in the definition of sporadic task is that its minimum interarrival time is known. This hypothesis enabled us to conservatively consider sporadic tasks as periodic tasks with a period equal to their minimum interarrival time. In some cases, this hypothesis is implicitly satisfied because devices often generate an interrupt only as a reaction to some software-issued command or inherently guarantee a minimum interarrival time anyway, like mechanical buttons if we ignore bounces.

As an example of the first category of devices, a hard-disk controller generates an end-of-transfer interrupt only after receiving and processing a data transfer command from its device driver. Moreover, it will not generate further interrupts of the same kind afterwards, unless the device driver issues another command to it.

As a consequence, the interrupt generation rate, which determines the minimum interarrival time of the interrupt handler, cannot exceed the rate at which the device driver sends commands to the device. If, for instance, a periodic task with period T_i is in charge of preparing and issuing one of those commands on each activation, we can safely use the same T_i as the minimum interarrival time of the interrupt handling sporadic pseudo-task. Besides being convenient for schedulability analysis, this approach is also useful to contain the maximum utilization a certain device can impose on the system, because we can cap the interrupt arrival rate by choosing a T_i as a suitable trade-off between data transfer bandwidth and the system load itself.

Unfortunately, this is not viable in all cases, mainly because some devices may generate interrupt requests independently of any software action. This is typical, for example, of most network controllers, which generate an interrupt request whenever they receive an incoming frame. In turn, unless the network is time-triggered or works according to a Time Division Multiplexed Access (TDMA) paradigm, the arrival time of incoming frames is basically unpredictable and uncorrelated with any local task activities.

Generally speaking, leaving the interrupt rate unchecked is inconvenient and often dangerous from several different points of view.

- As hinted previously, schedulability analysis becomes hardly possible. This is worrying not only from a theoretical point of view, but also from a practical standpoint, because worst-case task response times can then be assessed only by testing the system, while relying on the assumption that some test scenarios can indeed reproduce the worst-case interrupt load the system is going to face in practice.
- Even if some other physical characteristics of the system may also limit the maximum interrupt arrival rate as a side effect, the resulting load can nevertheless be too high to be sustainable. For instance, even a Controller Area Network interface running at a relatively modest rate of 1 Mb/s can still receive one frame every $47\,\mu$s, and potentially generate interrupt requests at the same rate in the worst case. This is because on that kind of network the minimum legal frame length, including inter-frame spacing, is 47 bits.
- It is also important to remember that interrupt handlers have a higher priority than any ordinary tasks in the system. Therefore, the execution time C_h of an interrupt handling pseudo-task τ_h has a direct, important impact on the response time of all ordinary tasks in the system. This can be clearly seen by referring back to the main response time analysis equation (4.5) and observing that C_h certainly contributes to the R_i of all ordinary tasks τ_i, because it surely is $h \in \mathrm{hp}(i)$ for all i.
- As a consequence, when a system is swamped with interrupts, it may have little time left to perform "real work." Moreover, interrupts have a negative impact on cache performance, which also slows other activities down. A related issue caused by high-speed devices that are able to perform direct memory access, as is typical of network interfaces, is memory bandwidth saturation, which slows down memory accesses issued by the processor.
- Last, but not least, unchecked interrupt handlers may open the door to denial of service attacks. For instance, if an attacker can get access to the CAN network mentioned previously, it can easily "flood" the system with minimum-size frames. Whether or not this leads to any service disruption then depends on how well the system is able to tolerate an interrupt arrival rate it potentially has not been designed and tested for.

The range of techniques to limit interrupt arrival rate can be divided into two broad categories: hardware and software-based.

- Depending on its sophistication, the device itself may offer a way to reduce and place an upper bound on its own interrupt generation rate. For instance, some CAN controllers implement rather sophisticated filters on the *identifier*—the part of a CAN message that uniquely identifies its contents—and can automatically store incoming frames into different device or memory-resident mailboxes, also chosen depending on their identifier.

  On one hand, this reduces the overall interrupt rate because the controller, when suitably programmed, can autonomously discard incoming frames that are of no interest to the software, without generating any interrupt. On the other hand, this also restricts the worst-case interrupt rate because, for each mailbox, these controllers put in place automatic message replacement and interrupt hold-off policies. For instance, they may keep only the most recent message destined to a certain mailbox when software is unable to process them all, and refrain from generating a new interrupt request until a programmable hold-off time has elapsed since the previons one.
- Similar mechanisms are also available on Ethernet controllers, especially the ones operating at or beyond 100 Mb/s. In general, faster network interfaces tend to be more sophisticated and offload the processor more. For instance, they are usually able to store incoming frames and fetch outgoing frames directly to and from memory-resident buffers they share with the device driver. In this case, available memory bus bandwidth becomes a factor to be taken into account. As an example, on a state-of-the-art embedded system RTEMS with its new protocol stack (see Chapter 10) can sustain 1 Gb/s at around 30-40% processor utilization.
- Most processors allow software to individually enable or disable each interrupt source. Therefore, the interrupt handler can disable its own interrupt source after setting a timer. In this way, no further interrupts from that source will be generated and handled until the timer expires. Upon timer expiration, interrupts are enabled again, thus imitating in software the hardware-based hold-off mechanism mentioned previously. With respect to the hardware-based approach, there are two shortcomings worth noting. Firstly, the software-based approach entails additional overhead, due to the work to be performed upon timer expiration. Depending on the way timers are implemented, this may also imply extra interrupts from the timer itself. Secondly, some devices may enter an error condition if their interrupt requests are not serviced timely. Depending on the device, the error condition may imply data loss—for instance, due to overflows of the receive buffers of a network interface—and, in extreme cases, the device may stop working altogether.

An alternative approach that addresses the second shortcoming is to keep device interrupts enabled at all times, but handle them differently depending on the circumstances. Namely, when the interrupt handler detects that an interrupt came "too close" to the previous one, it will perform only the minimal amount of house-keeping needed for the device to work correctly and nothing else, thus saving processor time and reducing the interference on other tasks. Actually, it turns out this is merely a special case of a more general technique, often called two-stage interrupt handling, to be discussed next.

Two-stage interrupt handling

In Section 3.2.3, we saw that a key part of a real-time scheduling algorithm is an appropriate priority assignment scheme. According to this scheme, the priority of a task shall depend on some properties of the task itself, for instance, its period for RM or its relative deadline for the deadline monotonic priority order.

This requirement is not of concern for ordinary tasks because software can set their priority at will. Instead, as described in Section 4.2.1, although interrupt handlers do have a software-assigned priority, they have nevertheless a priority higher than any ordinary task in the system. In other words, tasks are partitioned in two subsets: ordinary tasks and interrupt handlers. Although software can set task priorities at will *within* each subset, the second subset always has a higher priority than the first.

As a consequence, it may be impossible to fully adhere to the RM or deadline monotonic priority assignment, with two important side effects:

- Some schedulability analysis methods, like response time analysis (see Section 4.1.2), can still be used because they work with any priority assignment. Instead, simpler techniques, like utilization-based schedulability tests (see Section 4.1.1), are no longer applicable.
- The optimality theorems discussed in Section 4.1 obviously do not apply to priority assignments that do not satisfy their hypotheses. Moreover, by intuition, the further a given priority assignment is from an optimal one, the worse the system performs.

A simple workaround that alleviates these issues is two-stage interrupt handling, in which interrupt handling activities are split into two parts:

1. Time-critical activities are still performed in the interrupt handler, which can be modeled as before as a pseudo-task τ_k with worst-case execution time C_k. Afterwards, the interrupt handler wakes up an ordinary task τ_d.
2. Task τ_d runs at a lower priority than τ_k and takes care of less time-critical activities in a deferred way, with a worst-case execution time C_d.

Overall, two-stage interrupt handling can be modeled as two sporadic tasks τ_k and τ_d for each interrupt source. Both tasks have the same period $T_k = T_d$, equal to the minimum interrupt interarrival time, and known worst-case execution times C_k and C_d. Thus, their impact on the system can easily be calculated, for instance, with response time analysis.

The total interrupt handling time $C_k + C_d$ is still the same as for the one-stage monolithic interrupt handling approach discussed previously, and probably worse because of the additional synchronization overhead between the interrupt handler and the ordinary task. However, this approach is much more flexible because it enables programmers to choose (within certain limits) the most appropriate trade-off between code executed at high and low priorities, that is, between C_k and C_d.

Let us assume, as an example, that programmers would like to use the RM priority assignment but the minimum possible interrupt handling priority of τ_k is too high for the corresponding T_k. Although they cannot lower the priority of τ_k further, they can still set the priority of τ_d depending on T_d as RM requires. Then, they can make C_k as small as possible by moving most of the processing into C_d.

In this way, task τ_d is appropriately positioned in the RM priority hierarchy. Task τ_k is not, but its adverse impact on the system has been reduced by reducing its C_k. In many cases, it is possible to shrink the interrupt handler until it only invokes the synchronization primitives needed to wake τ_d up, leading to a very small C_k and bringing system behavior very close to the optimality guaranteed by RM.

4.3   SUMMARY

This chapter introduced a couple of popular schedulability analysis techniques, namely utilization-based tests and response time analysis (RTA). In the case of response time analysis, it was also shown how to refine and extend the analysis, starting from a basic periodic task model and then going towards a more complex model that incorporates task interactions, self-suspension, and interrupt handling.

Moving from theoretical to more practical considerations, the chapter discussed how interrupt handling and, more generally, exception handling is carried out in practice, using RTEMS as an example. Given its close analogy to exception handling, we also illustrated how RTEMS performs a context switch.

Part II

Task Management and Timekeeping

5 Task Management and Timekeeping, Classic API

CONTENTS

This chapter illustrates the general concepts of task management and timekeeping, focusing on the RTEMS scheduling algorithms for single-core systems, along with the facilities provided by RTEMS to manipulate tasks and account for the passage of time through its Classic Applications Programming Interface (API).

Chapter 6 will focus on how the same facilities are embodied in the POSIX API. A short comparison between these two APIs, also given in the present chapter, shows their relative advantages and disadvantages, and also highlights which facilities are available only in one API and not in the other.

The chapter ends with a description of some lower-level aspects of interrupt handling on single-core systems by means of the RTEMS Interrupt Manager. Although these aspects are generally outside the scope of general-purpose application programmers, they are often essential in embedded systems. Further information about multicore systems will be provided in Chapters 13.

5.1 TASK MANAGEMENT BASICS

The concept of task management encompasses all the operating system functions and interfaces that control a task through its entire lifetime, summarized in the task state diagram of Figure 3.4. Even more generally, we can say that task management represents the practical embodiment of the concurrent programming concepts discussed in Chapter 3 by RTEMS.

In a somewhat hierarchical view, at the top level there are operating system functions to create a new task and make it eligible for execution. These same functions

TABLE 5.1

Single-Core Scheduling Algorithms of RTEMS

Configuration macro (CONFIGURE_SCHEDULER_...)	Default name	Description
PRIORITY	"UPD "	Deterministic Priority Scheduler (default) [84]
SIMPLE	"UPS "	Simple Priority Scheduler [84]
EDF	"UEDF"	Earliest Deadline First Scheduler [84]
CBS	"UCBS"	Constant Bandwidth Server Scheduler [1]

also enable users to specify task scheduling parameters that, in turn, drive the operating system's scheduling algorithms, as described in Section 3.2. As in most other operating systems, another set of RTEMS functions is available to inspect and change the scheduling parameters of a task after its creation. Additional top-level functions exist to temporarily suspend and then resume a task, and to terminate it.

Below the surface, a number of RTEMS modules not directly accessible to end-users implement a variety of real-time scheduling algorithms [125]. Among them, the ones designed for single-core systems will be presented in Section 5.2 of this chapter, along with the *Scheduler Manager*, the RTEMS component responsible for managing schedulers and, on multicore systems, maintain the association of schedulers with the cores they manage and operate upon.

The algorithms are conceptually very close to the ones discussed in Chapter 3 and analyzed in Chapter 4, with additional features provided to enhance their practical applicability and usefulness. Instead, scheduling algorithms for multicore systems will be one of the topics of Chapter 13.

At the bottom of the hierarchy, a set of partly architecture-dependent RTEMS functions are responsible for putting scheduling decisions into practice by means of task dispatching and context switching. This is done according to the general scheme outlined, for instance, in Section 4.2 for the ARM Cortex-M architecture.

RTEMS provides access to the top level of the hierarchy just introduced by means of two primary application programming interfaces (APIs) named the *Classic* and *POSIX* interfaces and located at cpukit/rtems and cpukit/posix within the RTEMS source code, respectively. Section 5.3 summarizes the commonalities and differences between the two APIs. The rest of this chapter will focus on the classic API for task management and timekeeping, while Chapter 6 will be entirely devoted to the POSIX interface.

5.2 SCHEDULER MANAGER AND SINGLE-CORE SCHEDULING ALGORITHMS

A standard distribution of RTEMS provides the off-the-shelf single-core priority-based schedulers listed in Table 5.1. In addition, thanks to the modular, plugin-based

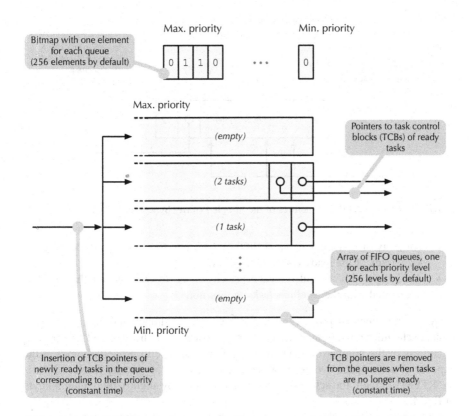

FIGURE 5.1 Abstract data structures of the Deterministic Priority Scheduler (DPS).

framework for schedulers of RTEMS, users can implement their own scheduling algorithms if necessary.

As it will be better described in Chapter 13, the same framework also supports multiple scheduler instances on a multicore system, each using a possibly different scheduling algorithm and governing a subset of the available cores.

On a single-core system, the specific scheduler to be used is selected by defining the macro listed in the leftmost column of the table in the RTEMS configuration, according to the general procedure described in Section 2.4.2. For instance, defining the configuration macro CONFIGURE_SCHEDULER_EDF selects the Earliest Deadline First (EDF) scheduler. No definitions are needed to select the Deterministic Priority Scheduler (DPS) because it is enabled by default.

The DPS is a preemptive, fixed-priority scheduler suitable to implement, for instance, the Rate Monotonic priority assignment (RM), Deadline Monotonic Priority Ordering (DMPO), and Optimal Priority Assignment (OPA), all described in Section 3.2.3.

For what concerns its practical implementation, it makes use of the abstract data structures depicted in Figure 5.1:

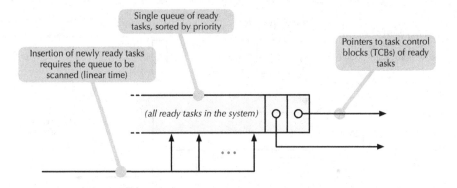

FIGURE 5.2   Abstract data structures of the Simple Priority Scheduler (SPS).

- An array of first-in, first-out (FIFO) *queues*, one for each priority level, to keep track of the ready tasks at that level.
- A *bitmap*, with one bit for each priority level, to tell whether the FIFO associated with the level has tasks in it or not.

When implemented properly, these data structures support the execution of all typical scheduler operations in a deterministic (that is, predictable and fixed) time, regardless of the number of tasks in the system. This useful property comes at the expense of some space overhead. When configured for 256 priority levels, which is the RTEMS default, the scheduler data structures occupy slightly more than 3 kbyte of RAM.

On small systems, which are anyway unable to support a large number of tasks, a more compact data structure may be preferable. This is provided by the Simple Priority Scheduler (SPS). It behaves the same as the DPS from the functional point of view but, as shown in Figure 5.2, it makes use of a single queue of ready tasks, implemented as a linked list. With respect to the DPS, most of the memory overhead due to queue headers clearly disappears.

However, since the scheduler must be able to pick the highest-priority ready task for execution efficiently, the queue must be kept sorted by task priority because, in this way, the selection of the highest-priority ready task can still be performed in constant time. On the contrary, the whole queue must be scanned whenever a task is inserted into it, with an execution time overhead that is proportional to the queue length, that is, the number of ready tasks.

In summary, the SPS is advantageous with respect to the DPS on small systems, in which the linear complexity of the SPS in the number of tasks is not an issue because the number of tasks is relatively small, whereas memory—especially the on-chip RAM of single-chip microcontrollers—often comes at a premium.

The EDF scheduler has clear advantages with respect to fixed-priority schedulers like the DPS and the SPS, from the point of view of schedulability analysis, as pointed out in Chapter 4. The downside comes from the necessity of keeping

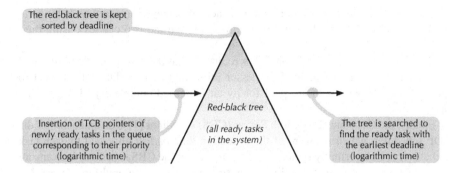

The red-black tree is kept sorted by deadline

Red-black tree

(all ready tasks in the system)

Insertion of TCB pointers of newly ready tasks in the queue corresponding to their priority (logarithmic time)

The tree is searched to find the ready task with the earliest deadline (logarithmic time)

FIGURE 5.3 Abstract data structures of the Earliest Deadline First (EDF) scheduler.

ready tasks ordered by *deadline* in an efficient manner, so that the scheduler can select for execution the task with the earliest deadline whenever the processor becomes available.

Although the problem might look similar to keeping ready tasks ordered by *priority*, there is a fundamental difference that makes the data structures used by the DPS, which are indeed very efficient, unsuitable for use. More specifically, as discussed previously, priorities have a limited range whereas deadlines in RTEMS are expressed in clock ticks as a 64-bit integer.

As a consequence, the memory overhead of any sorted data structure whose size is proportional to the range of the sort key, like the one of the DPS, may be appropriate for priorities but not for deadlines. For this reason, as shown in Figure 5.3, the EDF scheduler adopts a red-black tree [53] as its main data structure to keep track of all ready tasks in the system.

A red-black tree is similar to a binary search tree but insertion and deletion operations keep the tree balanced, so that the complexity of all main operations on the tree (search, insert and delete) is a logarithmic function of the number of nodes in the tree, which is the number of ready tasks in the system. Space overhead is proportional to the number of nodes.

The RTEMS implementation of the EDF scheduler supports two classes of tasks, the ones that declared a deadline (*foreground* tasks) and the ones that did not (*background* tasks). For background tasks, the EDF scheduler falls back to a fixed priority scheduler and executes them according to their priority, exactly like the DPS would do.

The scheduler enforces a strict hierarchy between the two classes, namely, all background tasks have a lower importance than any of the foreground tasks. In other words, background tasks are picked for execution, according to their priority, only if no foreground tasks are ready at the moment.

The deadlines used by this scheduler are declared by means of the RTEMS Rate Monotonic Manager, whose API will be described in Section 5.5, and are always *implicit*, that is, are assumed to be equal to the task period. A task belongs to the foreground class when it has a deadline declared using the Rate Monotonic

Manager. Instead, if a task's deadline is canceled or never declared, it belongs to the background class. Tasks can freely move between the two classes during their lifetime.

By contrasting Figures 5.1 and 5.3, we can see that the additional flexibility and the advantages in terms of total processor utilization of the EDF with respect to the DPS (see Section 4.1) are paid for by an increased complexity of scheduling operations.

Finally, the Constant Bandwidth Server (CBS) scheduler [1] is an extension of the EDF scheduler, which is aware of the execution time *budget* allocated to each task and enforces it. Informally speaking, the main goal of the CBS scheduler is to ensure that a task will not miss a deadline because other tasks consumed more processor time than they should.

Hence, it avoids the so-called *domino effect* that characterizes the EDF scheduler in a transient overload condition. A thorough description of the RTEMS API for this scheduler is beyond the scope of this book. Readers are referred to the RTEMS API manual [105] for more information about it.

Fine-grained scheduler control

Besides the coarse-grained choice among several priority-based scheduling algorithms and implementations with different trade-offs between time and space overheads, which affects all tasks managed by a certain scheduler, RTEMS also provides three finer-grained ways to control scheduler behavior, selected on a task-by-task basis.

The theoretical definitions given in Chapter 3 specify that all schedulers provided by RTEMS shall be *preemptive*, that is, they shall immediately take away a core from an executing task when a higher-priority task becomes ready for execution and no other suitable cores are available. Moreover, the scheduling analysis methods presented in Chapter 4 assume that the underlying algorithms accommodate preemption.

Unless otherwise specified, RTEMS schedulers indeed preempt tasks as needed. However, preemption can be disabled on a task-by-task basis by setting the RTEMS_NO_PREEMPT flag in the *task mode* when creating the task. Tasks can also inspect and modify their mode during their lifetime, by means of the RTEMS interfaces that will be outlined in Section 5.4.

In both cases, the end result is that, when a scheduler dispatches for execution a task with the RTEMS_NO_PREEMPT flag set, that task will not be preempted even if a higher-priority task under the control of the same scheduler is later released. In terms of the task state diagram depicted in Figure 3.4, this implies that transition *d.* cannot take place involuntarily. The task can still voluntarily relinquish its right to execute by blocking or yielding.

Although preemption control can be very useful from the practical point of view in specific cases, there are several downsides worth remarking about it:

1. The no-preemption flag only affects the execution of a task *once* it has started executing. It does not affect in any way the decision of whether or not to execute

the task and when, which is still totally controlled by the scheduler, according to the scheduling algorithm and the task priority.

2. Similarly, setting the no-preemption flag of a task cannot prevent other tasks from being executed if the task voluntarily blocks or yields after the scheduler picks it for execution. Programmers should be aware that this constraint also encompasses sources of blocking deeply hidden in libraries—for instance, to avoid race conditions—which might not be immediately evident.

3. Preemption control is sometimes used as a very efficient way to ensure mutual exclusion among tasks in single-core systems. In multicore systems, this approach usually does not work correctly, for the same reasons outlined in Section 12.3 while discussing hand-crafted priority elevation.

Another way to adjust the basic behavior of a scheduler is to enable *timeslicing* for some tasks. Timeslicing is ortogonal to preemption and is a way to fairly divide the available processing resources among ready tasks with the *same priority*.

By default timeslicing is disabled, and hence, a running task will continue execution—even if there are other ready tasks with the same priority—until it voluntarily blocks or yields, or is preempted by a higher-priority task if preemption is enabled. When a running task has timeslicing enabled, that is, the RTEMS_TIMESLICE flag in the task mode is set, RTEMS will instead enforce a limit on the maximum amount of time the task can execute before the core is given to another ready task with its same priority, if any.

The maximum amount of execution time is called *timeslice* and is measured in RTEMS ticks. If there are no other ready tasks at the same priority as the running task at the end of a timeslice, RTEMS grants the running task another timeslice and keeps executing it. Otherwise, it places the running task at the end of the queue of ready tasks with its same priority and dispatches another task, in a round-robin fashion. The timeslice length is globally set by means of the configuration macro CONFIGURE_TICKS_PER_TIMESLICE and applies to all tasks in the system.

Finally, the running task can explicitly ask the scheduler to reconsider its scheduling decision by *yielding*, that is, by explicitly giving up the core it is executing on. In this case, RTEMS places the yielding task at the end of the queue of tasks with its same priority, executes the scheduling algorithm anew, and acts accordingly to the result.

Unlike other operating systems that provide a dedicated interface for yielding, in RTEMS this is accomplished by means of the ordinary timed-wait interface rtems_task_wake_after with the special time interval RTEMS_YIELD_PROCESSOR, as described in Section 5.4.

Programmers must also be aware that, considering the way yielding works, the yielding task *may or may not* lose the core it is running on, depending on whether there are other ready tasks with the same priority or not.

TABLE 5.2

RTEMS Scheduler Manager API

Function	Purpose
`ident` [1]	Get a scheduler identifier given its name
`ident_by_processor`	Get a scheduler identifier given a processor index
`ident_by_processor_set`	Get a scheduler identifier given a processor set
`get_processor_maximum`	Get the total number of cores configured in the system
`get_processor_set`	Get the processor set managed by a scheduler
`get_maximum_priority`	Get the maximum task priority of a scheduler
`add_processor`	Add a processor to the set managed by a scheduler
`remove_processor`	Remove a processor from the set managed by a scheduler

[1] All API function names start with the `rtems_scheduler_` prefix.

Retrieving a scheduler identifier

As described previously, schedulers are primarily selected and configured statically, by means of the RTEMS configuration. In addition, the Scheduler Manager provides a set of primitives to alter the initial configuration at runtime. They are summarized in Table 5.2 and discussed in the rest of this section.

The first group of functions listed in the table provides users several ways to get a *scheduler identifier*. A scheduler identifier is an object of type `rtems_id` that uniquely identifies a scheduler in the system and must be used whenever it is necessary to refer to the scheduler. The function:

```
rtems_status_code rtems_scheduler_ident(
    rtems_name name,
    rtems_id *id
);
```

stores in the location pointed by `id` the scheduler identifier of the scheduler called `name`. By default, the name of a scheduler is the 4-character name given in the second column of Table 5.1 and encoded as a `rtems_name` data type, but it can be changed by the user in the RTEMS configuration. This is especially important in multicore systems, in which there may be multiple instances of the same scheduler, and each of them should have a unique name.

The function also returns a status code that indicates whether it completed successfully or not. In particular:

RTEMS_SUCCESSFUL means that the function was successful and the location referenced by `id` contains a valid scheduler identifier.

RTEMS_INVALID_ADDRESS indicates that the function failed because `id` was a `NULL` pointer.

RTEMS_INVALID_NAME indicates that the scheduler name passed as argument was invalid.

Another way to retrieve a scheduler identifier is by means of one of the cores it manages. The function:

```
rtems_status_code rtems_scheduler_ident_by_processor(
    uint32_t cpu_index,
    rtems_id *id
);
```

stores into the location pointed by id the identifier of the scheduler in charge of the core whose index is cpu_index. The index of a core, called *CPU index* in RTEMS terminology, is an integer that uniquely identifies a core in the system. It ranges from zero to the total number of cores minus one, included. In a single-core system, the only valid CPU index is therefore zero. The function rtems_scheduler_get_processor_maximum returns the total number of cores configured in the system.

The possible return values of rtems_scheduler_ident_by_processor are:

RTEMS_SUCCESSFUL means that the function was successful and the location referenced by id contains a valid scheduler identifier.

RTEMS_INVALID_ADDRESS indicates that the function failed because id was a NULL pointer.

RTEMS_INVALID_NAME indicates that the CPU index cpu_index passed as argument was invalid.

RTEMS_INCORRECT_STATE means that the CPU index cpu_index was valid, but the corresponding core was not currently assigned to any scheduler.

Similarly, the function:

```
rtems_status_code rtems_scheduler_ident_by_processor_set(
    size_t cpusetsize,
    const cpu_set_t *cpuset,
    rtems_id *id
);
```

stores into the location pointed by id the identifier of the scheduler in charge of the CPU set referenced by cpuset. The argument cpusetsize specifies the size of the CPU set in bytes. The highest-numbered online core in the set is taken as a reference to identify the scheduler.

A CPU set is represented by the data type cpu_set_t and is always passed by reference. The set of functions listed in Table 5.3 shall be used to manipulate a CPU set without delving into implementation details.

TABLE 5.3

RTEMS Functions to Manipulate CPU Sets

Macro	Purpose
CPU_ZERO	Initialize an empty CPU set
CPU_FILL	Initialize a CPU set that contains all cores in the system
CPU_COPY	Copy a CPU set into another
CPU_SET	Add a core to a set, given its CPU index
CPU_CLR	Remove a core from a set, given its CPU index
CPU_ISSET	Check whether a core belongs to a set, given its CPU index
CPU_COUNT	Return the number of cores that belong to a CPU set
CPU_EMPTY	Return true if (and only if) a CPU set is empty
CPU_EQUAL	Return true if (and only if) two CPU sets are identical
CPU_CMP	Alternative name of CPU_EQUAL
CPU_AND	Calculate the intersection of two CPU sets
CPU_OR	Calculate the union of two CPU sets
CPU_XOR	Calculate the set of cores that belong to one of two CPU sets, but not both [1] [2]
CPU_NAND	Calculate the set of cores that do not belong to the intersection of two CPU sets [2]

[1] This operation is called *disjunctive union* or *symmetric difference*.

[2] These two operations are not the same because the result of CPU_NAND includes cores that belong to neither of the two CPU sets, whereas the result of CPU_XOR does not.

Querying scheduler characteristics

The second group of functions enables the user to query certain characteristics of a scheduler, given its identifier. Namely, the function:

```
rtems_status_code rtems_scheduler_get_processor_set(
    rtems_id scheduler_id,
    size_t cpusetsize,
    cpu_set_t *cpuset
);
```

returns the set of cores currently associated with the scheduler identified by scheduler_id. The set is stored in an object of type cpu_set_t referenced by cpuset, whose size in bytes is specified by cpusetsize. Set contents can be inspected by means of the functions listed in Table 5.3.

The function:

```
rtems_status_code rtems_scheduler_get_maximum_priority(
    rtems_id scheduler_id,
    rtems_task_priority *priority
);
```

stores into the object pointed by priority the numerically maximum task priority of the scheduler scheduler_id. It should be noted that, according to the priority numbering scheme used by the RTEMS core, numerically higher priority values correspond to *lower* priorities, whereas in POSIX numerically higher priority values correspond to *higher* priorities.

Modifying the set of cores managed by a scheduler

The third and last group of functions listed in Table 5.2 provides a way to dynamically change the set of cores managed by a scheduler, thus changing the set originally specified in the static RTEMS configuration. The two functions in the group are:

```
rtems_status_code rtems_scheduler_add_processor(
    rtems_id scheduler_id,
    uint32_t cpu_index
);
```

and

```
rtems_status_code rtems_scheduler_remove_processor(
    rtems_id scheduler_id,
    uint32_t cpu_index
);
```

Both functions take as arguments the identifier of the scheduler they must operate upon, scheduler_id, and the CPU index of the core to be added or removed from the set of cores managed by that scheduler, cpu_index. They return the status code RTEMS_SUCCESSFUL when successful. Moreover, the function rtems_scheduler_add_processor may fail for the following reasons:

RTEMS_INVALID_ID The scheduler identifier scheduler_id is invalid.

RTEMS_NOT_CONFIGURED The core indicated by cpu_index is not configured for use by RTEMS.

RTEMS_INCORRECT_STATE The core is configured for use, but is not online.

RTEMS_RESOURCE_IN_USE The core is already assigned to a scheduler instance.

Similarly, rtems_scheduler_remove_processor may fail for the following reasons:

RTEMS_INVALID_ID The scheduler identifier scheduler_id is invalid.

RTEMS_INVALID_NUMBER The core indicated by cpu_index is not managed by the scheduler identified by scheduler_id.

RTEMS_RESOURCE_IN_USE There is at least one task that is using the scheduler identified by scheduler_id and it would become impossible to execute it if the core indicated by cpu_index is removed, for one of the following reasons:
  - The scheduler would be left with no cores at all on which to execute and dispatch the tasks under its control.
  - The task has an *affinity mask* (see Chapter 13) that restricts the cores it can execute on and it would become impossible to find a suitable core for it.

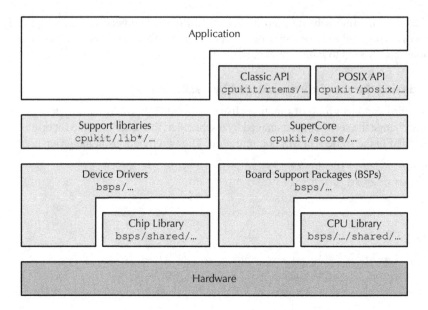

FIGURE 5.4 Layered structure of RTEMS components and their source code location.

5.3 RTEMS CLASSIC AND POSIX API

The RTEMS operating system offers two main Application Programming Interfaces (APIs) for many of its functions, including task management and timekeeping, called *classic* and *POSIX* API. These two APIs serve different yet overlapping purposes. The focus of the POSIX interface is compliance with the international standard ISO/IEC/IEEE 9945 [68] to support portable application development.

The Classic API aims to provide rich, expressive features useful to tailored real-time applications designed specifically to work with RTEMS. In addition, it allows access to operating system features, like the Scheduler Manager described in Section 5.2, currently not covered by the standard. Thus, application developers ought to choose between portability and expressiveness in deciding which API to use. Applications can use both APIs for the most expressive and powerful range of functionality, but any software that relies on Classic API services is non-portable to other operating systems.

The primary goal of RTOS interface standardization efforts is to facilitate portable real-time application development by defining the function-level interfaces and the behavior of a compliant implementation. An application that uses the interfaces correctly can be assured of proper execution in an RTOS with a standards compliant implementation. As shown in Figure 5.4, in RTEMS much of the implementation of RTOS services to satisfy these APIs is shared in an internal subsystem called the *SuperCore* located at cpukit/score in the RTEMS source code tree, which corresponds roughly with the notion of kernel in other operating systems. The SuperCore interface is not a public-facing API.

The Classic API is an evolution of the Real Time Executive Interface Definition (RTEID) [88] and the Open Real-Time Kernel Interface Definition (ORKID) [92] proposed RTOS interfaces that were never standardized, although they did inform POSIX 1003.1b-1993, the real-time extensions to the POSIX 1003.1 standard [65], which in turn converged into ISO/IEC/IEEE 9945 [68]. The POSIX API in RTEMS targets the real-time extensions plus a useful subset of POSIX.

The design and implementation of the Classic API followed and evolved naturally from the goals and constraints of real-time applications. Two distinguishing features of such applications are the need for accurate timekeeping to avoid deadline misses and multitasking to support responding to independent events in a timely and prioritized manner. The Classic API therefore provides a suite of function calls, or directives, that facilitate time and task management with other services provided to simplify application development. The full set of directives and in-depth documentation can be found in the RTEMS Classic API Guide [105].

The Classic API aims to maintain consistent behavior across the API to simplify the creation and management of system resources. Consistency is achieved by adopting an object-oriented approach to resource management, and dividing the API into managers that includes directives to create, delete, and use a single type of object. Each object created by a manager has a user-defined name and an RTEMS-defined identifier. Names are typically four ASCII characters, while identifiers are of an opaque `rtems_id` type that is representable by an unsigned integer.

A create directive associates the newly allocated object with an identifier that is returned to the user application, and an *ident* directive may be used to obtain the identifier associated with the object having a given object name. Most other directives of a manager take as first argument the identifier. A directive's return value usually is an opaque `rtems_status_code` type that indicates either success or a failure code. Directives can contain both input and output parameters, with inputs preceding outputs in the argument list. Parameters tend to be opaque rather than primitive types.

The rest of this chapter will discuss in more details Classic API primitives for task management and timekeeping. Chapter 6 contains a thorough discussion of the corresponding POSIX API.

5.4   TASK MANAGEMENT

The RTEMS functions (often called *directives*) for task management are implemented by the RTEMS Task Manager and summarized in Table 5.4. The main functions allow the caller to dynamically create new tasks and delete them. Additional functions allow the caller to query and modify the most important runtime properties of a task, such as its priority. The last two functions in this group is conceptually closer to the time management functions to be discussed in Section 5.6 and implements timed waits.

TABLE 5.4

RTEMS Task Manager

Function	Purpose
create [1]	Create a task, *without starting* it
start	Start a task, that is, make it ready or execution
restart	Restart a task and execute it again from its entry point
ident	Return the task identifier given its name
self	Return the identifier of the calling task
wake_after	Block the calling task, wake it up after a specified amount of time [2]
wake_when	Block the calling task until a specified time [2]
suspend	Suspend a task
resume	Resume a previously suspended task
is_suspended	Check whether a task is suspended or not
delete	Delete a task
exit	Delete the calling task
mode	Change the current task mode
set_priority	Set the priority of a task
get_priority	Get the current priority of a task
set_scheduler	Set the scheduler in charge of a task
get_scheduler	Get the scheduler currently in charge of a task
set_affinity	Set the affinity mask of a task [3]
get_affinity	Get the affinity mask of a task [3]
iterate	Invoke a user-specified function on all tasks in the system

[1] All API function names start with the `rtems_task_` prefix.

[2] In other words, `rtems_wake_after` performs a *relative* wait, with respect to the time of the call, whereas `rtems_wake_when` performs an *absolute* wait.

[3] Useful only on multicore systems, certain schedulers may impose restrictions on legal mask settings.

Creating and starting a task

The core calls used for creating a set of tasks in the Classic API are `rtems_task_create` and `rtems_task_start`. These API calls suffice to create a multitasking system amenable to real-time analysis. The `rtems_task_create` call allocates resources for a new task and initializes its task control block. The optional task create user extension is also called during `rtems_task_create` that can allocate additional application-defined resources for the newly created task.

The `rtems_task_create` directive is defined as:

```
rtems_status_code rtems_task_create(
    rtems_name           name,
    rtems_task_priority  initial_priority,
```

TABLE 5.5

RTEMS Task Attribute Constants

Constant	Meaning
RTEMS_NO_FLOATING_POINT	The task does not use the floating-point coprocessor
RTEMS_FLOATING_POINT	The task uses the floating-point coprocessor
RTEMS_LOCAL	The task is accessible only within the local node
RTEMS_GLOBAL	The task is accessible by other nodes of a multiprocessor system

```
    size_t              stack_size,
    rtems_mode          initial_modes,
    rtems_attribute     attribute_set,
    rtems_id            *id
);
```

The `name` is a user-defined four-character mnemonic for the task, which is rare to change after a task is created. The `initial_priority` is the priority value the scheduler uses for this task, which may change depending on the scheduling algorithm implemented by the scheduler in charge of the task.

By default, the newly created task inherits the scheduler from its creator. On multicore systems, the scheduler can be changed at a later time, preferably before starting the task, by means of the function `rtems_task_set_scheduler`, to be described later. On single-core systems there may be only one scheduler in charge of all tasks in the systems and there is no way of changing it.

A `stack_size` can be specified to increase or constrain the memory allocated to the task for its call stack, but not less than the minimum specified for the hardware processor architecture or the application-configured minimum. The `stack_size` may also assume two special values:

- RTEMS_MINIMUM_STACK_SIZE indicates that the stack size must *at least* the minimum, recommended stack size recommented by the RTEMS developers for the underlying processor architecture. If the system has been user-configured for a larger minimum stack size, the user configuration will prevail.
- RTEMS_CONFIGURED_MINIMUM_STACK_SIZE indicates that the stack size must be equal to the user-configured minimum stack size, regardless of whether it is larger or smaller than RTEMS_MINIMUM_STACK_SIZE.

In all cases, the stack size cannot be changed after task creation. The `initial_modes` define the execution mode the task should start with, explained further below. An `attribute_set` defines the task attributes, which cannot be changed later. The attribute set is built by means of the bitwise or of the attribute constants listed in Table 5.5 and include whether or not the task uses the floating point

TABLE 5.6

RTEMS Task Mode Masks and Values

Mask	Values	Aspect
RTEMS_PREEMPT_MASK	RTEMS_PREEMPT	Enable preemption
	RTEMS_NO_PREEMPT	Disable preemption [1]
RTEMS_TIMESLICE_MASK	RTEMS_TIMESLICE	Enable timeslicing
	RTEMS_NO_TIMESLICE	Disable timeslicing
RTEMS_ASR_MASK	RTEMS_ASR	Enable asynchronous signals
	RTEMS_NO_ASR	Disable asynchronous signals
RTEMS_INTERRUPT_MASK	RTEMS_INTERRUPT_LEVEL(0)	Enable all interrupts
	RTEMS_INTERRUPT_LEVEL(k)	Set interrupt level k [1] [2]

[1] Unsupported on multicore systems.

[2] Interrupt levels are architecture and platform-dependent, as described in Section 4.2.

co-processor (if one exists), and whether a task is *local* to the executing RTEMS kernel or is *global*, that is, it is accessible from remote nodes running a different RTEMS kernel in a multiprocessor setup. The default attributes are no floating point and local task.

The execution mode of a task controls whether or not it: can be preempted, uses timeslices, receives asynchronous signals, and what level of interrupts are enabled during its execution. Using the RTEMS_DEFAULT_MODES constant as initial_modes enables a default set of modes that enable preemption, disable timeslicing, enable asynchronous signals, and enable all interrupts. Any other setting of mode needs to be explicit in the call to create the task or can be changed later using the rtems_task_mode directive, which can also be used to obtain the current mode of the task.

A task mode is built by means of the bitwise or of several values that individually control a specific aspect of the mode. RTEMS defines a set of constants, listed in Table 5.6, which help users manipulate a task mode. More specifically, it provides:

- A *mask* to isolate a specific aspect from a task mode, for instance, preemption control.
- Within each mask, a set of valid *values* for that aspect. For instance, within the preemption control mask, two values are defined to enable and disable preemption during task execution.

When successful, the directive rtems_task_create returns RTEMS_SUCCESSFUL after storing the identifier of the newly created task into the location pointed by id. Otherwise, it returns a status code that provides more information about why it failed:

RTEMS_INVALID_ADDRESS indicates that the function failed because id was a NULL pointer.

RTEMS_INVALID_NAME indicates that the task `name` passed as argument was invalid.

RTEMS_INVALID_PRIORITY indicates that the `initial_priority` given to the task was invalid.

RTEMS_MP_NOT_CONFIGURED indicates that the task attributes asked for the creation of a global task on a single-node system.

RTEMS_TOO_MANY the maximum number of tasks in the system has been reached and no more tasks can currently be created.

RTEMS_TOO_MANY the maximum number of global objects in the system has been reached and no more global tasks can be created.

RTEMS_UNSATISFIED can be returned for two distinct reasons:

- There was not enough memory to allocate the task stack and its floating-point context, if requested.
- The task mode was illegal for a multicore system, namely, asking for non-preemption or setting a non-zero interrupt level is unsupported on such a system.

Although `rtems_task_create` allocates a new task, the caller does not get preempted because the new task is not yet added to the schedulable set of ready tasks, which is the job of the `rtems_task_start` directive, defined as:

```
rtems_status_code rtems_task_start(
    rtems_id              id,
    rtems_task_entry      entry_point,
    rtems_task_argument   argument
);
```

The directive operates on an existing task, specified by means of the task identifier `id`, which is an output from a previous call to `rtems_task_create`. The `entry_point` argument specifies the *entry point* of the task, that is, the address of the function from which task execution will begin.

Upon execution, the entry point function will receive a single argument `argument`, of type `rtems_task_argument`. This RTEMS-defined data type is opaque, but it is guaranteed to have the following useful properties:

- It is a numeric, unsigned integer type of unspecified width.
- Any valid pointer to `void` can be cast to this type and then back to a pointer to `void` without loss of information, that is, the result will be equal to the original pointer.

Importantly, the caller of `rtems_task_start` may be immediately preempted by the newly started task, depending on the number of cores in the system, the relative priorities of the two tasks, and whether or not preemption is enabled in the calling task. Besides RTEMS_SUCCESSFUL, which indicates that the task was started successfully, `rtems_task_start` may return one of the following status code to report a failure:

RTEMS_INVALID_ID The task id provided as argument was invalid.

RTEMS_INVALID_ADDRESS The task entry_point was invalid.

RTEMS_INCORRECT_STATE The task was not in the dormant state.

RTEMS_ILLEGAL_ON_REMOTE_OBJECT The task id referred to a remote task, but tasks can only be started locally—that is, from the same node they have been created on.

The rtems_task_restart directive, defined as:

```
rtems_status_code rtems_task_restart(
    rtems_id             id,
    rtems_task_argument  argument
);
```

restarts the task identified by id from any state except the dormant state, using the same entry_point specified when the task was originally started, possibly with a new argument. As for many other task-related primitives, the special task identifier RTEMS_SELF can be used to indicate that a directive shall operate on the calling task. For instance, in this case, a task can use RTEMS_SELF as id to restart itself, in which case rtems_task_restart will not return at all.

The possible non-normal status codes of rtems_task_restart are:

RTEMS_INVALID_ID The task id provided as argument was invalid.

RTEMS_INCORRECT_STATE The task was never started before, hence its entry point is unknown.

RTEMS_ILLEGAL_ON_REMOTE_OBJECT The task id referred to a remote task, but tasks can only be restarted locally—that is, from the same node they have been created on.

The final question that must be answered to completely address the topic of task creation is what are the tasks initially present in the system when it is first boot-strapped. An RTEMS application begins with two tasks named *init* and *idle*. The *idle* task has the lowest priority in the system and gets executed as a fallback only when no other tasks are ready. The *init* task is created by RTEMS and started at an application-configured function entry point, which is called Init by default.

Often a real-time application will initialize itself in Init by using a loop through a series of calls to create all the application's tasks, then another loop that starts them all, and then the *init* task will delete itself or self-suspend. Applications that are memory resource-constrained may re-use the *init* task as one of the application tasks. Barriers, described in Section 7.5, are very useful to ensure that the newly created tasks synchronize themselves at the very beginning to start executing in an orderly way.

Identifying a task

To operate on a task, most task management functions require an RTEMS task identifier, represented by the data type rtems_id. The task identifier is returned by the

TABLE 5.7

RTEMS Special Nodes

Node	Meaning
RTEMS_SEARCH_LOCAL_NODE	Search only the local (caller's) node
RTEMS_SEARCH_ALL_NODES	Search all nodes, starting from the local node
RTEMS_SEARCH_OTHER_NODES	Search all nodes except the local node

rtems_task_create directive upon task creation, but there are also other ways to retrieve it afterwards. The function:

```
rtems_status_code rtems_task_ident(
    rtems_name name,
    uint32_t node,
    rtems_id *id
);
```

returns the task identifier of a task, storing it into the location pointed by id, given its name and the node on which it resides. When searching other nodes besides the local node (the node on which the caller is executing) this function can find only global tasks, that is, tasks which have the RTEMS_GLOBAL flag set in their attributes.

Besides actual node numbers, the special nodes listed in Table 5.7 can be used to define the scope of the search. These special nodes can generally be used with any RTEMS function that looks up an object identifier given its name and node, and not only with rtems_task_ident.

If the task name is not unique within the search scope, the function returns the identifier of one of the tasks with that name. The special name RTEMS_SELF can be used by the calling task to retrieve its own identifier. When successful, the function returns RTEMS_SUCCESSFUL. Otherwise, it returns one of the following status codes:

RTEMS_INVALID_ADDRESS the function failed because id was a NULL pointer.
RTEMS_INVALID_NAME the task name passed as argument was invalid or no task with the given name was found.
RTEMS_INVALID_NODE the given node number was invalid.

A simpler way for a task to obtain its own identifier is to call:

```
rtems_id rtems_task_self(void);
```

This function can also be called from an interrupt handler, but takes a different meaning. In that case, it returns the identifier of the interrupted task.

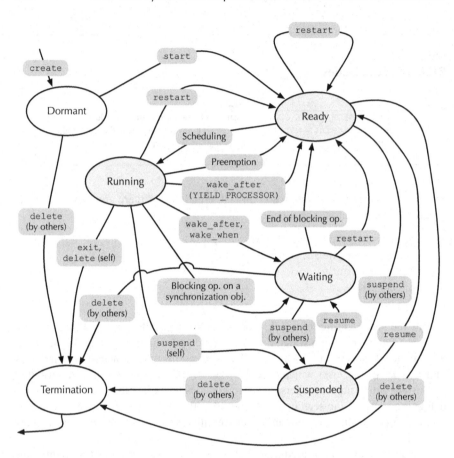

FIGURE 5.5 Simplified view of the RTEMS task state diagram (TSD).

Task states, timed delays, and suspension

Figure 5.5 provides a simplified view of the RTEMS task state diagram (TSD). This diagram is a specialization of the abstract diagram shown in Figure 3.4 and commented in Section 3.2. Although Figure 5.5 better captures the underlying complexity of a real operating system, it is still highly simplified and omits, for example, all the intricacies related to task termination and deletion.

In the figure, transitions whose label is written in typewriter font are triggered by the execution of the directive with the same name, described in this section. Transitions written in normal font are triggered by other operating system events.

After a task is started, it moves from the dormant to the ready state. Then, the scheduler may dispatch the task to run on a processor, which moves the task from the ready to the running state. While a task is running, it can voluntarily *block* itself in two possible ways:

TABLE 5.8

Fields of the `rtems_time_of_day` Data Type (Calendar Time)

Field	Contents	Range
`year`[1]	Year	≥ 1988
`month`	Month	$[1, 12]$
`day`	Day	$[1, 31]$
`hour`	Hour	$[0, 23]$
`minute`	Minute	$[0, 59]$
`second`	Second	$[0, 59]$
`ticks`	Fraction of second expressed in ticks	$[0, 99]$ [2]

[1] All fields are of type `uint32_t`, that is, unsigned, 32-bit integers.
[2] Assuming the default RTEMS clock tick length 10 ms is in use.

1. By calling the `rtems_task_wake_after` or `rtems_task_wake_when` directives, which specify a relative blocking interval in scheduling ticks or an absolute timeout interval in the `rtems_time_of_day` format, respectively.
2. By executing a blocking operation on a synchronization object. This form of self-blocking is needed whenever a task must interact with other tasks in an orderly way and is extremely articulate. It will be the subject of Chapters 7 and 9.

Passing a relative interval of `YIELD_PROCESSOR` to the *wake after* directive causes a task to yield the processor, thus moving the task into the ready state and allowing the scheduler to dispatch an equal or higher priority task. More specifically, the `rtems_task_wake_after` directive has the following interface:

```
rtems_status_code rtems_task_wake_after(
    rtems_interval ticks
);
```

Its only argument represent the number of `ticks` for which the calling task will wait. Although the function returns a status code, currently it always returns `RTEMS_SUCCESSFUL` because it cannot fail. The length of a tick, which also determines the resolution of this kind of wait, is determined by the configuration macro `CONFIGURE_MICROSECONDS_PER_TICK`.

Instead, the `rtems_task_wake_when` directive blocks the calling task until a certain time of day has been reached. The time of day is specified by reference, by means of the `time_buffer` argument:

```
rtems_status_code rtems_task_wake_when(
    rtems_time_of_day *time_buffer
);
```

The `rtems_time_of_day` data type is a structure with the fields listed in Table 5.8, which represents a calendar time. This function ignores the fractional part of the second held in the `ticks` field, and hence, the absolute wait it implements has a resolution of one second.

The `rtems_task_wake_when` function blocks the calling task and then returns `RTEMS_SUCCESSFUL` when successful. Otherwise, it immediately returns one of the following status codes:

RTEMS_INVALID_ADDRESS the `time_buffer` pointer was null.

RTEMS_INVALID_TIME_OF_DAY the contents of the structure pointed by `time_buffer` were invalid.

RTEMS_NOT_DEFINED the system data and time have not been set as specified in Section 5.6.

The caller of the *wake after* and *wake when* directives is placed in the waiting state until the interval elapses or the timeout is reached, respectively. Changing the system's current date and time does not affect *wake after*, but it does affect *wake when* and may cause the waiting task to unblock if the modified time exceeds the timeout. It is also important to remark that these two functions are related and conceptually close to the other timekeeping functions presented in Section 5.6.

A task can also *suspend* itself or another task by using the `rtems_task_suspend` directive, which is undone by the `rtems_task_resume` directive. As shown in Figure 5.5, the suspended state is complementary to other blocking, for example, a prior self-suspension via `rtems_task_wake_after` or blocking on a synchronization object, and a task that is resumed may remain in a waiting state in case it is blocked. The *suspend* and *resume* directives both take a single argument, the identifier `id` of the task whose state is changed: ʹ

```
rtems_status_code rtems_task_suspend(
    rtems_id id
);
```

```
rtems_status_code rtems_task_resume(
    rtems_id id
);
```

Both functions return one of the following status codes:

RTEMS_SUCCESSFUL The target task was successfully suspended or resumed.

RTEMS_INVALID_ID The task identifier `id` passed as argument was invalid.

RTEMS_ALREADY_SUSPENDED An attempt was made to suspend a task that was already suspended.

RTEMS_INCORRECT_STATE An attempt was made to resume a task that was not currently suspended.

Finally, the function:

```
rtems_status_code rtems_task_is_suspended(
    rtems_id id
);
```

checks whether the task identified by id is suspended or not. It returns RTEMS_SUCCESSFUL if the task is not currently suspended and RTEMS_ALREADY_ SUSPENDED if it is. Moreover, it may fail and return one of the following status codes:

RTEMS_INVALID_ID The task identifier id passed as argument was invalid.

RTEMS_ILLEGAL_ON_REMOTE_OBJECT The function cannot be used on a remote task.

Deleting a task

A task is deleted by calling the directive:

```
rtems_status_code rtems_task_delete(
    rtems_id id
);
```

passing its task identifier id as argument. A task can delete itself by using RTEMS_SELF as task identifier. In this case, rtems_task_delete does not return and the system selects another task to execute.

If a task is deleting another, rtems_task_delete returns RTEMS_ SUCCESSFUL if it successfully deleted the target task, or one of the following status codes upon failure:

RTEMS_INVALID_ID The task identifier id passed as argument was invalid.

RTEMS_ILLEGAL_ON_REMOTE_OBJECT The function cannot be used on a remote task. In other words, tasks must be deleted by a task residing on their same node.

When begin deleted, a task goes through a sequence of states in which most, but not all, of the resources allocated to it are reclaimed. All these states are subsumed under the termination state depicted in Figure 5.5. Two main aspects of this process are very important from the programmer's point of view:

1. RTEMS, like most other operating systems, is unable to reclaim *all* resources somewhat associated with or owned by the task. This is especially true for what concerns synchronization devices to be discussed in Chapters 7 and 9.

   From this point of view, unless the task to be deleted is unable to execute or experienced other kinds of critical failure, it is therefore better to ask the task to delete itself (after releasing all its resources) rather than deleting it abruptly from another task.

2. After deleting a task, RTEMS may reuse the same task identifier for a newly created task. Similarly, if a new instance of the same task is created after deleting it, the operating system will likely give it a different identifier. Again, this is consistent with what other operating systems do and is analogous, for instance, to the way Unix-like operating systems reuse process identifiers.

   Special care must be taken in the application code to handle this scenario correctly because, if the code merely keeps using a previously cached task identifier, it may easily refer to the wrong task.

Moreover, as it will be better discussed in Chapter 5, part of the context of a task created with the POSIX API may survive its deletion if the task was *joinable*, until another task performs a join operation on it.

A more concise way for a task to delete itself is to invoke the directive:

```
void rtems_task_exit(void);
```

which never returns to the caller. Since it is an error for a task created with the Classic API to return from the function specified as its entry_point, often the last line of code in such entry points is indeed a call to rtems_task_exit. Also in this case, it is worth remarking that the rules for tasks created with the POSIX API are different. In that case, returning from the entry point terminates the task normally.

Changing the task mode and other properties

As described previously, tasks are given an initial set of attributes and an execution mode upon creation. Attributes cannot be changed during the lifetime of a task. Instead, a task can inspect and change its execution mode by means of the directive rtems_task_mode:

```
rtems_status_code rtems_task_mode(
    rtems_mode mode_set,
    rtems_mode mask,
    rtems_mode *previous_mode_set
);
```

This directive implicitly targets the calling task, and hence, it does not take any task identifier as argument. It can be used in two different ways:

- When mode_set is the special value RTEMS_CURRENT_MODE, the directive only stores into the location pointed by previous_mode_set the current task mode, without changing it.
- Otherwise, the mask argument identifies which aspects of the execution mode are to be changed, and mode_set specifies their new value. Aspects not covered by the given mask are left unaltered. The mask argument is formed as the bitwise or of the masks listed in the leftmost column of Table 5.6, while mode_set is the bitwise or of the corresponding values listed in the middle column of the same table. As in the previous case, the directive also stores into the location pointed by previous_mode_set the task mode as it was before changing it.

Besides RTEMS_SUCCESSFUL, which indicates a successful completion, the directive may return one of the following status codes:

RTEMS_INVALID_ADDRESS the previous_mode_set pointer was null.
RTEMS_NOT_IMPLEMENTED the new task mode is not supported by the system. For instance, disabling preemption is not supported on multicore systems.

Two functions allow the caller to inspect and modify the current priority of a task. They are:

```
rtems_status_code rtems_task_set_priority(
    rtems_id id,
    rtems_task_priority new_priority,
    rtems_task_priority *old_priority
);

rtems_status_code rtems_task_get_priority(
    rtems_id task_id,
    rtems_id scheduler_id,
    rtems_task_priority *priority
);
```

The `rtems_task_set_priority` function sets the priority of a task, targeted by the task identifier `id`, to the value `new_priority`. It also stores the task's previous priority into the location pointed by `old_priority`.

By default, legal priority values range from 1 (which represents the highest possible priority) to 255 (the lowest). The number of priorities that RTEMS supports can be changed by setting the `CONFIGURE_MAXIMUM_PRIORITY` configuration macro and, on multicore systems, also on a scheduler-by-scheduler basis.

Due to the way resource locking protocols work—for instance, the priority inheritance protocol to be discussed in Section 8.1—lowering the priority of a task may be postponed until the task does not hold any resources managed by those protocols.

The function returns `RTEMS_SUCCESSFUL` when successful, or one of the following status code to indicate failure:

RTEMS_INVALID_ID The task identifier `id` passed as argument was invalid.

RTEMS_INVALID_ADDRESS the `old_priority` pointer was null.

RTEMS_INVALID_PRIORITY the priority specified by `new_priority` was invalid.

The `rtems_task_get_priority` function stores into the location pointed by `priority` the current priority of the task `id` with respect to the scheduler instance `scheduler_id`. Unlike `rtems_task_set_priority`, it can be used only on local tasks. The function may fail for the following reasons:

RTEMS_INVALID_ID The task identifier `id` or the scheduler instance identifier `scheduler_id` were invalid.

RTEMS_INVALID_ADDRESS The `priority` pointer was null.

RTEMS_ILLEGAL_ON_REMOTE_OBJECT The directive cannot be used on a remote task.

RTEMS_NOT_DEFINED The task has no priority defined with respect to the given scheduler instance.

In both cases, the priority returned by the directives reflects the actual, current task priority, which at times may be different than the priority specified by the user, due to the temporary priority adjustments that RTEMS may perform to implement the locking protocols mentioned previously.

By default, a newly created task inherits the scheduler from the task that created it. If necessary, it can be changed at a later time, by means of the function:

```
rtems_status_code rtems_task_set_scheduler(
    rtems_id id,
    rtems_id scheduler_id,
    rtems_task_priority priority
);
```

This function sets the scheduler of task `id` to `scheduler_id` and gives to the task a new `priority`. It is generally legal to change the scheduler of a task before starting it. Depending on the specific task state, attempting to change the scheduler in other scenarios may lead to `RTEMS_NOT_SATISFIED` errors. On multicore systems, the scheduler instance inherited by the task upon creation or set by means of this function is the *home* scheduler instance of the task, which is normally in charge of its execution. However, when a task acquires certain kinds of shared synchronization device, to be described in Chapter 7, the operating system can temporarily place it under the control of other scheduler instances.

The `rtems_task_set_scheduler` returns `RTEMS_SUCCESSFUL` when successful. Otherwise, it returns one of the following status codes:

RTEMS_INVALID_ID The task identifier `id` or the scheduler instance identifier `scheduler_id` were invalid.

RTEMS_INVALID_PRIORITY The given `priority` was invalid.

RTEMS_UNSATISFIED The scheduler indicated in the call has no cores assigned to it and performing the requested scheduler change would have rendered the task unable to execute.

RTEMS_RESOURCE_IN_USE The current task state did not allow a change of scheduler.

RTEMS_ILLEGAL_ON_REMOTE_OBJECT The directive cannot be used on a remote task.

The directive:

```
rtems_status_code rtems_task_get_scheduler(
    rtems_id id,
    rtems_id *scheduler_id
);
```

stores the scheduler currently in charge of task `id` into the location pointed by `scheduler_id`. Besides `RTEMS_SUCCESSFUL`, the status codes that the function can possibly return are:

RTEMS_INVALID_ID The task identifier `id` was invalid.

RTEMS_INVALID_ADDRESS the `scheduler_id` pointer was null.

By default, on a multicore system a task is eligible for execution on all online cores and the scheduler can freely choose among them. As better detailed in Chapter 13, some multicore schedulers give users finer control on this set of cores by means of an *affinity mask*. The affinity mask of a task is a CPU set represented by the data type `cpu_set_t`. The scheduler is allowed to use a certan core to execute the task if and only if the core belongs to the set. As in other cases, users can manipulate CPU masks by means of the functions listed in Table 5.3. The function:

```
rtems_status_code rtems_task_set_affinity(
    rtems_id id,
    size_t cpusetsize,
    const cpu_set_t *cpuset
);
```

sets the affinity mask of the task `id`. The `cpuset` argument points to an instance of the `cpu_set_t` data type that holds the CPU set and `cpusetsize` indicates its size in bytes.

The function does not change the scheduler in charge of the task and the affinity mask may contain cores that are not assigned to that scheduler. In any case, the final set of cores on which the task may be executed is the intersection between the set of cores indicated in the affinity mask and the set of cores assigned to the scheduler. For the task to actually be executed, it is therefore crucial that this intersection is not empty.

Users should also take into account that some multicore locking protocols to be described in Chapter 13 may temporarily run a task on cores not specified in its affinity mask. Finally, if the underlying scheduler does not support task affinity, it is important that the affinity mask is kept to the widest possible setting, that is, the mask should include all cores in the system.

The `rtems_task_set_affinity` primitive returns RTEMS_SUCCESSFUL when successful. Otherwise, it returns one of the following status codes without touching the affinity mask:

RTEMS_INVALID_ID The task identifier `id` was invalid.

RTEMS_INVALID_ADDRESS the `cpuset` pointer was null.

RTEMS_INVALID_NUMBER the affinity mask was invalid.

RTEMS_ILLEGAL_ON_REMOTE_OBJECT The directive cannot be used on a remote task.

Finally, the directive:

```
rtems_status_code rtems_task_get_affinity(
    rtems_id id,
    size_t cpusetsize,
    cpu_set_t *cpuset
);
```

enables the caller to retrieve the affinity mask of task `id`. The `cpuset` argument must point to the location in which the directive will store the mask and `cpusetsize` must indicate its size, in bytes. Besides `RTEMS_SUCCESSFUL` the directive may return one of the following status codes:

RTEMS_INVALID_ID The task identifier `id` was invalid.

RTEMS_INVALID_ADDRESS the `cpuset` pointer was null.

RTEMS_INVALID_NUMBER the size specified in `cpusetsize` is too small for the affinity mask of the task.

RTEMS_ILLEGAL_ON_REMOTE_OBJECT The directive cannot be used on a remote task.

Operating on all tasks

The last Task Manager directive to be discussed here, `rtems_task_iterate`, allows the caller to apply a user-defined function to the task control block (TCB) of all tasks currently present in the system:

```
void rtems_task_iterate(
   rtems_task_visitor visitor,
   void *arg
);
```

The directive takes two arguments:

- `visitor` is a pointer to the user-defined function, and
- `arg` is an additional argument to be passed to `visitor`, besides the task control block.

The `visitor` function is invoked once for each task currently in the system and also receives two arguments:

- a pointer to the task control block (data type `rtems_tcb`), and
- the user-defined `arg` initially passed to `rtems_task_iterate` that can be used, for instance, to pass contextual information to the `visitor` function and propagate it from one invocation to the next.

The `visitor` function returns a Boolean value (data type `bool`) that, if `true`, aborts the iteration prematurely. This function must be designed with care because RTEMS task control blocks (TCBs) are not normally user-visible and altering them can easily cause problems. A full description of the internals of RTEMS task control blocks is beyond the scope of this book.

5.5 THE RATE MONOTONIC MANAGER

The Task Manager presented in Section 5.4 allows users to dynamically create new "generic" tasks, whose internal behavior is completely opaque to the operating system. In particular, no information is provided to the system about the timing patterns and requirements of those tasks. This is similar to the limited level of knowledge that general-purpose operating systems have about their processes, which are merely seen as entities that are sometimes ready to execute and consume CPU time, to somehow fulfill a useful purpose.

However, in Chapters 3 and 4 we saw that a whole wealth of theoretical work has been devoted to analyze specific kinds of task—most importantly, *periodic* tasks—in order to accurately predict their worst-case behavior in terms of response time as well as the system's ability to fulfill their timing constraints, represented by *deadlines*, thus guaranteeing their *schedulability*. This was done because these kinds of task indeed have a special relevance for many real-time systems.

As described in Chapter 3 and shown in Figure 3.5, a periodic task consists of an infinite sequence of *jobs*. Each job is *released*, that is, it becomes ready for execution, at regular time intervals and more specifically at the beginning of each task period.

In the simplest case, a job stays ready for execution and runs, that is, consumes the CPU time that the operating system scheduler allocates to it, until it attains a certain total execution time. The job then concludes and the periodic task as a whole is no longer ready for execution until the next job is released, at the beginning of the next period. This very specific timing pattern continues for the whole task's life. For what concerns timing constraints, the simplest and often most natural way of proceeding is to establish an *implicit* deadline, that is, stipulate that a job must necessarily conclude before the next job of the same task is released, in order to be considered schedulable.

In more complex scenarios, a job can also *block*—for instance, to synchronize and communicate with other tasks, and to perform input–output operations—but the top-level articulation of a task in an infinite sequence of jobs released at regular time intervals stays the same. In some circumstances, it may also be convenient to consider *constrained* deadlines, which are shorter than the task period, especially for tasks dealing with abnormal situations that must be handled urgently as they arise.

The consequence of these more complex task behaviors and constraints have been discussed in Section 4.1.3 for what concerns schedulability analysis. Then, Chapters 7 and 8 will give more information about how the inter-task synchronization and communication mechanisms work.

For what concerns the practical implementation of periodic tasks, existing real-time operating systems and APIs may endorse two complementary approaches:

1. Stay at the level of abstraction seen in the RTEMS Task Manager and leave all the burden to programmers. In return, programmers get absolute freedom to implement these tasks in the way that most effectively suits their application, but also to make mistakes along the way. This is the approach taken by the POSIX API [68] and described in Chapter 6.

   Programmers who do not wish to implement their own periodic tasks from scratch can of course resort to additional libraries, like [30] for Linux, which provide the higher-level notion of periodic task, with varying degrees of sophistication and complexity, starting from what the underlying operating system API makes available.

TABLE 5.9

RTEMS Rate Monotonic Manager API

Function	Purpose
create [1]	Create a period object
ident	Get the identifier of a period object given its name
delete	Delete a period object
period	Mark the conclusion of a period and wait for the next
cancel	Cancel a period
get_status	Get the current status of a period object
get_statistics	Retrieve the execution statistics of a period object
reset_statistics	Reset the execution statistics of a specific period object
reset_all_statistics	Reset the execution statistics of all existing period objects
report_statistics	Print the execution statistics of all existing period objects [2]

[1] All API function names start with the rtems_rate_monotonic_ prefix.

[2] Except period objects that have not executed at least one period since they were created or their statistics were reset.

2. Supply a general framework that helps programmers to implement periodic tasks with less effort in an easier (and hopefully more correct) way, even though they may or may not grasp all the subtleties of the framework internals. At the same time, this approach enables the operating system to automatically gather more information about periodic tasks and drive its scheduling algorithms with it. This is the approach taken by RTEMS with its Rate Monotonic Manager, which will be the subject of this section.

For example, knowing that a certain task is periodic and has an implicit deadline rather than being a generic task of unknown characteristics, allows RTEMS to feed its Earliest Deadline First (EDF) scheduling algorithm described in Section 5.2 with accurate data without any further programmers' assistance.

Table 5.9 summarizes the user-visible functions of the Rate Monotonic Manager.

Creating and deleting period objects

As explained previously, the two main top-level phases of a periodic task are an *active* phase, in which one of its job has been released but has not been concluded yet, and an *inactive* phase, in which the current job has been concluded and the task is blocked waiting for the release of another job, which will take place at the beginning of the next period.

RTEMS drives a task through these two phases repeatedly by means of a *period object*, also called Period Control Block (PCB) in the RTEMS documentation [105]. The period control block of RTEMS should not be confused with the term Process

Control Block, also abbreviated PCB and used by other operating systems (together with many theoretical textbooks) to indicate the information that an operating system must manage and maintain to represent a process. A task can create a new period object by means of the function:

```
rtems_status_code rtems_rate_monotonic_create(
    rtems_name name,
    rtems_id *id
);
```

Like many other kinds of RTEMS object, period objects have a 4-character, possibly human-readable name. When successful, this function stores the unique identifier of the period object just created into the location pointed by id and returns RTEMS_SUCCESSFUL. The period object identifier must be used whenever it is necessary to refer to the period object iself. Upon failure, the function returns one of the following status codes:

RTEMS_INVALID_NAME the task name passed as argument was invalid.

RTEMS_INVALID_ADDRESS the id pointer was null.

RTEMS_TOO_MANY the maximum number of period objects in the system has been reached and no more can currently be created.

While creating a period object, the rtems_rate_monotonic_create function records the identity of the calling task into it. This is because a period object can be used for timing purposes only by the task that created it. Any attempt to use a period object belonging to a different task, except for retrieving its status or statistics, or deleting it, leads to the error RTEMS_NOT_OWNER_OF_RESOURCE. For similar reasons, period objects cannot be global.

The function:

```
rtems_status_code rtems_rate_monotonic_ident(
    rtems_name name,
    rtems_id *id
);
```

stores in the location pointed by id the identifier of the period object identified by name. Since object names are not guaranteed to be unique, if there are multiple period objects with the same name the function returns one of them, but exactly which one is unspecified. The function returns RTEMS_SUCCESSFUL when it completed successfully. Otherwise, it returns one of the following status codes:

RTEMS_INVALID_ADDRESS the id pointer was null.

RTEMS_INVALID_NAME No period object with the given name was found.

When no longer in use, a period object can be deleted by passing its identifier id to the function:

```
rtems_status_code rtems_rate_monotonic_delete(
    rtems_id id
);
```

The function returns RTEMS_SUCCESSFUL when successful, or the status code RTEMS_INVALID_ID to indicate that the period object identifier was invalid, for instance, because it was not found in the system. After a period object has been deleted, the storage allocated to it, all its system resources, and also its identifier can be reused by RTEMS.

As for other RTEMS objects, care must be taken not to use an object identifier after the corresponding object has been deleted because the identifier may become invalid or, even worse, refer to the wrong object. A period object can be deleted by the task that created it, or also by another task. The second option is useful to forcefully reclaim all the resources allocated to a task that has been aborted and is therefore unable to execute anymore.

Implementing periodic tasks

In its simplest form, an RTEMS periodic task is built according to the template shown in Figure 5.6. In the template most checks on the return value of RTEMS directives, except for the ones related to the timings of the periodic task, and the related error handling have been omitted. This is obviously contrary to good programming practice, but in this context has the advantage of keeping the listing short. The code can be divided into three parts:

1. In the first part the periodic task initializes itself, allocates all the resources it needs, and prepares for periodic execution. This is represented in the template by the pseudo-function init(). Then, it makes use of the Rate Monotonic Manager to create the period object, which will control its timings during the second part.
2. The second part consists of a while loop in which the task repeatedly waits until the right time arrives for the release of its next job, by means of the directive rtems_rate_monotonic_period, and then performs it. The job code is represented by the pseudo-function job().

   The task moves into its third part only when the Rate Monotonic Manager reports a timing failure by means of the RTEMS_TIMEOUT status code of rtems_rate_monotonic_period. As it will be explained in further details below, this status code signifies that the job missed its implicit deadline, an event often called *overrun* in literature.
3. During the third part, the task deletes the period object, releases all other resources it owns, and eventually deletes itself. Resource release is represented in the template in an abstract way, by means of the pseudo-function fini().

The key directive for the periodic execution part is:

```
rtems_status_code rtems_rate_monotonic_period(
    rtems_id id,
    rtems_interval length
);
```

which makes use of the period object id to establish and maintain a periodic job execution pattern of period length ticks.

```
#define PERIOD_IN_TICKS 500
rtems_task Periodic_task (rtems_task_argument arg)
{
    rtems_id period;

    /* init (); */
    rtems_rate_monotonic_create (
        rtems_build_name ( 'P', 'E', 'R', ' ' ), &period);

    while (1)
    {
        if (rtems_rate_monotonic_period (
            period, PERIOD_IN_TICKS) == RTEMS_TIMEOUT)
            break;

        /* job (); */
    }

    rtems_rate_monotonic_delete (period);
    /* fini (); */
    rtems_task_delete (RTEMS_SELF);
}
```

FIGURE 5.6 Typical structure of an RTEMS periodic task (status checks and error handling, except for period overrun, omitted for brevity).

As shown in Figure 5.7, when this function is invoked on an *inactive* period object, that is, a period object that has not yet been used after having been created, it immediately returns to the caller a RTEMS_SUCCESSFUL status code, thus allowing it to proceed and perform its first job.

At the same time, the function also makes the period object *active*, records the instant of the call, and sets a time mark k ticks in the future, where k is the value of the length argument. The width of this time interval is the nominal period of the calling task and the interval itself represents the maximum amount of wall time allotted to the first job to complete.

In the normal case, that is, when the periodic task is not going to miss its deadline, the job will conclude its execution before the interval ends, the task will start another iteration of the while loop, and eventually call rtems_rate_monotonic_period again. This call will find that the period object is active and block the caller until k ticks have elapsed since the previous call. Then, it will set a new time mark k ticks further in future and return a RTEMS_SUCCESSFUL status code.

This ensures that, regardless of how much time the previous job took to complete, the next job will be released exactly k ticks after the release of the previous one.

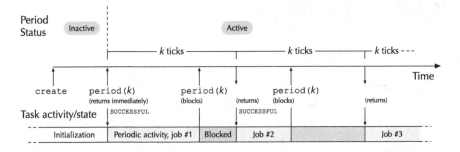

FIGURE 5.7 Normal timeline of an RTEMS Rate Monotonic task.

Figure 5.7 exemplifies the case in which, although the second job took less time than the first to complete, all three jobs shown are released exactly k ticks apart.

It is also worth noting that the timing mechanism is insensitive to the cause of these variations that, as we saw in the previous chapters, may be due to several reasons, like:

- The inherent variability of the execution time of the job, for instance, because the job contains data-dependent code paths with different execution times.
- Interference from other, higher-priority tasks or interrupt handlers. Although interference does not affect the job execution time in itself, it does increase its response time.
- Different amounts of blocking time, if the job makes access to shared resources, synchronizes with other tasks, or performs synchronous input–output operations.

Let us now discuss how the period object reacts to a *deadline miss* scenario, that is, when k ticks since the release of the current job elapse without rtems_rate_monotonic_period having being called again. If the job has been properly designed—so that its worst-case execution and blocking time do not lead it to exceed its period—this is typically caused by a transient system *overload*, that is, the amount of interference the job endured exceeded what was foreseen during schedulability analysis.

As shown in Figure 5.8, the period object performs two distinct actions in this scenario:

- When reaching the deadline, the state of the period object becomes *expired*, to mark the fact that the current job has missed its deadline.
- At the same time, the period object sets a new time mark k ticks in the future, which indicates where the deadline of the next job (job #2 in the figure) would be, assuming it had been released at the expected time.

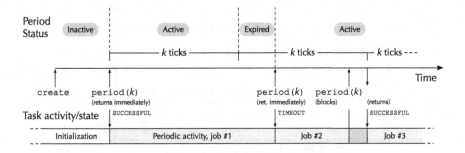

FIGURE 5.8   Timeline of an RTEMS Rate Monotonic task upon a deadline miss.

When the task eventually calls `rtems_rate_monotonic_period` (in the figure, after job #1 finished late) the period object is already in the expired state. Hence, the directive immediately returns to the caller, to let job #2 start as soon as possible, and the state of the period object goes back to active. The caller is also given the abnormal status code `RTEMS_TIMEOUT` to signal that the job just completed (job #1 in the figure) missed its deadline.

If the deadline miss was due to an occasional overload rather than a systemic timing issue, it is possible that the periodic task recovers. That is, as shown in the figure, job #2 may complete before its deadline although it started late, and `rtems_rate_monotonic_period` may be called again before reaching the next time mark. If this happens, the period object starts behaving again as in the normal case, namely:

- It blocks the caller until reaching the time mark.
- It sets a new time mark k ticks in the future.
- It returns to the caller to release job #3.
- Its status code is `RTEMS_SUCCESSFUL` to reflect the fact job that #2 did *not* miss its deadline.

By comparing Figures 5.7 and 5.8, we can observe that a very useful result of this approach to deadline misses is that, after recovering from the transient overload, job #3 is still released at exactly the same instant as it would have been if the overload did not take place at all and job #1 did not finish late.

So far, we examined the simplest case of recovery, in which the period tasks is able to go back to its normal timings immediately after an overload occurs, that is, during the execution of the job that immediately follows the one affected by the overload. This is not always possible and, for this reason, the Rate Monotonic Manager adopts are more complex strategy to keep track of deadline misses than what we described so far.

As shown in Figure 5.9, each period object has a *postponement counter* associated with it. The value of the counter is set to zero when the period object is initialized and stays at zero if there are no deadline misses. Its value changes in two circumstances:

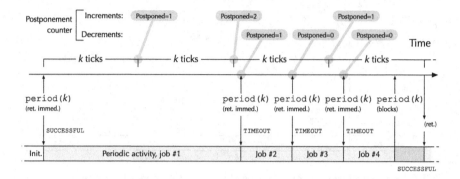

FIGURE 5.9   Effect of the postponement counter on a Rate Monotonic task timeline.

- When a time mark is reached before the task called the directive rtems_rate_monotonic_period to wait for the next period, the value of the postponement counter is incremented by one. In the figure, this happens k, $2k$, and $3k$ time units after the initial call to rtems_rate_monotonic_period, which activates the period object.
- When rtems_rate_monotonic_period returns RTEMS_TIMEOUT immediately in order to release a new job as soon as possible, due to a deadline miss, the postponement counter is decremented by one. In the figure, this happens when jobs #2, #3, and #4 are released.

The value of the postponement counter represents the number of jobs that have currently been postponed due to a deadline miss. As Figure 5.9 shows, the operating system uses it to determine how many jobs should be released immediately while recovering from a deadline miss.

In our example, after job #2 is released immediately using the logic depicted in Figure 5.8, jobs #3 and #4 are also released immediately after calling rtems_rate_monotonic_period because it is already late for them to start. The period object reverts to its normal behavior, as in Figure 5.7, when rtems_rate_monotonic_period is called at the conclusion of job #4, before its deadline expired.

As we can see, even in the more complex and general case of multiple deadline misses, periodic tasks managed by period objects still have two very useful properties:

- After one or more deadline misses occurs, new jobs are released as soon as possible to "help" the periodic task recover.
- When recovery is complete, jobs are released at exactly the same time as they would had been if the deadline misses did not occur.

At the same time, the periodic task is informed about deadline misses by means of the RTEMS_TIMEOUT status code of rtems_rate_monotonic_period. In this way, the task may speed up recovery by reducing the execution time demand of jobs released while recovery is in progress, or even skip them completely if this is not detrimental to the task purpose. Interested readers may refer to [35] for further details about the method that RTEMS uses to handle deadline misses and more thorough information about its properties.

Besides being used to delimit the jobs of a periodic tasks and control their release as we saw so far, `rtems_rate_monotonic_period` can also be called at any time with the special value `RTEMS_PERIOD_STATUS` as `length`. In this case, the directive never blocks the caller, and just returns to the caller a summary of the current state of period object `id`, without altering it in any way. In particular, the directive uses the following status codes to represent the state:

RTEMS_NOT_DEFINED The period object is inactive.
RTEMS_SUCCESS The period object is active.
RTEMS_TIMEOUT The period object has expired.

More detailed information about the state of a period object can be gathered by means of the `rtems_rate_monotonic_get_status` function, to be discussed in the following. Regardless of the way in which `rtems_rate_monotonic_period` is used, it may also fail and return the status code `RTEMS_NOT_OWNER_OF_RESOURCE` when the calling task is not the same task that created the period object, or `RTEMS_INVALID_ID` to indicate that the period object identifier was invalid.

Last, but not least, it is worth remarking two more important aspects of period objects:

- Although in the previous examples the `length` of the task period was kept constant at k for all calls to `rtems_rate_monotonic_period` its value can indeed be changed, for instance, to implement variable-period tasks. In this case, the new period takes effect from the first job released after the call to `rtems_rate_monotonic_period`.
- A task can also create and use more than one period object when its job is structured as a sequence of sub-jobs and each sub-job must be released at a certain fixed temporal offset from the release of the main job, regardless of how much time the previous sub-jobs took to conclude. The RTEMS documentation [105] contains more information about this aspect.

A period object can be canceled and brought back to the inactive state, as it was immediately after creation, by means of the function:

```
rtems_status_code rtems_rate_monotonic_cancel(
    rtems_id id
);
```

which takes as argument the `id` of the period object to be canceled. Then, the period object may be activated again by means of `rtems_rate_monotonic_period`.

TABLE 5.10
Fields of RTEMS Rate Monotonic Period Status

Field	Meaning
owner	Identifier of the task that owns the period object
state	Current state of the period object [1]
postponed_jobs_count	Number of jobs that could not be released timely
since_last_period	Wall time elapsed since the beginning of the current period [2]
executed_since_last_period	CPU time consumed since the beginning of the current period

[1] Possible values are: RATE_MONOTONIC_INACTIVE, RATE_MONOTONIC_ACTIVE, and RATE_MONOTONIC_EXPIRED.

[2] The beginning of the current period is marked by the most recent execution of the rtems_rate_monotonic_period directive.

Deleting a period object with rtems_rate_monotonic_delete implicitly cancels it, if it is running. As usual, rtems_rate_monotonic_cancel returns RTEMS_SUCCESSFUL when successful, or one of the following status codes upon failure:

RTEMS_INVALID_ID The period identifier id was invalid.

RTEMS_NOT_OWNER_OF_RESOURCE The calling task was not the same task that created the period object.

Users can acquire more detailed information about the status of a period object than what rtems_rate_monotonic_period provides by means of the function:

```
rtems_status_code rtems_rate_monotonic_get_status(
    rtems_id id,
    rtems_rate_monotonic_period_status *status
);
```

The directive takes as arguments the identifier of the period object id and status, a pointer to a data structure that it will fill with status information upon successful completion. Table 5.10 lists the fields of the data structure. Among them:

- state contains essentially the same information as the directive rtems_rate_monotonic_period returns when called with a period length equal to RTEMS_PERIOD_STATUS.
- postponed_jobs_count is the postponement counter discussed previously, represented as a uint32_t, that is, a 32-bit unsigned integer. The counter is incremented with saturating arithmetic, so that it stays at the maximum possible uint32_t value and never wraps around even in the (fairly unlikely) case in which there are more than about 4 billion postponed jobs.

In addition, if the period object is not inactive, the fields `since_last_period` and `executed_since_last_period` provide the amount of wall time elapsed and the amount of CPU time consumed by the job since the beginning of the current period. When `rtems_rate_monotonic_get_status` is used at the conclusion of each job, this information can be particularly useful to understand how close (or how far) the actual behavior of a periodic task is to its design parameters. Moreover, it also gives valuable insights on how much time margin was left before the job missed its deadline, besides the simpler "yes or no" answer provided by the next invocation of `rtems_rate_monotonic_period`.

In the latest version of RTEMS both the wall time and the CPU time are expressed as an integer number of seconds, plus a fractional part in nanoseconds, and stored in a `struct timespec`, a POSIX data type that will also be discussed in Section 6.6.

The function returns RTEMS_SUCCESSFUL when it completed successfully. Otherwise, it returns one of the following status codes:

RTEMS_INVALID_ID The period identifier `id` was invalid.

RTEMS_INVALID_ADDRESS the `status` pointer was null.

Gathering execution statistics

Even if the user does not collect status information about a period object explicitly, it is used anyway to calculate execution statistics. These statistics can be retrieved by means of the function:

```
rtems_status_code rtems_rate_monotonic_get_statistics(
    rtems_id id,
    rtems_rate_monotonic_period_statistics *statistics
);
```

The function has an interface very similar to `rtems_rate_monotonic_get_status`. It takes as arguments the identifier `id` of the period object on which it should operate, and a pointer `statistics` to a data structure that the function will fill with statistics and whose fields are listed in Table 5.11. Statistics are updated at the end of each period and consist of:

- The total number of jobs executed so far (`count`) and the number of jobs that missed their deadline (`missed_count`).
- The minimum, maximum, and total CPU time consumed by all jobs executed so far (`min_cpu_time`, `max_cpu_time`, and `total_cpu_time`).
- The minimum, maximum, and total wall time consumed by all jobs executed so far (`min_wall_time`, `max_wall_time`, and `total_wall_time`).

The minimum and maximum wall time estimate the best and worst-case response time of the periodic task, and the minimum and maximum CPU time estimate its best and worst-case execution time. Therefore, in the long run, they provide useful

TABLE 5.11

Fields of RTEMS Rate Monotonic Period Statistics

Field	Meaning
count	Number of jobs executed so far
missed_count	Number of jobs that missed their deadline
min_cpu_time	Minimum CPU time consumed by a job
max_cpu_time	Maximum CPU time consumed by a job
total_cpu_time	Total CPU time consumed by all jobs
min_wall_time	Minimum wall time consumed by a job
max_wall_time	Maximum wall time consumed by a job
total_wall_time	Total wall time consumed by all jobs

insights on the accuracy of the task parameters used for scheduling analysis. Moreover, when compared with the deadline, the maximum wall time gives an idea of how much time margin is left before jobs possibly miss their deadline. Last, but not the least, the ratio of the maximum CPU time to the period represents the fraction of CPU time the periodic task consumes, and gives a valuable starting point to determine which tasks are promising candidates for optimization.

The function rtems_rate_monotonic_get_statistics returns RTEMS_SUCCESSFUL when successful. Otherwise, it returns one of the following status codes:

RTEMS_INVALID_ID The period identifier id was invalid.
RTEMS_INVALID_ADDRESS the statistics pointer was null.

Statistics can be reset by invoking either of the following two functions:

```
rtems_status_code rtems_rate_monotonic_reset_statistics(
    rtems_id id
);
```

```
void rtems_rate_monotonic_reset_all_statistics(void);
```

The difference between them is that the first function resets the statistics of a single period object, given its identifier id, whereas the second resets all period objects that currently exist in the system. The first function fails and returns RTEMS_INVALID_ID when the period identifier is invalid, whereas the second always succeeds. Finally, the function:

```
void rtems_rate_monotonic_report_statistics(void);
```

prints out the statistics of all existing period objects on the system console, in human-readable form. Since this function is meant to be used mainly during testing and debugging, there is no way to change its output device and format. However, the same underlying information can be retrieved, on an object-by-object basis, by means of `rtems_rate_monotonic_get_statistics` and used as a starting point to generate custom printouts.

5.6 TIMEKEEPING: CLOCKS AND TIMERS

Timekeeping in RTEMS revolves around two related concepts, *clocks* and *timers*.

- RTEMS supports a single *clock* that maintains the notion of time across the whole system. It has a known resolution, which coincides with the *tick interval*. The value of the clock—that is, its notion of current time of day often called *calendar time*—is updated whenever a tick interval elapses. In order to do this, RTEMS requires a device able to generate a periodic interrupt on every tick, which usually consists of a hardware timer.

  As shown in the top-left part of Figure 5.10, the RTEMS Board Support Package (BSP) is responsible for initializing and starting the hardware timer upon system initialization. The BSP also installs a software handler for the periodic interrupt that the timer generates. From it, it invokes the `rtems_clock_tick` primitive, which is the entry point for RTEMS timekeeping as a whole.

- As also shown in the figure, the clock acts as a time base for many crucial internal time-related operating system activities, of which the main ones are:

  - Scheduler timeslicing, discussed in Section 5.4.
  - Service of the internal per-task timers used to implement the directives `rtems_task_wake_after` and `rtems_task_wake_when` also discussed in Section 5.4.
  - Similarly, the clock also services the timers used by the Rate Monotonic Manager described in Section 5.5, internal to its period objects.
  - Most operating system primitives for inter-task synchronization and communications, to be discussed in Chapters 7 and 8 *time out* and return to the caller a failure indication if they cannot complete their work within a certain amount of time. Also in this case, internal timers are used to keep track of elapsed time.

  Last, but not least, the clock also acts as a time reference for user-level *timers*, to be described next.

A timer is an object that *fires*, that is, triggers a call to a user-defined timer service routine either when a certain amount of time elapses since its initiation (relative time) or when a specified time of day has been reached (absolute time), according to the notion of time maintained by the aforementioned clock. The timer service routine can be called in two different ways, at the user's choice on an initiation-by-initiation basis:

- Directly from the clock tick handler, that is, with the processor executing in an *interrupt context*.

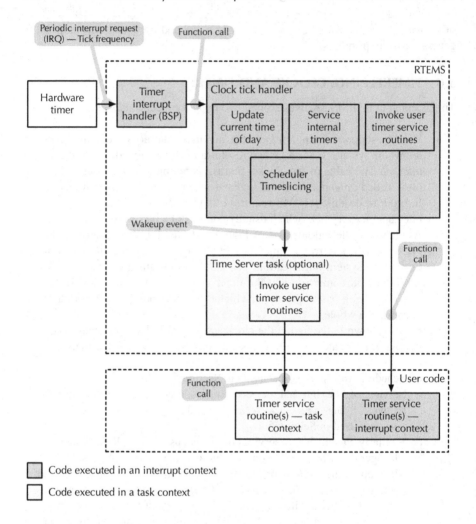

FIGURE 5.10 General outline of the RTEMS clock and timer chain.

- Through the interposition of a *timer server task* internal to RTEMS. In this case, the execution of the timer service routine takes place in a *task context*.

In Figure 5.10, modules executed in an interrupt context are shown as dark grey rectangles, while modules executed in a task context are painted light gray. There are several differences between these two approaches to execute the timer service routine that are worth remarking:

- There are significant limitations on what a timer service routine that runs in an interrupt context can do in two important areas. Firstly, many operating system directives cannot be invoked from such a context. Secondly, any

TABLE 5.12
RTEMS Clock Manager API

Function	Purpose
set [1]	Set the time of day
get_tod	Get the time of day as a rtems_time_of_day object
get_tod_timeval	Get the time of day as a struct timeval object
get_seconds_since_epoch	Get the number of seconds elapsed since the RTEMS epoch
get_ticks_per_second	Retrieve the clock resolution in ticks per second
get_ticks_since_boot	Get the number of clock ticks since boot
get_uptime	Get the time since boot as a struct timespec object
get_uptime_timeval	Get the time since boot as a struct timeval object
get_uptime_seconds	Get the time since boot in seconds
get_uptime_nanoseconds	Get the time since boot in nanoseconds
tick_later	Return a future tick counter (time interval in ticks)
tick_later_usec	Return a future tick counter (time interval in μs)
tick_before	Return true if the given tick counter is in the future

[1] All API function names start with the rtems_clock_ prefix.

kind of blocking is forbidden and any time delay must be necessarily be implemented by means of an active waiting loop.

- Interrupt handlers are implicitly given a priority higher than any task in the system. For the reasons outlined in Section 4.2, the amount of time spent in an interrupt context must necessarily be kept to a minimum, due to the significant impact it has on the overall schedulability of the system.

- Instead, the priority of the timer server task can freely be chosen by the user when it is started, in order to protect higher-priority tasks from any interference from timer service routines.

- The price to be paid for this additional flexibility is that the latency of timer service routines, with respect to the time they should nominally be invoked, will generally be higher if they are called indirectly by the timer server task, rather than directly by the clock tick handler.

- In order to reduce operating system overheads, RTEMS does not start the timer server task automatically. Therefore, users must explicitly call a dedicated operating system primitive to start the timer server task before initiating any timer that will make use of it.

Both the clock and timers are under the control of the RTEMS Clock Manager. Table 5.12 summarizes its directives.

Setting and getting the time of day

When RTEMS starts, its notion of time of day must be initialized before use. The initialization may be performed at boot time by the Board Support Package (BSP) if it manages a hardware device, often called Real Time Clock (RTC), able to keep track of the time of day even while the operating system is not running and, possibly, most other hardware components are powered off. Time of day initialization can also be performed at any time by user code, by means of the function:

```
rtems_status_code rtems_clock_set(
    rtems_time_of_day *time_buffer
);
```

The argument `time_buffer` points to a `rtems_time_of_day` data structure, whose fields are listed in Table 5.8 and contains the time of day to be set. RTEMS only supports times of day after its *epoch*, a well-defined point in the past defined as January 1st, 1988 at midnight. To avoid confusion, it is worth remarking that the POSIX API also defines the concept of epoch, as explained in Section 6.6, but the reference point is different.

However, in most cases this difference is kept hidden from users as long as they do not mix times of day held in data structures with differing data types. This is because, on one hand, RTEMS functions that load or store a time of day from an RTEMS-specific data structure, like `rtems_time_of_day`, use the RTEMS's own notion of epoch. On the other hand, functions that use a POSIX data structure, like `struct timeval` (Chapter 11) and `struct timespec` (Chapter 6), adhere to the POSIX's definition of epoch, the only restriction being that times prior to the RTEMS epoch cannot be represented.

The `rtems_clock_set` returns `RTEMS_SUCCESSFUL` when it successfully set the time of day, or one of the following status code upon failure:

RTEMS_INVALID_ADDRESS the `time_buffer` pointer was null.

RTEMS_INVALID_CLOCK the contents of the data structure referenced by `time_buffer` were invalid, for instance, because they represented a time before the RTEMS epoch.

The following three functions retrieve the current time of day in different formats:

```
rtems_status_code rtems_clock_get_tod(
    rtems_time_of_day *time_buffer
);

rtems_status_code rtems_clock_get_tod_timeval(
    struct timeval *time_buffer
);

rtems_status_code rtems_clock_get_seconds_since_epoch(
    rtems_interval *time_buffer
);
```

All functions take as argument a pointer `time_buffer` to a user-allocated buffer that they will fill with time-of-day information. For the first function, `rtems_clock_get_tod`, the data type of the buffer is the RTEMS-specific data structure `rtems_time_of_day`, whose fields have been described in Table 5.8.

For the second function, `rtems_clock_get_tod_timeval`, the argument refers to a POSIX-defined `struct timeval`. This structure expressed time as two quantities, namely, an integral number of seconds and a fractional part in microseconds elapsed since the POSIX epoch, defined as January 1st, 1970 Coordinated Universal Time (UTC) at midnight. Chapter 11 provides more information about the `struct timeval` and its usage in the POSIX API.

For the third function, `rtems_clock_get_seconds_since_epoch`, the argument is a pointer to an integer of type `rtems_interval`, which will be set to the number of seconds elapsed since the RTEMS epoch. We already saw that the RTEMS data type `rtems_interval` is used to express a time interval in ticks, for instance, as an argument of `rtems_task_wake_after`. In this case and in several others to be discussed next, the same data type is overloaded to hold other kinds of information.

All three functions return `RTEMS_SUCCESSFUL` when they were able to retrieve the requested information. Otherwise, they return one of the following status codes:

RTEMS_INVALID_ADDRESS the `time_buffer` pointer was null.
RTEMS_NOT_DEFINED the time of day has not been set yet.

The last function in this group enables users to obtain the resolution of the RTEMS clock. More specifically, the function:

```
rtems_interval rtems_clock_get_ticks_per_second(void);
```

returns the *frequency* at which the clock is updated, that is, the number of ticks per second, as an integer.

Elapsed time since boot

Besides the time of day, another significant reference for many embedded systems is the boot time. The function:

```
rtems_interval rtems_clock_get_ticks_since_boot(void);
```

returns the current RTEMS *tick counter*, which represents the number of ticks elapsed since the system was booted. When using this information, users must be aware that `rtems_interval` is currently defined as a `uint32_t` (a 32-bit unsigned integer), and hence, it will wrap around after a relatively short amount of time. For instance, with the default clock tick period of 10 ms, the tick counter will wrap around after about 497 days, that is, about 1.38 years. This may be of concern for long-lived missions, also because it becomes proportionally shorter if the tick period is shortened to improve clock resolution.

The next set of functions provides the same information, but with reduced risks related to wrap-arounds and potentially higher resolution. More specifically, the function:

```
rtems_status_code rtems_clock_get_uptime(
    struct timespec *uptime
);
```

stores the time elapsed since the system was booted (also called *uptime*) in the data structure pointed by `uptime`, whose data type is `struct timespec`. The resolution of the returned information does not depend on the tick frequency, but only on the resolution of the hardware timers present on the platform and supported by the RTEMS Board Support Package (BSP).

The POSIX-defined `struct timespec` is similar to `struct timeval`. The main difference between the two—and also a common source confusion and programming errors—is that the fractional part of a second is expressed in *nanoseconds* in a `struct timespec`, whereas it is expressed in *microseconds* in a `struct timeval`. Chapter 6 contains more information on how the `struct timespec` data type is used in the POSIX API.

The `rtems_clock_get_uptime` function returns `RTEMS_SUCCESSFUL` when successful, or `RTEMS_INVALID_ADDRESS` if the `uptime` pointer was null. The function:

```
void rtems_clock_get_uptime_timeval(
    struct timeval *uptime
);
```

is very similar to the previous one. The main differences are:

- It requires a `struct timeval` instead of a `struct timespec` buffer.
- It does not check if the `uptime` pointer is null.

The directives:

```
time_t rtems_clock_get_uptime_seconds(void);
uint64_t rtems_clock_get_uptime_nanoseconds(void);
```

also return the time elapsed since boot, but expressed in seconds and nanoseconds, respectively. With these functions, the likelihood of a wrap-around becomes extremely remote, considering the data types involved. For instance, the number of nanoseconds since boot is expressed as a `uint64_t` (a 64-bit unsigned integer). Therefore, the returned value will wrap around only about 584 years after the system was booted.

Future time markers and busy waiting

The functions `rtems_clock_tick_later` and `rtems_clock_tick_later_usec` calculate and return the tick counter that corresponds to a time instant located at a specified distance in the future from current time:

```
rtems_interval rtems_clock_tick_later(
    rtems_interval delta
);
```

```
rtems_interval rtems_clock_tick_later_usec(
    rtems_interval delta_in_usec
);
```

They differ in the unit of measurement of time distance, which is expressed in ticks for `rtems_clock_tick_later` and in microseconds for `rtems_clock_tick_later_usec`.

Their return value can then be compared with the current time by passing it as argument to the function:

```
bool rtems_clock_tick_before(
    rtems_interval tick
);
```

This function compares the `tick` value provided as argument with the current tick counter and returns true if `tick` is still in the future. For instance, the following fragment of code:

```
rtems_interval future = rtems_clock_tick_later(10);
while(rtems_clock_tick_before(future));
```

traps a task in the busy-waiting `while` loop for 10 ticks.

Creating and deleting timers

A task can create a new timer by invoking the directive:

```
rtems_status_code rtems_timer_create(
    rtems_name name,
    rtems_id *id
);
```

The directive takes as argument the `name` of the timer to be created and stores into the location pointed by `id` its unique identifier, which must be used to operate on the timer with other Timer Manager functions. Timers are local objects, they can be used only on the RTEMS node they have been created on. In addition, on multicore systems, the core on which the calling task is currently running determines which core the operating system will use to manage the timer internally for its whole lifetime. Table 5.13 shows the directives available in the Timer Manager.

The returns value is RTEMS_SUCCESSFUL when the function completes successfully, or one of the following status codes:

RTEMS_INVALID_ADDRESS the id pointer was null.

RTEMS_INVALID_NAME the timer name passed as argument was invalid.

RTEMS_TOO_MANY the maximum number of timers in the system has been reached and no more can currently be created.

The identifier of a timer can also be obtained at a later time, given the timer's name, by means of the function:

TABLE 5.13
RTEMS Timer Manager API

Function	Purpose
create [1]	Create a timer
ident	Get the identifier of a timer given its name
delete	Delete a timer
fire_after	Execute a service routine after a time interval
fire_when	Execute a service routine at a specified time
cancel	Cancel an active timer
reset	Reset a relative interval timer
initiate_server	Start the RTEMS Timer Server
server_fire_after	Like fire_after, but execute through the Timer Server
server_fire_when	Like fire_when, but execute through the Timer Server

[1] All API function names start with the rtems_timer_ prefix.

```
rtems_status_code rtems_timer_ident(
    rtems_name name,
    rtems_id *id
);
```

This function takes as argument the name of an existing timer and stores into the location pointed by id its identifier. As for other kinds of RTEMS objects, timer names are not guaranteed to be unique (although timer identifiers are). For this reason, if there are multiple timers with the same name, this function will surely provide the identifier of one of them, but exactly which one is unspecified. The function returns RTEMS_SUCCESSFUL upon successful completion, or one of the following status codes, which describe the reason why it failed:

RTEMS_INVALID_ADDRESS the id pointer was null.
RTEMS_INVALID_NAME the timer name passed as argument was invalid or no timer was found with that name.

When a timer is no longer needed, it can be deleted by means of the function:

```
rtems_status_code rtems_timer_delete(
    rtems_id id
);
```

The only argument of this function is the identifier id of the timer to be deleted. If the timer is currently active, the function automatically deactivates it. A timer created by a task may be deleted by a different task residing on the same node. Deleting unused timers is recommended because the resources allocated to them are returned to the system and reused for other, new timers. After a successful call to `rtems_timer_delete` the timer identifier passed as argument becomes invalid and shall no longer be used, also because the system may also reuse the same identifier and for a new timer.

The function returns `RTEMS_SUCCESSFUL` when successful, or `RTEMS_INVALID_ID` if `id` was invalid, for instance, because it did not correspond to any existing timer.

Starting, cancelling, and resetting a timer

A timer can be activated by means of several different directives, depending on whether the expiration time is expressed in a *relative* or *absolute* way, and whether the timer service routine must be invoked *directly* from the clock tick handler or *indirectly* through the Timer Server task.

Staying with direct timer service routine invocation for now, the directive:

```
rtems_status_code rtems_timer_fire_after(
    rtems_id id,
    rtems_interval ticks,
    rtems_timer_service_routine_entry routine,
    void *user_data
);
```

activates timer `id` so that it will expire, or *fire*, after `ticks` clock ticks have elapsed. In other words, `ticks` represents a relative time interval with respect to the time of the call to `rtems_timer_fire_after`. If the timer is already active, the previous activation is automatically canceled before activating the timer again. The `routine` argument indicates the user-specified timer service routine, with signature:

```
void (* routine)(
    rtems_id id,
    void *user_data
);
```

that will be invoked upon timer expiration, with the expired timer `id` and the user-provided `user_data` pointer as arguments.

The function returns `RTEMS_SUCCESSFUL` when successful, or one of the following status codes:

RTEMS_INVALID_ID The timer identifier `id` was invalid.
RTEMS_INVALID_ADDRESS the `routine` pointer was null.
RTEMS_INVALID_NUMBER the time interval `ticks` was invalid.

Another directive enables the caller to activate a time so that it fires at a specified time of day, that is, when the time of day reaches a given absolute point in time:

```
rtems_status_code rtems_timer_fire_when(
    rtems_id id,
    rtems_time_of_day *wall_time,
```

```
    rtems_timer_service_routine_entry routine,
    void *user_data
);
```

Besides the arguments id, routine, and user_data, which have the same meaning as for rtems_timer_fire_after, this directive takes a pointer wall_time to a rtems_time_of_day data structure. It holds the absolute time at which the timer will fire. The directive returns RTEMS_SUCCESSFUL when successful, or one of the following status codes:

RTEMS_INVALID_ID The timer identifier id was invalid.

RTEMS_INVALID_ADDRESS the routine or wall_time pointer was null.

RTEMS_NOT_DEFINED the time of day has not been set yet.

RTEMS_INVALID_CLOCK the contents of the data structure referenced by wall_time were invalid, for instance, because they represented a time before the RTEMS epoch.

An timer can be canceled by means of the function:

```
rtems_status_code rtems_timer_cancel(
    rtems_id id
);
```

whose only argument is the timer id. A canceled timer will no longer expire, and hence, its timer service will no longer be called unless it is activated again. It is not an error to cancel an inactive timer. The function returns RTEMS_SUCCESSFUL when successful or RTEMS_INVALID_ID when given an invalid timer id.

Finally, the function:

```
rtems_status_code rtems_timer_reset(
    rtems_id id
);
```

provides a convenient way to reset and reactivate a relative timer multiple times with the same time interval and timer service routine, without having to specify them explicitly every time. If the timer indicated by id is active, it is canceled. Then, it is immediately activated again using the same interval and timer service function specified in the last call to rtems_timer_fire_after.

It is not an error to use this function on a timer that is not active, in this case it simply activates the timer. The function returns RTEMS_SUCCESSFUL when successful, or one of the following status codes:

RTEMS_INVALID_ID The timer identifier id was invalid.

RTEMS_NOT_DEFINED The timer has never been activated before, or it was activated last time with the rtems_timer_fire_when directive instead of rtems_timer_fire_after. This is an error because there is no valid interval or timer service function to refer to and reuse.

Both `rtems_timer_reset` and `rtems_timer_cancel` can also be used on a timer whose timer service routine is to be invoked by the Timer Server, to be discussed next. Therefore, for instance, it is legal to use `rtems_timer_reset` after `rtems_timer_server_fire_after`.

Using the timer server

In order to benefit from the indirect invocation of timer service routines through the RTEMS Timer Server task, it must be first of all created and started by means of the directive:

```
rtems_status_code rtems_timer_initiate_server(
    uint32_t priority,
    uint32_t stack_size,
    rtems_attribute attribute_set
);
```

The directive allows the caller to specify several crucial parameters of the Timer Server task, namely, its `priority`, `stack_size`, and `attribute_set`. These values are passed directly to the `rtems_task_create` directive described in Section 5.4.

The function returns `RTEMS_SUCCESSFUL` upon successful completion. Otherwise, it may return any of the status codes defined for `rtems_task_create`. Priority can be the value `RTEMS_TIMER_SERVER_DEFAULT_PRIORITY`. This is the default value of the Timer Server priority, which corresponds to the highest possible task priority in the system.

After starting the Timer Server, users may use the functions:

```
rtems_status_code rtems_timer_server_fire_after(
    rtems_id id,
    rtems_interval ticks,
    rtems_timer_service_routine_entry routine,
    void *user_data
);
```

```
rtems_status_code rtems_timer_server_fire_when(
    rtems_id id,
    rtems_time_of_day *wall_time,
    rtems_timer_service_routine_entry routine,
    void *user_data
);
```

They have the same arguments as `rtems_timer_fire_after` and `rtems_timer_fire_when`, respectively, but the timer service `routine` will be invoked by the Timer Server instead of the clock tick handler. These functions return the same status codes as their direct-invocation counterparts, with the same meaning, plus:

TABLE 5.14

RTEMS Interrupt Manager API, Basic Functions

Function	Purpose
catch [1, 2]	Associate an handler to an interrupt source
disable	Disable interrupts on a single-core system
enable	Restore interrupt level on a single-core system
flash	Restore interrupt level
local_disable	Disable interrupts locally, on the calling core
local_enable	Enable interrupts locally, on the calling core

[1] The spinlock-related Interrupt Manager functions are listed in Table 13.3.
[2] All Interrupt Manager functions start with the rtems_interrupt_ prefix.

RTEMS_INCORRECT_STATE The Timer Server has not been initiated, and hence, these functions cannot be used.

5.7   PREEMPTION AND INTERRUPT MANAGEMENT

In Section 5.4, we saw that on single-core systems it is possible to disable task preemption by setting the task mode appropriately. This method can be used as a very efficient way to ensure mutual exclusion *among tasks* in this kind of system.

The mutual exclusion domain can be extended to *interrupt handlers* by means of a set of functions provided by the RTEMS Interrupt Manager and listed in Table 5.14, together with the other main Interrupt Manager functions. However, before describing them, it is important to remark once more that this mutual exclusion method only works on single-core systems, for the reasons outlined in Section 12.3. The Interrupt Manager provides additional synchronization methods to guarantee the mutual exclusion between tasks and interrupt handlers also on multicore systems. They are listed in Table 13.3 and will be discussed in Chapter 13.

The directive:

```
rtems_status_code rtems_interrupt_catch(
    rtems_isr_entry new_isr_handler,
    rtems_vector_number vector,
    rtems_isr_entry *old_isr_handler
);
```

installs an interrupt handler—called Interrupt Service Routine (ISR) in RTEMS—for an interrupt source, with the interposition of the RTEMS Interrupt Manager. The interrupt source is identified by means of an interrupt vector number vector. The correspondence between vector numbers and interrupt sources is platform-specific and is established through the platform's Board Support Package (BSP).

The directive installs the new interrupt handler `new_isr_handler` for the given `vector` and stores the previous interrupt handler for the same vector into the location pointed by `old_isr_handler`. Interrupt handlers are C functions of type `rtems_isr_entry` and must adhere to the following prototype:

```
rtems_isr new_isr_handler(
    rtems_vector_number vector
);
```

They receive as argument the vector number, so that they can identify the interrupt source. In this way, the same handler can handle interrupts coming from different sources appropriately. On some architectures, interrupt handlers may receive additional architecture-dependent arguments, like a pointer to the exception frame created to handle the interrupt. The return type `rtems_isr` is currently defined as `void`.

The interposition of the RTEMS Interrupt Manager, both before and after invoking the user-specified interrupt handler, plays a crucial role to ensure that the interrupt itself and processor scheduling are both handled appropriately. In particular:

- Before invoking the interrupt handler, RTEMS takes care of saving all registers of the interrupted task that are not automatically saved by the hardware or the C calling conventions, as explained in Section 4.2 using the ARM Cortex-M processor as an example. This ensures that the whole context of the interrupted task can be restored when the task is dispatched again for execution and, at the same time, allows the interrupt handler to be written as a normal C function.
- After the interrupt handler returns, RTEMS performs task scheduling and dispatching, because the interrupt handler may have readied a task whose priority is higher than the interrupted task. On architectures that support interrupt nesting, this is done only when the outermost interrupt handler returns, as an optimization.

The use of the Interrupt Manager is not mandatory for all interrupt sources. It may be bypassed, for efficiency, by installing interrupt handlers directly in the processor's interrupt vectoring data structures. However, for the reasons outlined previously, if a certain interrupt handler invokes any RTEMS primitive, the Interrupt Manager must be used to install it, as well as all interrupt handlers it may interrupt.

Instead, interrupt handlers that do not invoke any RTEMS primitive and cannot be interrupted by a handler that does may be installed directly, without passing through the Interrupt Manager. This rule also extends to Non-Maskable Interrupt (NMI) handlers, which cannot be masked by definition. They can be installed directly, but cannot invoke any RTEMS primitive because there would be no way for RTEMS to protect its critical sections against them.

In any case, there are restrictions on exactly which RTEMS directives an interrupt handler may legally invoke. The general rule of thumb is that any directive that can possibly block the caller is forbidden. The RTEMS documentation—for instance,

Reference [105] for the Classic API—provides a complete list of permissible directives.

On a single-core system, the directive:

```
void rtems_interrupt_disable(rtems_interrupt_level level);
```

stores the current interrupt level into `level` and disables all maskable interrupts. This directive can actually *write* into its argument although it is apparently passed by value because it is implemented as a C preprocessor macro.

Symmetrically, on a single-core system, the directive:

```
void rtems_interrupt_enable(rtems_interrupt_level level);
```

restores the processor interrupt level to the value specified by its argument `level`.

Together, these two directives can be used to bracket a critical section that must be protected from preemption by other tasks and interrupt handlers. This kind of critical section is very strong, and also quite disruptive for system timings, because it is entirely executed with interrupts disabled.

It is important to remark that, although the directive to be placed at the end of the critical section is named *interrupt enable*, it does not necessarily enables interrupts when called. For instance, when two or more critical sections are properly nested, this mechanism ensures that interrupts are not actually re-enabled until the execution flows out of the outermost critical section.

All invocations of `rtems_interrupt_enable` except the last one do not re-enable interrupts because the matching `rtems_interrupt_disable` was invoked while interrupts were already disabled and recorded this fact into its `level` argument. Only the very last `rtems_interrupt_enable` may re-enable interrupts because it restores the interrupt level as it was when the matching outermost `rtems_interrupt_disable` was invoked.

The directive:

```
void rtems_interrupt_flash(rtems_interrupt_level level);
```

is available only on single-core systems for backward compatibility. It is also implemented as a macro and is equivalent to the sequence of calls:

```
rtems_interrupt_enable(level);
rtems_interrupt_disable(level);
```

Its argument must come from a previous call to `rtems_interrupt_disable` or `rtems_interrupt_flash` itself. When called from a non-nested critical section delimited by `rtems_interrupt_disable` and `rtems_interrupt_enable`, its effect is to re-enable interrupts—thus allowing any pending interrupt request to interrupt the calling task and be serviced—and immediately disable them again before resuming execution in the critical section itself. It is typically invoked at regular intervals within a relatively long critical section, when it is safe to do so, to reduce interrupt servicing latency. Instead, it has no effect when called from a nested critical section.

Disabling interrupts globally is often a complex and time-consuming affair on multicore architectures and, in many cases, it cannot be done in an atomic way across all cores in the system. For this reason, the directives `rtems_interrupt_disable` and `rtems_interrupt_enable` are not available on multicores and are replaced by the following weaker variants. They are available on both single-core and multicore systems, and only disable interrupts on the processor on which the calling task is currently running:

```
void rtems_interrupt_local_disable(
    rtems_interrupt_level level
);
```

```
void rtems_interrupt_local_enable(
    rtems_interrupt_level level
);
```

On multicore systems, these directives are not generally suitable for mutual exclusion because:

- They do not protect the critical section against other tasks. Although those other tasks are prevented from running on the same core, because they inhibit preemption anyway, nothing prevents them from running on other cores, unless affinity masks are set very carefully when possible.
- They do not protect the critical section against interrupt handlers either, unless it can somehow be guaranteed that interrupt requests will be accepted and handled only by the same core that is executing within the critical section, which can be difficult.

As remarked previously, both issues can be solved by means of more sophisticated mutual exclusion mechanisms suitable for multicore systems, to be described in Chapter 13.

5.8 SUMMARY

This chapter introduced readers to the RTEMS scheduling algorithms and its task management and timekeeping directives made available by the RTEMS Classic API. Sections 5.1 and 5.2 provided a general description of basic task management concepts and a short description of the RTEMS single-core scheduling algorithms and the Scheduler Manager functions to control and manage them.

Then, after a short comparison between the Classic and POSIX APIs, given in Section 5.3, the three main RTEMS components involved in task management and timekeeping—namely, the Task Manager, the Rate Monotonic Manager, the Clock Manager, and the Timer Manager—were discussed in Sections 5.4 to 5.6.

Last, but not least, Section 5.7 gave an overview of interrupt management on single-core systems through the RTEMS Interrupt Manager.

6 Task Management and Timekeeping, POSIX API

CONTENTS

The POSIX Application Programming Interface (API) for task management and timekeeping is more complex than its RTEMS Classic counterpart presented in Chapter 5. For this reason, its description has been divided into several parts:

- The chapter starts by discussing *attribute objects*, the main mechanism used by POSIX to configure and customize the behavior of threads and other entities it manages.
- Then, we present the main *thread creation* and *termination* functions, which are used in virtually all POSIX concurrent multithreaded programs and we look at the POSIX scheduling model and its policies, which define how threads are executed. The discussion of the functions typical of multicore scheduling is deferred to Chapter 13.
- Another important topic addressed in this chapter is how POSIX supports thread *cancellation*, that is, the forced termination of a thread upon request from another thread.
- *Signals* are the way POSIX supports the software equivalent of an interrupt directed to a thread. This is the main mechanism used to notify threads of external asynchronous events, like timer expiration and message arrival.
- A discussion of the main POSIX *timekeeping* functions, realized by means of clocks and timers, concludes the chapter.

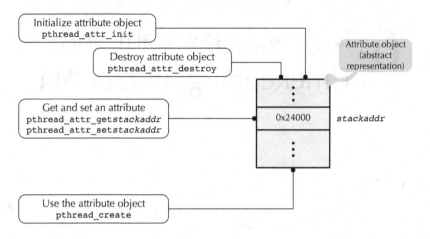

FIGURE 6.1   Attribute objects in POSIX.

6.1   ATTRIBUTE OBJECTS

Before describing any other POSIX object in detail, it is important to discuss *attribute objects*, which are used to determine some aspects of the behavior of threads and other POSIX objects. By means of attribute objects it becomes possible to support future extensions of POSIX objects in a portable way and, in most cases, without changing the API of the functions that operate on them.

In addition, they provide a clean isolation of the configurable, and sometimes inherently non-portable, aspects of a POSIX object. For example, the *stack address* of a thread (that is, the address of the area of memory to be used as the thread stack) is an important characteristic of a thread because using a kind of memory or another can significantly affect the thread execution speed. However, it can rarely be expressed in a portable way, and hence, must be adjusted when the software is ported from one system to another. Therefore, it makes sense to represent it as an attribute of the thread and store it in an attribute object to keep it isolated from the portable portions of the software.

Since the same attribute object can be used to create multiple objects of the same kind, like threads, setting up such an attribute object also provides a way to define some characteristics common to a whole class of threads in a centralized way.

Figure 6.1 shows how attributes and attribute objects are managed, still using the stack address attribute as an example. The description that follows refers to thread attribute objects, but the same concepts also apply to other kinds of attribute object, only data types and function names differ.

Attribute objects must be initialized before use. For thread attribute objects, the initialization function is:

```
int pthread_attr_init(pthread_attr_t *attr);
```

The only argument of this function is a pointer to an object of type `pthread_attr_t` that represents the attribute object. When successful, the function initializes the object, setting all attributes it contains to their default value, and returns zero. Upon failure, it returns a non-zero status code. The only status code mandated by the standard is:

ENOMEM The attribute object could not be initialized due to lack of memory.

Initializing the same attribute object twice in a row leads to undefined behavior, although the standard recommends that `pthread_attr_init` detects this error and returns the status code:

EBUSY The attribute object pointed by `attr` is already initialized.

When an attribute object is no longer needed, it should be destroyed by means of the function:

```
int pthread_attr_destroy(pthread_attr_t *attr);
```

to release the dynamic memory possibly allocated for it by `pthread_attr_init`. Although it is not mandatory, this function may return the following non-zero status code when it detects that `attr` does not point to an initialized attribute object:

EINVAL The `attr` argument does not point to an initialized attribute object.

Attribute objects shall not be referenced after they have been destroyed but the same attribute object can be reinitialized with a new call to `pthread_attr_init` and then used again.

As shown in Figure 6.1, an attribute object can be seen as a "container" of attributes. Each individual attribute can be retrieved from and stored into the attribute object by means of a pair of functions whose name generally is `pthread_attr_get` and `pthread_attr_set`, respectively, followed by the name of the attribute. For instance, the functions `pthread_attr_getstackaddr` and `pthread_attr_setstackaddr` get and set the stack address attribute. There are some exceptions, though. For example, the functions to get and set the `contentionscope` thread attribute are called `pthread_attr_getscope` and `pthread_attr_setscope`.

These functions invariably take as first argument a pointer to the attribute object they must operate upon, followed by additional arguments whose number and type depend on the specific attribute. All of them return zero upon success and a non-zero status code when they fail.

Morever, some convenience functions exist, which are able to get and set multiple, related attributes at a time. These functions will be discussed in more detail along with the attributes they operate upon.

TABLE 6.1

RTEMS Thread Creation and Termination Primitives, POSIX API

Function	Purpose
pthread_create	Create and start a new thread
pthread_exit	Voluntary terminate a thread
pthread_join	Wait until a thread terminates
pthread_detach	Detach a thread
pthread_equal	Check whether two thread identifiers are equal
pthread_self	Return the calling thread identifier

6.2   THREAD CREATION AND TERMINATION

The main functions related to the creation of a thread and its voluntary termination are summarized in Table 6.1. In POSIX, threads are represented by a thread descriptor, that is, an object of type pthread_t. They are created by means of the function:

```
int pthread_create(pthread_t *restrict thread,
    const pthread_attr_t *restrict attr,
    void *(*start_routine)(void*), void *restrict arg);
```

The first two arguments of this function indicate where it must store the newly created thread descriptor (in the location pointed by thread) and the thread attributes (held in the thread attribute descriptor pointed by attr).

The argument start_routine is the name of a C function that the new thread must execute, that is, its entry point, and arg is a pointer to be passed to it as argument. Unlike Classic tasks, tasks created through the POSIX API are ready for execution upon creation and might start executing immediately if their priority is sufficiently high, possibly preempting the task that created them.

The function pthread_create returns zero when it succeeded to create the new thread. It shall return a non-zero error number to indicate failure:

EGAIN The system does not have enough resources to create the new thread or the maximum number of threads specified by PTHREAD_THREADS_MAX would be exceeded.
EPERM The calling thread does not have sufficient privileges to set the thread attributes indicated by attr, most notably the scheduling policy and priority.

The standard also recommends that, if the function detects that attr does not point to a correctly initialized thread attribute object, it should fail and return the EINVAL error number.

Table 6.2 summarizes the thread attribute manipulation functions, while Table 6.3 lists the default attribute values in the current version of RTEMS. The functions to create and destroy an attribute object have already been described in Section 6.1.

TABLE 6.2
RTEMS Thread Attributes, POSIX API

Function	Purpose
pthread_attr_init	Initialize a thread attribute object
pthread_attr_destroy	Destroy a thread attribute object
pthread_attr_getdetachstate	Get the detachstate attribute
pthread_attr_getschedpolicy	Get the schedpolicy attribute
pthread_attr_getschedparam	Get the schedparam attribute
pthread_attr_getinheritsched	Get the inheritsched attribute
pthread_attr_getscope	Get the contentionscope attribute
pthread_attr_getstackaddr	Get the stackaddr attribute
pthread_attr_getstacksize	Get the stacksize attribute
pthread_attr_getstack	Get both stackaddr and stacksize together
pthread_attr_getguardsize	Get the guardsize attribute
pthread_attr_getaffinity_np	Get the (non-portable) affinity_np attribute
pthread_attr_setdetachstate	Set the detachstate attribute
pthread_attr_setschedpolicy	Set the schedpolicy attribute
pthread_attr_setschedparam	Set the schedparam attribute
pthread_attr_setinheritsched	Set the inheritsched attribute
pthread_attr_setscope	Set the contentionscope attribute
pthread_attr_setstackaddr	Set the stackaddr attribute
pthread_attr_setstacksize	Set the stacksize attribute
pthread_attr_setstack	Set both stackaddr and stacksize together
pthread_attr_setguardsize	Set the guardsize attribute
pthread_attr_setaffinity_np	Set the (non-portable) affinity_np attribute

TABLE 6.3
Default Values of Thread Attributes, POSIX API

Attribute	Value
detachstate	PTHREAD_CREATE_JOINABLE
schedpolicy	SCHED_FIFO
schedparam	Scheduling priority 2
inheritsched	PTHREAD_INHERIT_SCHED
contentionscope	PTHREAD_SCOPE_PROCESS
stackaddr	Set upon thread creation
stacksize	Set upon thread creation to the minimum possible value
guardsize	0
affinity_np	All cores enabled

The other functions follow the general naming scheme also discussed in that section, with some exceptions that will be better pointed out in the following. To keep the discussion short, individual functions will not be discussed in detail. Interested readers may refer to the POSIX standard [68] for more information. The thread attributes are:

detachstate This attribute is an `int` and determines whether or not the newly created thread can be the target of a *join* operation, performed by means of the `pthread_join` function to be discussed in the following. Its two possible values are `PTHREAD_CREATE_JOINABLE` (the thread supports the join operation) and `PTHREAD_CREATE_DETACHED` (the thread runs as a *detached* thread and does not support the join operation).

schedpolicy This attribute is an `int` that establishes the scheduling policy for the new thread, that is, the set of rules that the system applies to govern the transition of the thread between the *ready* and *running* states. The related attribute `schedparam` holds the scheduling parameters of the thread, which is a piece of information attached to the thread and used by the policy to take scheduling decisions about it. The algorithms implemented by the policies defined by the POSIX standard will be presented in more detail in Section 6.3.

schedparam This is a composite attribute of type `struct sched_param`. It contains the scheduling parameters of the thread. For most scheduling policies, they coincide with an integer scheduling priority, although the underlying attribute mechanism supports a more complex set of parameters when needed.

Unlike other attributes, both `schedpolicy` and `schedparam` can also be modified after thread creation, by means of the functions to be discussed in Section 6.3.

inheritsched This attribute is an `int` that controls whether the new thread shall get its scheduling policy and parameters from the attributes `schedpolicy` and `schedparam` (when its value is `PTHREAD_EXPLICIT_SCHED`), or shall inherit them from the calling thread (when it is `PTHREAD_INHERIT_SCHED`).

contentionscope This attribute is an `int` that determines against which other threads the new thread will compete to reach the *running* state. In summary, if the contention scope is `PTHREAD_SCOPE_SYSTEM`, the thread competes, on the base of its scheduling policy and parameters, directly against all the other threads with the same scope.

If the scope is `PTHREAD_SCOPE_PROCESS`, the new thread competes directly against the other threads within the same process, whereas the way all these threads compete with threads belonging to other processes in the system depends on the scheduling policy and parameters set at the process level. Since the distinction is important only for operating systems that, unlike RTEMS, support multiple processes, it will not be discussed further.

stackaddr This attribute, of type `void *`, allows the caller to set up a so-called *application-managed thread stack*. More specifically, the attribute value points to the lowest address of the memory area to be used as a stack, while `stacksize` denotes its size in bytes.

With this approach, the memory that the new thread will use as a stack is allocated and managed manually by the calling thread, rather than automatically by the operating system. In this way, it becomes possible to accurately determine the location of the stack within the address space and, for instance, make use of a specific kind of memory for it.

This comes at the expense of a higher software complexity and dubious portability. Those mainly come from the fact that, for architectural reasons, implementations may impose alignment requirements on `stackaddr` that are stricter that what `malloc` guarantees and are left unspecified by the standard. As a consequence, a pointer returned by `malloc` may not be used directly to set the `stackaddr` or a thread.

stacksize This attribute is of type `size_t` and indicates the size of the new thread stack, in bytes. The value must be higher than or equal to `PTHREAD_STACK_MIN`. Besides the ordinary functions that get and set the `stackaddr` and `stacksize` attributes separately, the two functions `pthread_getstack` and `pthread_setstack` conveniently get and set them together.

guardsize This attribute, also of type `size_t`, indicates the size of the *guard area* of the thread stack. A guard area is a reserved area of memory that is located at the far end of the stack (that is, beyond the locations that are filled last when the processor pushes items onto the stack). It is used by the system to detect stack overflow, usually with the assistance of a memory management unit (MMU), and signal an error.

The special value zero means that no guard area shall be provided. This attribute is ignored when an application-managed thread stack is in use. Moreover, the current version of RTEMS does not implement guard areas although it provides the functions to get and set the `guardsize` attribute.

affinity_np The standard gives implementations the freedom of defining *non-portable* features and entities, identified by the _np or _NP suffix in their name and the functions related to them. This is an example of non-portable attribute because it is not specified by the POSIX standard and represents the set of cores on which the newly created thread will be allowed to run. Its type is `cpu_set_t`. This attribute is the POSIX counterpart of the Classic API directives that manipulate the affinity mask of a task, described in Section 5.4.

A thread *implicitly* terminates when it returns from its entry point, that is, from the `start_routine` passed to `pthread_create`. It can also terminate *expicitly*, by calling the following function from anywhere in the code:

```
void pthread_exit(void *value_ptr);
```

In both cases, the terminating thread can summarize its final status and the reason for its termination in a value of type `void *`, which is the return value of `start_routine` or the argument given to `pthread_exit`, respectively. These two ways of termination are equivalent because returning from `start_routine` has the same effect as calling `pthread_exit` with the return value as argument.

The function `pthread_exit` never returns to the caller by definition, and hence, no error handling is necessary for it.

If the attribute `detachstate` of the terminating thread is set to the value `PTHREAD_CREATE_JOINABLE`, another thread can wait for its termination by means of a join operation, to be discussed in the following, and get access to its final status. Otherwise, all the information associated to the thread, including the final status, is discarded upon termination and is no longer accessible elsewhere.

In all cases, the thread still executes two kinds of *cleanup* action immediately before terminating. In particular:

- It executes the *cleanup handlers* that have been registered by means of the function `pthread_cleanup_push`, to be discussed in Section 6.4. The execution takes place in last-in, first-out order, that is, the cleanup handler pushed last is executed first.
- It invokes the *finalizers* of all thread-specific data items associated with the thread and whose pointer is not `NULL`. Thread-specific data, not discussed in this book due to lack of space, are per-thread data structures accessed by means of a global key, whose memory may have to be freed when the thread terminates.

Both ways of terminating a thread presented so far are *voluntary*, that is, the thread itself takes the initiative of terminating. The standard also specifies an *involuntary* termination mechanism, in which a thread requests the termination of another thread. This mechanism is called *cancellation* and will be discussed in more detail in Section 6.4.

The function:

```
int pthread_join(pthread_t thread, void **value_ptr);
```

waits for the termination of the thread represented by the argument `thread`. The thread must have been created with the attribute `detachstate` set to `PTHREAD_CREATE_JOINABLE`. If the argument `value_ptr` is not `NULL` the function also stores the final status code of `thread` into the location referenced by `value_ptr`.

The final status code is the one provided by the thread itself upon voluntary termination, or the special value `PTHREAD_CANCELED` if the thread was canceled. After a successful join, the system can freely destroy all the information associated with the terminated thread. For this reason, multiple joins targeting the same thread are not allowed.

The function returns zero after the target `thread` has terminated if it succeeds. Otherwise, the standard recommends that the function detects the following error conditions and returns the corresponding non-zero error codes:

EINVAL The thread descriptor `thread` does not refer to a *joinable* thread, that is, to a thread whose `detachstate` attribute is `PTHREAD_CREATE_JOINABLE`.

ESRCH The thread descriptor `thread` refers to a terminated thread that has already been the target of a join operation.

EDEADLK The thread descriptor `thread` refers to the calling thread. Therefore, executing the join would result in a deadlock.

The definition of the join mechanism just described implies that the system must preserve some information about a joinable thread—namely, a data structure reachable from its thread descriptor that contains, at least, its final status code—after thread termination and until the thread is targeted by a `pthread_join`. If a thread has been created as joinable, but this feature is no longer needed or desired, it is highly advisable to *detach* the thread to avoid wasting the system resources just mentioned by keeping them allocated for an undetermined amount of time.

Moreover, systematically omitting join operations may also put some systems at risk of exceeding `PTHREAD_THREADS_MAX` due to a high number of "terminated but not yet joined" threads, because the standard leaves unspecified whether or not these threads count against that limit. A thread can be detached by calling the function `pthread_detach` with the corresponding thread descriptor `thread` as argument:

int pthread_detach(pthread_t thread);

As many other POSIX functions, `pthread_detach` returns zero when successful. Otherwise, the standard recommends that the function detects and reports the following error conditions:

ESRCH The thread descriptor `thread` is unknown to the system.

If upon thread creation it is already known for certain that a thread will not need to be joined, a simpler alternative is to create the thread with the `detachstate` attribute set to `PTHREAD_CREATE_DETACHED`. This action cannot be undone at a later time, though. In other words, a joinable thread can be detached after being created, but a detached thread cannot be changed into a joinable thread.

The function:

int pthread_equal(pthread_t t1, pthread_t t2);

checks whether two thread descriptors, `t1` and `t2`, are equal or not. It returns a non-zero value (interpreted as true in a C-language logical expression) if they are, and zero otherwise. The behavior of the function is undefined if one or both thread descriptors are invalid.

This function must exist because the standard leaves the definition of the `pthread_t` data type to the implementation. In some cases, the choice may fall on a data type (for instance, a `struct`) for which the C language does not support direct comparison, thus making logical expressions like `t1 == t2` illegal.

The thread descriptor is not automatically made available to the thread itself upon creation, but can be retrieved if needed, by means of the function:

pthread_t pthread_self(**void**);

This function simply returns to the calling thread its thread descriptor. No error conditions are defined for it and no special return values are reserved to report an error.

6.3   THREAD SCHEDULING

In its most general form, the POSIX standard specifies that a software system is organized hierarchically as a set of *processes*, each containing one or more *threads* of execution. Each process has its own address space and encompasses most system resources needed for the execution of the threads it contains, most notably memory. Instead, threads are the units of scheduling, as described in Section 6.2. Therefore, they comprise the system resources needed to support an autonomous flow of control, like the processor context to be saved and restored in a context switch, discussed in Section 4.2.3.

System resources allocated at the process level are implicitly shared among all its threads but do not go beyond process boundaries. Therefore, for instance, an area of memory allocated by a thread with `malloc()` is automatically shared with, and accessible to, all other threads belonging to, and executing within, the same process— provided they get to know its address somehow. On the contrary, threads belonging to different processes live in disjoint address spaces and, unless special actions are taken, do not share memory.

However, the efficient implementation of multiple processes each having its own, protected address space requires hardware assistance in the form of a full-fledged memory management unit (MMU) or, at the very least, a memory protection unit (MPU). Since these resources might be unavailable or difficult to use without worsening execution determinism in many real-time systems, the standard allows implementations to still support multitasking by means of the single-user, single-process (SUSP) approach. This concept was first introduced with the POSIX real-time oriented application environment profiles defined in [64].

In the SUSP execution environment that RTEMS supports, a single process is implicitly created when the system starts up, and one of its threads goes on to execute the `main()` C-language entry point. This initial user thread can then dynamically create other threads as described in Section 6.2, but all these threads always live within the same initial process boundary. For this reason, in the following we will mainly focus on thread-level scheduling and keep the description of process-level aspects to a minimum. These will be further mentioned only when the hierarchical relationship between processes and threads becomes essential for the discussion.

The abstract POSIX scheduling model defines a range of *global* priority levels, common to the whole system, and an ordered list of threads for each level. The list contains all the ready threads having that priority at the moment. The global priority range is represented by the tall rectangle on the right of Figure 6.2.

It is worth remarking that all scheduling-related data structures and algorithms defined in the POSIX standard are part of a *conceptual* scheduling model. They are purposefully kept as simple as possible to make the specification clearer and do not imply any recommendations for the actual implementation of the model. Implementations are free to replace them with others, as they often do for the sake of efficiency, provided they preserve the model semantics.

The algorithm used to assign an idle core to a ready thread is very straightforward. The scheduler simply selects the thread at the head of the highest-priority, non-empty

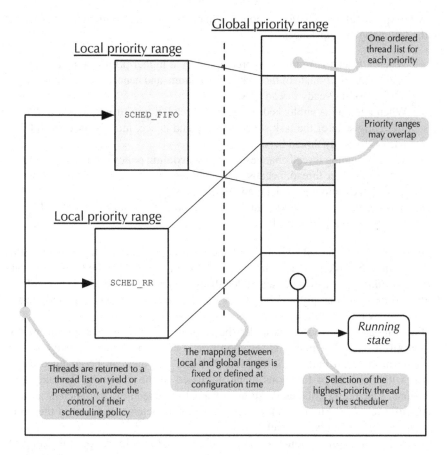

FIGURE 6.2 Summary of the POSIX scheduling model and policies.

thread list. The selected thread is removed from the list and moved to the *running* state of the task state diagram. In the figure, this action is depicted by the arrow near the bottom right corner, which goes from the global thread lists to the *running* state.

Instead, other components called scheduling *policies* are responsible for deciding whether a running thread shall be preempted, that is, forcibly removed from the *running* state and brought back into one of the thread lists. Moreover, scheduling policies also determine how a thread that becomes ready again—for instance, after temporarily transitioning to the *blocked* state—is inserted into the thread lists, and whether and how threads are moved among lists, thus dynamically changing their priority.

In some cases, operations that threads perform on synchronization objects also trigger priority changes, in order to preserve some useful real-time execution properties of the system. An example will be given in Chapter 8, while describing the priority inheritance and ceiling protocols.

More specifically, a scheduling policy is in charge of determining thread ordering in the following circumstances:

- When a thread is running, is preempted by a higher-priority thread, goes back to the *ready* state of the task state diagram, and hence, must be inserted into one of the ready thread lists.
- When a thread is unblocked by another thread, it goes from the *blocked* to the *ready* state of the task state diagram, and hence, must be inserted into one of the ready thread lists.
- When a running thread changes its own scheduling policy or priority, or the ones of another thread, because this may require the running thread to be moved back into one of the ready thread lists.
- In other circumstances (for instance, at regular time intervals) defined by the scheduling policy itself.

In other words, the scheduling policy of a thread affects all transitions of the thread within the task state diagram shown in Figure 3.4, except the transition from *ready* to *running* (for which the scheduler always selects the thread at the head of the highest-priority non-empty list) and from *running* to *blocked* (which is undertaken voluntarily by the thread itself).

Each thread in the system is under the control of one single scheduling policy at any given time. The policy is set upon thread creation, by means of the thread attribute mechanism outlined in Sections 6.1 and 6.2, and can be changed afterwards with the help of functions to be discussed in the following. To take their decisions, scheduling policies make use of a set of scheduling *parameters* associated with each thread under their control. These parameters can be set in the same way as the scheduling policy itself.

Individual scheduling policies work within a *local* priority range, consisting of at least 32 distinct priority levels. The mapping of local priority ranges within the global range is either fixed or established at system configuration time, but cannot be altered afterwards. Instead, suitable primitives exist to retrieve mapping information at runtime. More specifically, the functions sched_get_priority_min and sched_get_priority_max take a scheduling policy identifier as argument and return the minimum and maximum possible priority for that policy, respectively.

As shown in Figure 6.2, priority ranges belonging to different scheduling priorities may overlap. Where ranges overlap, multiple scheduling policies compete for inserting ready threads into the same lists. The fixed-priority thread selection mechanism presented at the beginning of this section is eventually responsible to resolve the competition and choose which threads will actually run, regardless of the policy they are controlled by.

The standard explicitly specifies the scheduling policies listed in Table 6.4. It also leaves implementations free to offer additional policies with the proviso that any application that makes use of them will not be portable from one POSIX system to another.

TABLE 6.4

RTEMS Thread Scheduling Policies, POSIX API

Policy	Description
SCHED_FIFO	First-in, first-out scheduling (without time sharing)
SCHED_RR	Round-robin scheduling (FIFO plus time sharing within a priority level)
SCHED_SPORADIC	Sporadic server scheduling [115] (optional, not provided by RTEMS)
SCHED_OTHER	Implementation-defined scheduling (non necessarily real-time)

Of them, we will not further discuss the SCHED_OTHER policy because it is a generic policy and the standard leaves the definition of its behavior completely up to the implementation. This policy may coincide with either SCHED_FIFO or SCHED_RR if the implementation so decides, but it must not necessarily provide any real-time execution guarantee.

The main purpose of it being explicitly defined in the standard is to enable programmers to specify, in a portable way, that a certain thread does not require any specific real-time scheduling policy.

The SCHED_SPORADIC policy is optional and implements a variant of the *sporadic server* algorithm described in [115]. It must be present only on implementations that support the "Process sporadic server" or "Thread sporadic server" options of the POSIX standard, which RTEMS currently does not. For this reason, it will no longer be discussed in the following, either.

Of the two remaining policies listed in Table 6.4, the SCHED_FIFO is the most basic one. Its only scheduling parameter is an integer-valued, non-negative priority held in the sched_prio field of the struct sched_param. Contrary to the convention of many real-time operating systems, higher values correspond to higher priorities. The policy operates as follows:

- When a running thread is preempted by a higher-priority thread, it is returned to the *head* of the thread list it belongs to, according to its priority.
- When a blocked thread becomes ready again, having been unblocked by another thread, it is inserted at the *rear* of the thread list it belongs to, according to its priority. This may trigger a preemption in favor of the thread just unblocked if its priority is higher than the priority of a running thread.
- When the priority of a running or ready thread is modified, either by itself or by another thread, it is placed at the *rear* of the ready thread list that corresponds to its new priority, except when a specific function pthread_setschedprio, has been used to request the change.
- In the last case, the placement depends on the sign of the priority change, namely:
  - If the priority has been increased, it is placed at the *rear*.
  - If the priority is unchanged, the thread does not change place.

- If the priority has been decreased, it is placed at the *head*.
- When a running thread voluntarily relinquishes the core by means of a *yield* operation, it is placed at the *rear* of the thread list it belongs to.

The rules concerning pthread_setschedprio may seem rather convoluted at first sight, but they are indeed necessary to enable the user-level implementation of strategies to avoid unbounded priority inversion. Examples of these strategies are the priority inheritance and priority ceiling protocols—to be discussed in Chapter 8. They may already be available for some built-in POSIX synchronization objects, but a user-level replacement or extension may be needed to apply them to other, more sophisticated objects or implement other variants of these algorithms.

The SCHED_RR policy has much in common with the SCHED_FIFO policy, plus an additional mechanism that prevents a long-running thread at a certain priority from monopolizing processing resources at the expense of other ready threads with its same priority. Considering, for simplicity, a single-processor single-core system, when threads are scheduled in a SCHED_FIFO manner and a certain thread is running, it may exit the running state only if it voluntarily blocks or yields, or is preempted by a higher-priority thread. If none of those conditions apply, it may keep running indefinitely although there are other ready threads at its same priority.

The SCHED_RR policy addresses this shortcoming by means of the *round-robin* rotation mechanism summarized in the following rules, to be applied in addition to the rules of the SCHED_FIFO policy:

- If a thread keeps running for the amount of time called *time quantum*, it is returned to the *rear* of the ready thread list corresponding to its priority, and another thread is chosen for execution.
- If a thread loses the CPU due to a preemption and resumes execution at a later time, it is allowed to execute for the portion of time quantum it has not consumed yet.

Informally speaking, the additional rules above force ready threads with the same priority to "rotate" in and out of the running state at a pace determined by the time quantum (the shorter the quantum, the faster the pace). In this way, each of them obtains on average its fair share of execution time when the SCHED_RR policy is used in isolation. If the policy is used together with others, due to the local–global priority mapping scheme depicted in Figure 6.2, this may no longer be true. For instance, nothing prevents same-priority threads scheduled by other policies from monopolizing the available cores, since they may not be subject to any time quantum limit.

The length of the time quantum, also called *round-robin time interval*, is a per-process item of information and is either fixed or part of the system configuration. It cannot be changed at run-time, although the function sched_rr_get_interval can be used to retrieve it. In the case of RTEMS, it is given by the ticks_per_timeslice item of the operating system configuration table. It is expressed in ticks and its default value is 50. Given that the default

TABLE 6.5

RTEMS Thread Scheduling Primitives, POSIX API

Function	Purpose
`sched_get_priority_min`	Return the minimum priority allowed for a given policy
`sched_get_priority_max`	Return the maximum priority allowed for a given policy
`sched_rr_get_interval`	Return the time quantum of the `SCHED_RR` policy
`pthread_getschedparam`	Retrieve the scheduling policy and parameters of a thread
`pthread_setschedparam`	Change the scheduling policy and parameters of a thread
`pthread_setschedprio`	Change the priority of a thread
`sched_yield`	Relinquish the CPU in favor of other threads

tick length is 10 ms, the default RTEMS time quantum is therefore 500 ms in the current operating system version.

After this general overview of the POSIX scheduling model, we can now give a deeper look at some additional functions related to it, besides the thread creation and termination functions discussed in Section 6.2. They are listed in Table 6.5. The first three functions enable the software to query, at run time, how the scheduling policies have been configured. More specifically, the functions:

```
int sched_get_priority_min(int policy);
int sched_get_priority_max(int policy);
```

return the minimum and maximum priority allowed for the scheduling policy indicated by `policy`. Both functions shall instead return −1, a negative value that cannot represent a valid priority, upon failure. In this case, they shall also set the `errno` variable to an error number that provides more information about the reason for the failure. The standard specifies the following error number:

EINVAL The value of the `policy` parameter does not represent a valid scheduling policy defined in the system.

The function:

```
int sched_rr_get_interval(pid_t pid,
    struct timespec *interval);
```

stores into the data structure referenced by `interval` the time quantum assigned to process `pid` and all the threads it contains. It returns zero upon successful completion. Upon failure, it shall return −1 and set `errno` to the value:

ESRCH The process indicated by `pid` was not found in the system.

The `struct timespec` is a POSIX data type specifically used to represent time and time intervals. It will be described in more detail in Section 6.6.

As shown in Section 6.2, the scheduling policy and parameters of a thread can be set when the thread itself is created. They can be retrieved at a later time by means of the function:

```
int pthread_getschedparam(pthread_t thread,
    int *restrict policy,
    struct sched_param *restrict param);
```

This function, given a thread descriptor `thread`, stores the scheduling policy of the thread into the location pointed to by `policy` and its scheduling parameters into the location pointed to by `param`. The function `pthread_getschedparam` returns zero when successful. Otherwise, it shall return a non-zero error code. The standard recommends that the function should detect and report the error:

ESRCH The argument `thread` refers to a thread that does not exist in the system.

The scheduling policy of a thread and its scheduling parameters can also be changed at run time by calling the function:

```
int pthread_setschedparam(pthread_t thread,
    int policy, const struct sched_param *param);
```

Given a thread descriptor `thread`, this function changes the scheduling policy of the thread to the value of the argument `policy` and its scheduling parameters to the contents of the data structure referenced by `param`. The function `pthread_setschedparam` returns zero when successful. Otherwise, it shall fail for the following reasons:

EINVAL The scheduling policy indicated by `policy` or one of its scheduling parameters in the data structure pointed by `param` are invalid.
ENOTSUP The scheduling policy indicated by `policy` or one of its scheduling parameters in the data structure pointed by `param` are valid, but the implementation does not support them.
EPERM The calling thread does not have the permission to change the scheduling policy and/or the scheduling parameters of the target thread.

As for the `pthread_getschedparam` function just described, the standard recommends that the `pthread_setschedparam` should detect and report the additional error:

ESRCH The argument `thread` refers to a thread that does not exist in the system.

The function:

```
int pthread_setschedprio(pthread_t thread, int prio);
```

has a more limited scope than `pthread_setschedparam` because it enables the caller to modify the priority of the thread whose descriptor is `thread` to the value `prio`, but not its scheduling policy. However, it triggers a more sophisticated handling of the priority change at the scheduling policy level, as outlined previously, while describing individual policies.

The `pthread_setschedprio` returns zero upon successful completion. Otherwise, it shall return a non-zero error code. The function may fail for the following reasons:

EINVAL The priority indicated by `prio` is invalid for the scheduling policy of `thread`.

ENOTSUP The priority indicated by `prio` is valid for the scheduling policy of `thread`, but the implementation does not support it.

EPERM The calling thread does not have the permission to change the priority of the target thread.

The last function listed in Table 6.5 has a very simple signature:

```
int sched_yield(void);
```

By invoking this function, the calling thread signals its will to relinquish the core it is running on in favor of other threads. As a consequence, the calling thread is returned to one of the ready thread lists, as determined by its scheduling policy, and the scheduler picks a new ready thread for execution.

Calling `sched_yield` does not necessarily imply that the calling thread will stop executing, though. This may happen even in very simple scenarios, for instance, when the calling thread is the highest-priority ready thread in the system at the moment and there are no other ready threads with the same priority.

The `sched_yield` function shall return zero when successful and a non-zero status code upon failure. However, the standard does not define any error condition for it.

6.4 FORCED THREAD TERMINATION (CANCELLATION)

The *cancellation* mechanism allows a thread to demand the *forced involuntary termination* of another thread and complements the two *voluntary* termination methods described in Section 6.2.

However, terminating a thread at an arbitrary point of its execution can lead to unexpected side effects that may easily undermine the stability of the whole system. For instance, if a thread is forcibly terminated while it is updating a complex, shared data structure, it will likely leave the data structure in an inconsistent state. The consequences of the inconsistency will likely become evident only when another thread tries to access the data structure at a later time, but may be severe.

Similar issues may also occur even if the thread is not necessarily sharing data with others. If a thread is terminated abruptly while it holds some process-level resources, like an area of memory it dynamically allocated with `malloc()`, these resources will become permanently unavailable until the process as a whole terminates unless they can be accessed in other ways, because they will not be reclaimed automatically. In the case of `malloc()`, the only way to free the memory would be for

TABLE 6.6

RTEMS Thread Cancellation Primitives, POSIX API

Function	Purpose
pthread_cancel	Forcibly terminate a thread
pthread_setcancelstate	Set thread cancellation state
pthread_setcanceltype	Set thread cancellation type
pthread_testcancel	Test for a pending cancellation request
pthread_cleanup_push	Push a thread cancellation handler
pthread_cleanup_pop	Pop and possibly execute a thread cancellation handler

another thread to somehow know the address of the memory area and call `free()` on it.

For these reasons, threads are given the ability to determine how they will respond to any cancellation request directed to them. The possible responses to a cancellation request range from ignoring it completely to terminating immediately. Moreover, threads are also given the opportunity to execute some final cleanup actions before terminating, by installing appropriate *cancellation handlers*. The functions used to send a cancellation request and set up the way a thread is going to respond to it are summarized in Table 6.6.

A cancellation request is submitted by calling the function `pthread_cancel` using as argument the thread descriptor of the target thread:

```
int pthread_cancel(pthread_t thread);
```

The function returns zero when executed successfully. Otherwise, it shall return a non-zero value to report the error back. The standard does not mandate any specific error check, but recommends that `pthread_cancel` should fail if the thread descriptor given as argument is invalid, and return:

ESRCH The argument `thread` refers to a thread that does not exist in the system.

The way the target thread reacts to a cancellation request directed to it depends on its *cancelability*. This information is set by the thread itself to one of the three settings listed in Table 6.7 by means of the following *cancellation control* functions:

```
int pthread_setcancelstate(int state, int *oldstate);
int pthread_setcanceltype(int type, int *oldtype);
```

They set the cancelability *state* and *type* of the calling thread to the values given by the arguments `state` and `type`, respectively. At the same time, they store the previous value of state and type into the locations pointed by `oldstate` and `oldtype`, respectively. The valid values of the cancelability state and type are:

- `PTHREAD_CANCEL_DISABLE` or `PTHREAD_CANCEL_ENABLE` for the cancelability state, and

TABLE 6.7

Thread Cancelability States and Types, POSIX API

Thread cancelability state/type (all macros begin with PTHREAD_CANCEL_)		Meaning
State	**Type**	
DISABLE	*ignored*	Cancellation disabled
ENABLE	DEFERRED	Enabled upon reaching a cancellation point
ENABLE	ASYNCHRONOUS	Enabled, immediate (asynchronous)

- PTHREAD_CANCEL_DEFERRED or PTHREAD_CANCEL_ASYNCHRONOUS for the cancelability type.

Both functions return zero upon successful completion. They shall return a non-zero error code upon detecting the following errors:

EINVAL The value of state (for pthread_setcancelstate) or type (for pthread_setcanceltype) is invalid.

The three possible cancelability settings of a thread naturally affect how it reacts to a cancellation request directed to it:

1. If the cancelability state is PTHREAD_CANCEL_DISABLE, the cancelability type setting has no effect. Any cancellation request sent to the thread has no immediate effect, but is held pending and will be reconsidered once the thread changes its cancelability state to PTHREAD_CANCEL_ENABLE.

   This memory effect should be carefully taken into account while programming because it may lead to the unexpected cancellation of a thread immediately after it changes its cancelability state, due to a cancellation request sent to it a long time before, and also because the standard currently does not specify any way to retract a cancellation request after issuing it.

   Programmers should also take into account that, while a thread cancelability state is PTHREAD_CANCEL_DISABLE there is no way to terminate it, unless it does so voluntarily. This makes the thread safe from all the data consistency and resource loss issues discussed previously, but also makes its execution unmanageable from the outside, especially if it gets trapped in a processor-intensive loop.

2. If the cancelability state is PTHREAD_CANCEL_ENABLE and the type is PTHREAD_CANCEL_DEFERRED the thread honors cancellation requests in a *deferred* fashion, that is, a cancellation request directed to it stays pending and takes effect as soon as the thread reaches a *cancellation point*.

   In other words, this setting enables a thread to honor cancellation requests, which is a desirable feature, but in a controlled way to counteract the possibility of data inconsistency and resource loss, because programmers know in advance where in

TABLE 6.8
Cancellation Points, POSIX API

Task management (Chapter 5)	
pthread_join	pthread_testcancel

Signal handling (Chapter 5)	
sigwait	sigwaitinfo
sigsuspend	

Inter-task communication based on shared memory (Chapter 7)	
sem_wait	sem_timedwait
pthread_cond_wait	pthread_cond_timedwait

Inter-task communication based on message passing (Chapter 9)	
mq_send	mq_receive
mq_timedsend	mq_timedreceive

the code a certain thread could be canceled, regardless of its exact timing relationship with respect to the thread requesting the cancellation.

Several of the functions described in this book are cancellation point, Table 6.8 summarizes them. In addition, many other POSIX functions not explicitly discussed here (for instance, input–output functions like open and printf) shall or may also be cancellation points.

Readers should refer to the standard [68] for a full list with the caveat that, if the standard says a function *may* be a cancellation point, portable code shall be written as if it actually *were* a cancellation point. This is to prevent thread cancellation from occurring at unexpected places on some implementations rather than others, which could easily lead to nasty, platform-dependent issues.

Generally speaking, and with some exceptions, the rationale behind Table 6.8 is that any function that could possibly block the calling thread for a long or unbounded amount of time shall be a cancellation point. In this way, a thread in deferred cancelability mode honors any pending cancellation request *before* engaging in such a wait.

A cancellation point may also be explicitly introduced in the code where needed by calling the function:

void pthread_testcancel(**void**);

3. If the cancelability state is PTHREAD_CANCEL_ENABLE and the type is PTHREAD_CANCEL_ASYNCHRONOUS the thread honors cancellation requests *immediately* as they arrive, that is, *asynchronously*.

Although the advantage of executing a thread in asynchronous cancellation mode—a prompt response to cancellation requests—may sound appealing, it must be carefully weighted against the disadvantages. Besides the risk of data inconsistency and resource loss already recalled previously, while a thread is

TABLE 6.9

Functions Safe Against Asynchronous Cancellation, POSIX API

Function
pthread_cancel
pthread_setcancelstate
pthread_setcanceltype

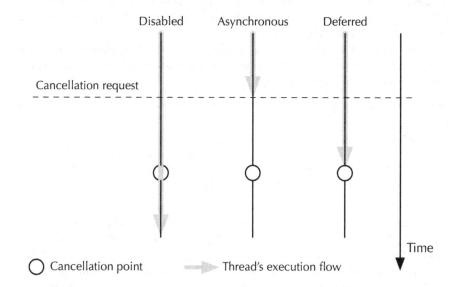

FIGURE 6.3   Possible responses to POSIX cancellation requests.

executing in this mode it must not call any POSIX functions, except the ones explicitly defined to be *async-cancel safe* in the standard. These functions are in rather scarce supply and are summarized in Table 6.9. Calling any other function, directly or indirectly, leads to undefined behavior.

Newly created threads, including the thread that first executes main() when the process starts, have their cancelability state set to PTHREAD_CANCEL_ENABLE and their cancelability type to PTHREAD_CANCEL_DEFERRED, which puts them in deferred cancelability mode. Unlike other thread characteristics, this default cannot be set by means of thread attributes while creating the thread. The new thread must explicitly call pthread_setcancelstate and/or pthread_setcanceltype itself.

Figure 6.3 depicts and summarizes the three possible ways of responding to a cancellation request that a thread may set up as previously described. Since they represent a trade-off between reacting quickly, without wasting execution resources, and not placing additional burdens on programmers, it is unlikely that a moderately complex thread keeps the same cancelability settings for its whole lifetime.

The definition of the functions pthread_setcancelstate and pthread_ setcanceltype—in particular, their ability to return the previous cancelability state and type while setting them to new values—helps to define well-delimited regions of code, or *scopes*, in which the thread cancelability state is certainly set to the desired value.

At the same time, a proper use of these functions ensures that, at the end of a scope, the thread cancelability state and type are restored to the values they had upon entering the scope itself. In this way, scopes can be nested at will without side effects, at least as long as inner scopes set the cancelability state and type to more conservative values than outer scopes.

For instance, let us imagine that a function f() needs to disable cancellations while it is executing a fragment of code. It can be coded in the following way, neglecting error handling for simplicity:

```
void f(void)
{
    int ps;
    pthread_setcancelstate(PTHREAD_CANCEL_DISABLE, &ps);
        ... code executed with cancellations disabled ...
    pthread_setcancelstate(ps, &ps);
}
```

The first call to pthread_setcancelstate sets the thread cancelability state to completely disable cancellations and, at the same time, stores the previous state into ps. The second call to pthread_setcancelstate restores the cancelability state to the value it had upon entry into f(). The second argument in the second call to pthread_setcancelstate is &ps rather than NULL, although the value that the function stores into ps is actually unused, because the POSIX standard does not require pthread_setcancelstate to accept NULL pointers, although many implementations do.

By delimiting the scope in this way, we can safely call function f() from another function g(), which needs cancellations to be handled in a deferred way:

```
void g(void)
{
    int pt;
    pthread_setcanceltype(PTHREAD_CANCEL_DEFERRED, &pt);
        ... code executed with deferred cancellations ...
        f();
        ... code executed with deferred cancellations ...
    pthread_setcanceltype(pt, &pt);
}
```

Scope nesting still works if, for instance, g() is called by an outer function h(), which already runs with cancellations disabled. In this case, neither the calls to `pthread_setcanceltype` performed by `g()`, nor the calls to `pthread_setcancelstate` performed by `f()` have any effect and the whole code is executed with cancellations disabled. This is correct because this is the most conservative cancelability state and type that `h()`, `g()`, and `f()` call for.

Before terminating voluntarily, a thread has all the means to perform any cleanup operation it may need in a very straightforward way. However, this is not the case when the thread terminates involuntarily due to a cancellation. For this reason, the POSIX standard specifies a way for a thread to register a set of *cleanup handlers* that will automatically be executed immediately before the thread terminates. For consistency, cleanup handlers are also executed when the thread terminates voluntarily, by calling `pthread_exit` or by returning from its entry point.

A cleanup handler consists of a function that receives a pointer of type `void *` as argument and returns `void`. The function `pthread_cleanup_push` pushes the cleanup function `routine` and its argument `arg` onto a *stack*:

```
void pthread_cleanup_push(void (*routine)(void*),
    void *arg);
```

Upon thread termination, cleanup handlers are popped from the stack and executed one after another, that is, the handler that was pushed last is executed first. Each handler receives as argument the same `arg` that was given to `pthread_cleanup_push` when the handler was pushed. Cleanup handlers can also be manually popped, and optionally executed, by calling the function:

```
void pthread_cleanup_pop(int execute);
```

Its only argument, `execute`, is a Boolean flag that, when not zero, indicates that the cleanup handler shall be executed besides being popped. To ensure that `pthread_cleanup_push` and `pthread_cleanup_pop` have well-defined semantics and enable their implementation as function-like macros, they are subject to several usage restrictions. The most important ones are:

- Syntactically, they must appear as statements.
- They must appear in pairs, in the same lexical scope, and be used to properly bracket a block of code that needs a cleanup handler.
- Abandoning the bracketed block of code prematurely, for instance, by means of `return`, `longjmp`, or `goto`, leads to undefined behavior.

Programmers should refer to the standard [68] as a reference to be aware of all these restrictions while they write their code, because compilers may be unable to check whether they have been compiled with or not.

6.5 SIGNAL HANDLING

Signals are a facility that was already specified by the ISO C standard [69] in 1989 and is widely available on most general-purpose operating systems. Signals provide a mechanism to convey information to a process or thread when it is not necessarily waiting for input.

The signal mechanism has a significant historical heritage. For instance, it was first designed when multithreading was not yet in widespread use. Therefore, its interface and semantics underwent many adjustments and extensions since their inception. Originally, the main goal of signals was to give processes the ability to react properly to external and usually fatal events that occurred asynchronously with respect to the notified process, like:

- Some hardware-detected errors occurring during the execution of the process, for example, a memory reference through an invalid pointer.
- Various system and hardware failures, such as an impending power failure.
- The explicit generation of a signal, often with the purpose of forcing the target process to terminate.

For instance, when a user types the `intr` control character (Control-C by default) on a terminal controlled by a POSIX-compliant shell, what the shell does is to send a signal (called `SIGINT`) to the process currently running in the foreground. Unless the foreground process has set up some special way of handling the signal, this causes it to terminate, so that the user can regain control of the terminal.

Currently, the signal handling mechanism owes most of its complexity to the need of maintaining compatibility with the historical implementations of the mechanism made, for example, by the various flavors of the influential Unix operating systems.

In this book, for the sake of clarity and conciseness, the compatibility interfaces will not be described and the discussion will be limited as much as possible to how signals work in the single-user, single-process (SUSP) execution environment that RTEMS supports. Readers should refer to the POSIX standard [68] for a more comprehensive description of the mechanism as a whole.

The POSIX standard further extends the signal mechanism to make it suitable for real-time asynchronous event handling. In particular, in the rest of this book we are going to discuss how signals are used to convey:

- Time-related events, that is, events that make a process or thread aware of the passing of time. These events are generated by POSIX *timers*, to be discussed in Section 6.6.
- Data availability in one of the inter-process communication mechanisms foreseen by the POSIX standard, the *message queue*, to be described in Chapter 9.

With respect to the ISO C signal behavior, the POSIX standard introduces two main enhancements of interest to real-time programmers:

1. In the ISO C standard, the various kinds of signals are identified by an integer number that, in most cases, should be denoted by a symbolic constant, like `SIGINT`, to make the application code portable. When multiple signals of different kinds are pending, they are serviced in an unspecified order.
   The POSIX standard continues to use signal numbers for backward compatibility, but specifies that, in a subset of their allowable range, between the values

SIGRTMIN and SIGRTMAX, a priority hierarchy among signals is in effect, so that the lowest-numbered signal has the highest priority of service.

The range must include RTSIG_MAX distinct signal numbers. This value is a runtime invariant and must be defined as a symbolic constant in the header limits.h when it is known at compile time. If the value is indeterminate at compile time (for instance, because it is part of the operating system configuration) the definition is omitted and processes can retrieve the value by means of the sysconf function. In any case, RTSIG_MAX shall be 8 as as minimum.

2. In the ISO C standard, there is no provision for signal queues. When multiple signals of the same kind are raised before the target process had a chance of handling them, all signals but the first are lost. Instead, POSIX specifies that the system must be able to keep track of multiple signals with the same number. Signals are always enqueued and serviced in the same order as they have been generated. This policy cannot be changed by the user.

   Moreover, POSIX also adds the capability of attaching a limited amount of information to each signal request, so that multiple signals with the same signal number can be distinguished from each other. The information must fit in an object of type union sigval, which is capable of holding either an integer (in its sival_int field) or a pointer (in sival_ptr).

As outlined previously, each signal has a signal number associated to it, which uniquely identifies its kind. For example, the interrupt signal sent by the shell to a process with the user types the intr character on the terminal has the number SIGINT associated to it. The object of type union sigval, which is associated to each signal and whose contents are chosen by the application upon signal generation, helps to further distinguish among different signals of the same kind.

For some kinds of events, like the time-related events to be described in Section 6.6, the POSIX standard specifies that the notification may also be carried out by the execution of an event-handling function in a separate thread, if the application so chooses. This mechanism is apparently more straightforward than signal-based notification, but requires more operating system resources and entails higher overheads. Moreover, since it involves the dynamic creation of a new thread at unpredictable times, it may introduce other shortcomings in a real-time execution environment, which may potentially escalate to a denial of service if the system is flooded by events. For these reasons, this notification method is currently unavailable on RTEMS.

Figure 6.4 depicts the life of a signal from its generation up to its delivery. Depending on their kind and source, signals may be directed to either a specific thread in a process, or to the process as a whole. In the latter case, every thread belonging to the process is a candidate for the delivery of the signal, by the rules described in the following.

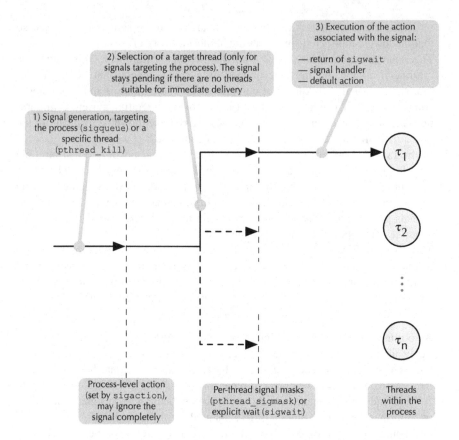

3) Execution of the action associated with the signal:

— return of `sigwait`
— signal handler
— default action

2) Selection of a target thread (only for signals targeting the process). The signal stays pending if there are no threads suitable for immediate delivery

1) Signal generation, targeting the process (`sigqueue`) or a specific thread (`pthread_kill`)

τ_1

τ_2

τ_n

Process-level action (set by `sigaction`), may ignore the signal completely

Per-thread signal masks (`pthread_sigmask`) or explicit wait (`sigwait`)

Threads within the process

FIGURE 6.4   Depiction of the POSIX signal handling mechanism.

Generation of a signal

The decision to target a specific thread or a whole process with a signal must be taken upon signal generation. For signals generated by the system, rather than by an explicit action performed by a thread, the POSIX standard specifies that the decision of whether the signal must be directed to the process as a whole or to a specific thread within a process must indicate the source of the signal as closely as possible.

Namely, if a signal is attributable to an action carried out by a specific thread, like a memory access violation, the signal shall be directed to that thread and not to the process. If such an attribution is either impossible or not meaningful, the signal shall be directed to the process. This is the case, for example, of the SIGINT signal. In addition, signals related to real-time asynchronous event handling shall always be directed to the process.

Threads also have the ability of synthesizing signals by means of two main interfaces, depending on the target of the signal:

- The `sigqueue` function, given a process identifier and a signal number, generates a signal directed to that process. An additional argument, of type `union sigval`, allows the caller to associate a limited amount of information to the signal, provided that the `SA_SIGINFO` flag is set for that signal number.
- The `pthread_kill` function generates a signal directed to a specific thread within the process to which the calling thread belongs. It is not possible to generate a signal directed to a specific thread of *another* process.

Additional interfaces exist to generate a signal directed to a group of processes, for example, the `killpg` function. However, they will not be further discussed here because they have not been extended for real time and lack the ability of associating any additional information to the signal. Moreover, they are of limited usefulness in a SUSP environment, in which there is only one process in the system.

Process-level action

For each kind of signal defined in the system, that is, for each valid signal number, processes may set up an action by means of the `sigaction` function. In general, the action may be:

- Ignore the signal completely. In this case, the delivery of the signal has no effect whatsoever on the process. If the signal is already pending when `sigaction` is called, it shall be discarded.
- A default action performed by the operating system on behalf of the process, and possibly with process-level side effects, such as the termination of the process itself.
- The execution of a signal handling function specified by the programmer.

However, some special signals whose intent is to enable the unconditional termination of a malfunctioning process, like `SIGKILL`, cannot be caught or ignored. Moreover, ignoring a signal generated as a consequence of a fatal error (like a memory access violation) and trying to continue leads the process to an undefined behavior.

In addition, the `sigaction` function allows the caller to set zero or more *flags* associated with the signal number. Of the rather large set of flags specified by the POSIX standard, the following ones are of particular interest to real-time programmers:

- The `SA_SIGINFO` flag, when set, enables the association of a limited amount of information to each signal, which will then be conveyed to the signaled process or thread. As previously mentioned, the information is stored in an object of type `union sigval`.
  If the action associated with the signal is the execution of a user-specified signal handler, setting this flag also extends the list of arguments passed to the signal handler to include additional information about the reason why

the signal was generated and about the receiving thread's context that was interrupted when the signal was delivered.

- The SA_RESTART flag, when set, enables the automatic, transparent restart of interruptible system calls when they are interrupted by the signal. If this flag is clear, system calls that were interrupted by a signal fail with the error indication EINTR and must be explicitly performed again by the application, if appropriate.

- The SA_ONSTACK flag, when set, commands the switch to an alternate stack, set up by means of the sigaltstack function, for the execution of the signal handler. If this flag is not set, the signal handler executes on the regular stack.

It should be noted that the setting of the action associated with each kind of signal takes place at the *process* level, that is, all threads within a process share the same set of actions. Hence, for example, it is impossible to arrange for two different signal handling functions (for two different threads) to be executed in response to the same kind of signal.

Another important aspect to be taken into account is that there are considerable limitations on what a signal-handling function can do. More specifically, a portable signal-handling function can invoke, directly or indirectly, only functions that are explicitly marked as *async-signal-safe* in the POSIX standard [68]. A signal-handling function can, of course, also invoke other functions if they are known to be safe on the operating system in use, but this makes it not portable to other POSIX systems.

Very importantly, neither malloc nor free have to be async-signal-safe, thus ruling out dynamic memory allocation within a signal-handling function. Also, neither setjmp nor longjmp have to be async-signal-safe either, and hence, non-local transfer of control from a signal-handling function is not portable across POSIX systems.

Moreover, even async-signal-safe POSIX functions can still change the value of the errno variable of the calling thread, and this property holds even when they are called from a signal-handling function. This may lead to an inconsistent behavior if a signal interrupts the main code of a thread after it detected that a POSIX function that it invoked has failed, but before it could check errno to gather more information about the failure. This is because, when the interrupted thread will eventually check errno after signal handling has been completed, it will find the error code set by the signal-handling function rather than the one set by the POSIX function that failed in the first place.

Going back to the lifetime of a signal depicted in Figure 6.4, the system checks the process-level action associated with the signal in the target process immediately after signal generation, and discards the signal if the action is set to ignore it. Otherwise, it proceeds to check whether the signal can also be acted on immediately, or the signal should remain *pending*.

Signal delivery and acceptance

Provided that the action associated to the signal at the process level does not specify to ignore the signal, a signal can be either *delivered to* or *accepted by* a thread within the process.

Unlike the action associated to each kind of signal, which is common to all threads in a process, each thread has its own *signal mask*. By means of the signal mask, each thread can selectively *block* some kinds of signal from being delivered to it, based on their signal number. The pthread_sigmask function allows the calling thread to examine and change its signal mask, represented as a signal set of type sigset_t. It can be set up and manipulated by means of the functions:

sigemptyset: Initializes a signal mask so that all signals are *excluded* from the mask.
sigfillset: Initializes a signal mask so that all signals are *included* in the mask.
sigaddset: Given a signal mask and a signal number, adds the specified signal to the signal mask. It has no effect if the signal was already in the mask.
sigdelset: Given a signal mask and a signal number, removes the specified signal from the signal mask. It has no effect if the signal was not in the mask.
sigismember: Given a signal mask and a signal number, checks whether the signal belongs to the signal mask or not.

A signal can be *delivered* to a thread if and only if that thread does not block the signal. When a signal is successfully delivered to a thread, that thread executes the process-level action associated with the signal. Alternatively, a thread may explicit wait for one or more kinds of signal, by means of the sigwait function. This function blocks the calling thread until one of the signals passed to sigwait, by means of an argument of type sigset_t, is conveyed to the thread. When this occurs, the thread *accepts* the signal and continues past the sigwait function.

Since the standard specifies that signals in the range from SIGRTMIN to SIGRTMAX are subject to a priority hierarchy, when multiple signals in this range are pending, the sigwait shall consume the lowest-numbered one. It it important to remark that, for this mechanism to work correctly, the thread must block the signals it wishes to accept by means of sigwait, through the signal mask passed to pthread_sigmask. Otherwise, signal delivery takes precedence over acceptance.

Two, more powerful, variants of the sigwait function exist: sigwaitinfo has an argument used to return additional information about the signal just accepted, including the union sigval associated to the signal when it was generated. Furthermore, sigtimedwait also allows the caller to specify the maximum amount of time that shall be spent waiting for a signal to arrive.

Selection of the target thread

The way in which the system selects a thread within a process to convey a signal depends first of all on where the signal is directed:

TABLE 6.10
RTEMS Signal-Related Primitives, POSIX API

Function	Purpose
sigqueue	Generate a signal directed to a process
pthread_kill	Generate a signal directed to a specific thread
sigaction	Set the process-level action associated with a signal
sigaltstack	Set up an alternate stack to be used for signal handling
pthread_sigmask	Set the per-thread signal mask of the calling thread
sigwait	Wait to accept a signal belonging to a given signal set
sigwaitinfo	Like sigwait, but returns additional information about the signal
sigtimedwait	Like sigwaitinfo, but with an upper limit on the waiting time
sigemptyset	Initialize a signal set with no signal numbers in it
sigfillset	Initialize a signal set with all signal numbers in it
sigaddset	Add a signal number to a signal set
sigdelset	Delete a signal number from a signal set
sigismember	Check whether or not a signal number belongs to a signal set

- If the signal is directed toward a specific thread, only that thread is a candidate for delivery or acceptance.
- If the signal is directed to a process, all threads belonging to that process are candidates to receive the signal.

In the second case the system looks for one thread within the process, which has an appropriate signal mask (for delivery), or which is performing a suitable sigwait (for acceptance). Three cases are possible:

- If there is *no* thread waiting to accept the signal or available for its delivery, the signal remains *pending* until its delivery or acceptance becomes possible. If the process-level action associated to that kind of signal is set to ignore it in the meantime, the system forgets everything about the pending signal, and all the other signals of the same kind.
- If there is exactly *one* thread waiting to accept the signal, or available for delivery, the signal goes to that thread.
- If there are *multiple* threads for which signal acceptance or delivery is possible, the system picks exactly one of them, using an unspecified algorithm.

Signal-related POSIX primitives

Table 6.10 lists the main signal-related POSIX primitives, divided into four groups. The first three groups contain the functions that can be used to generate a signal, set up how a process responds to a signal, and determine how individual threads within a process accept signals or wait for them, respectively. The fourth group at the bottom of the table contains some utility functions used to initialize and manipulate signal sets.

Going back to the first group, the function:

```
int sigqueue(pid_t pid, int signo,
    const union sigval value);
```

sends signal number `signo` to process `pid`. Moreover, depending on how the target process has chosen to react to the signal (by means of the `sigaction` function to be discussed later) it may also attach to the signal the information `value`, sometimes called the *signal value*. Even upon successful completion, the function returns to the caller immediately after queuing the signal. In other words, it does not synchronize the calling and the target processes in any way.

To avoid consuming a large and possibly unbounded amount of system resources in queuing signals, the standard specifies that threads belonging to a certain process can queue at most `SIGQUEUE_MAX` signals in total, even if they target different processes. The value of `SIGQUEUE_MAX` must be at least 32. As for `RTSIG_MAX`, `SIGQUEUE_MAX` must be defined as a symbolic constant when it is known at compile time, or made available by means of the `sysconf` function if it is not.

The `sigqueue` function returns zero upon successful completion. Upon failure, it shall return −1 and set `errno` to one of the following codes:

EINVAL The signal number `signo` is invalid or corresponds to a signal that the system does not support.
ESRCH The process `pid` does not exist.
EAGAIN There are insufficient resources to queue the signal, either because the process tried to exceed `SIGQUEUE_MAX`, the maximum number of currently queued signals, or a system-wide limit has been exceeded.
EPERM The operation was not performed due to lack of appropriate permissions.

For what concerns permission checks and the `EPERM` error code, the rules that determine whether a process may send a signal to another are fairly complex. However, they will not be further discussed in the following because they do not apply to a SUSP execution environment. In fact, in this kind of environment there is only one process and a process is always allowed to send a signal to itself. The function:

```
int pthread_kill(pthread_t thread, int signo);
```

has the same purpose as `sigqueue`, but is more specific and allows a thread belonging to a certain process to send a signal to another thread belonging to *the same* process. The target thread is identified by means of its thread descriptor `thread`. Signals sent in this way skip the target thread selection phase within the destination process. Unlike `sigqueue`, this function allows the caller to specify the signal number `signo`, but not a signal value.

Another difference with respect to `sigqueue` is that, upon error, `sigqueue` returns −1 and uses `errno` to provide more information about the error. Instead, `pthread_kill` returns the error code directly. The function `pthread_kill` shall fail for the following reasons:

EINVAL The signal number `signo` is invalid or corresponds to a signal that the system does not support.

Moreover, the function may detect an attempt to use a thread descriptor that corresponds to a non-existent thread, for instance, because it terminated, and report this error with the code:

ESRCH The given `thread` does not exist.

In order to examine and/or set the process-level action associated with a signal, any threads in the process may call the function:

```
int sigaction(int signo,
    const struct sigaction *restrict act,
    struct sigaction *restrict oact);
```

This function performs two distinct operations involving the process-level action associated to signal number `signo`:

- If the `oact` pointer is not NULL, it stores the current process-level action associated to `signo` into the data structure it references.
- If the `act` pointer is not NULL, it changes the process-level action associated to `signo` as instructed by the data structure it references.

The function returns zero upon successful completion. Otherwise, it shall return −1 and set `errno` to an error code without changing the process-level action associated with the signal `signo`. The `sigaction` function shall fail if:

EINVAL The signal `signo` is invalid or the process-level action indicated by `act` cannot be set for that specific signal. For instance, as previously mentioned, some signals cannot be caught or ignored.

ENOTSUP The SA_SIGINFO flag is set in the `sa_flags` field of the structure referenced by `act`, but this option is not supported for the given signal number `signo`. This may happen if `signo` is not in the range from SIGRTMIN to SIGRTMAX.

Both `act` and `oact` point to a data structure of type `struct sigaction`, whose fields are listed in Table 6.11. Besides the `flags` field, whose contents have already been discussed in the general description of the process-level action, the main fields of this data structure are `sa_handler` and `sa_sigaction`, which indicate the signal-handling function, that is, the function to be executed when a signal is delivered.

- The `sa_handler` field points to the signal-handling function to be used when the SA_SIGINFO flag is not set. It receives one single integer argument, the signal number.
- Instead, the `sa_sigaction` field points to the function to be used when the SA_SIGINFO flag is set. Besides the signal number, it also receives a pointer to an object of type `siginfo_t`, which provides more information about why and how the signal was generated, and a `void *` pointer.

TABLE 6.11

Fields of a `struct sigaction`, POSIX API

Field (prefixed by sa_)	Purpose
handler	Pointer to a "simple" signal-handling function with signature `void (*)(int)`, or one of the reserved values `SIG_IGN` (ignore the signal completely) or `SIG_DFL` (use the default handler).
sigaction	Pointer to an "enhanced" signal-handling function with signature `void (*)(int, siginfo_t *, void *)`, which is used instead of the simple signal-handling function `handler` when `SA_SIGINFO` is set in the `flags` field. With respect to `handler`, it receives two additional arguments. The first one is a pointer to an object of type `siginfo_t` and the second one is a pointer to `void`. Both arguments are discussed in more detail in the main text.
flags	This field contains the bitwise or of a set of flags, which modify the behavior of the signal. Among them, `SA_SIGINFO`, `SA_RESTART`, and `SA_ONSTACK` are discussed in the main text, while `SA_NODEFER` is described below.
mask	This field is a signal set (of type `sigset_t`). It contains the additional set of signal numbers that must be blocked during the whole execution of the signal-handling function. The signal number that triggered the execution of the signal-handling function is automatically blocked unless `flags` contains `SA_NODEFER`.

The `sa_handler` field, besides pointing to a signal-handling function, can also be set to one of the following reserved values:

SIG_IGN Ignore the signal completely, although ignoring some signals related to fatal error conditions leads to undefined behavior.

SIG_DFL Revert to the default handling action for the signal, which in most cases terminates the process.

Table 6.12 lists the contents of the main fields of a `siginfo_t` object. The third argument of the signal handling function pointed by `sa_sigaction` is actually a pointer to an object of type `ucontext_t`, which contains the saved context of the thread that was interrupted when the signal was delivered.

When the flag `SA_ONSTACK` is set, signal handling is performed on an alternate, dedicated stack instead of the regular stack of the thread to which the signal is delivered. The function:

```
int sigaltstack(const stack_t *restrict ss,
    stack_t *restrict oss);
```

can be used to set up the alternate stack and/or retrieve the current settings. The function `sigaltstack` uses the same underlying logic as `sigaction`. It can retrieve

TABLE 6.12

Main Fields of a `siginfo_t`, POSIX API

Field (prefixed by si_)	Purpose
signo	Signal number. This information is sometimes redundant because it coincides with the first argument passed to a signal-handling function (for signal delivery) and with the return value of `sigwaitinfo` (for signal acceptance).
code	Signal code that provides more information about the source of the signal. Codes of particular interest to real-time applications include: `SI_QUEUE`, which indicates that the signal was sent by `sigqueue`; `SI_TIMER`, denoting that the signal was sent by a timer expiration (see Section 6.6); and `SI_MESGQ`, which indicates that the signal was sent by a message queue (see Section 9.4).
value	The `union sigval` that contains the signal value, that is, the additional information associated with the signal when it was generated.

the current signal handling stack information (and store it into the data structure referenced by `oss`, if it is not a `NULL` pointer), and/or set up a new signal handling stack (according to the contents of the data structure referenced by `ss`, if it is not a `NULL` pointer). This function must be present only if the operating system supports the XSI (Single UNIX Specification) POSIX option. It is not implemented by RTEMS and will not be further discussed here for this reason.

The function:

```
int pthread_sigmask(int how, const sigset_t *restrict set,
    sigset_t *restrict oset);
```

enables the calling thread to examine and/or manipulate its own signal mask. Also in this case, the function uses the same underlying logic as `sigaction`. More specifically:

- If the pointer `oset` is not `NULL`, the function stores the current signal mask of the thread into the signal set referenced by it.
- If the pointer `set` is not `NULL`, the function also modifies the signal mask after possibly storing it using `oset` as just described.

The specific action that the function performs when `set` is not `NULL` depends on the value of the `how` argument:

- If it is `SIG_SETMASK`, the new signal mask is set as specified by the signal set referenced by `set` and the current mask is simply discarded.
- If it is `SIG_BLOCK`, the new signal mask is set to the union of the current mask and the signal set referenced by `set`. In other words, the signal

numbers indicated by `set` are *added* to the set of signals blocked by the calling thread.

- If it is `SIG_UNBLOCK`, the new signal mask is set to the intersection between the current mask and the complement of the signal set referenced by `set`. In other words, the signal numbers indicated by `set` are *removed* from the set of signals blocked by the calling thread.

The function `pthread_sigmask` returns zero upon successful completion. Otherwise, it shall not modify the signal mask and return an error code stating the reason for the failure. The standard specifies the following error code:

EINVAL The `how` argument is invalid, that is, it does not have any of the values previously mentioned.

A thread can explicitly announce it is willing to *accept* a certain set of signals, and possibly wait until one of them is sent to the process it belongs to, by calling the function:

```
int sigwait(const sigset_t *restrict set,
    int *restrict sig);
```

The argument `set` represents the set of signal numbers the thread is willing to accept. The function shall wait until one of these signals becomes pending, accept it, store its signal number into the variable referenced by `sig`, and finally return zero to the caller. If one of the signals is already pending at the time of the call, the function accepts it without waiting.

Upon failure, the `sigwait` function must return −1 and store an error code into `errno`. The function may fail for the following reason:

EINVAL The signal set referenced by `set` contains an invalid or unsupported signal number.

It is worth highlighting once more the peculiar relationship between the signal set passed to the `sigwait` function and the signal mask of the calling thread. In order to effectively wait for a set of signals, the corresponding signal numbers must be in the calling thread's signal mask, thus *blocking* these signals from being delivered to it. Although this may seen counterintuitive at first, it is indeed necessary because, otherwise, signal delivery would take precedence over acceptance.

Similarly, if a thread waits for a certain signal number by means of `sigwait`, it is essential that all the other threads in the process block that signal number and none of them also calls `sigwait` with a signal set that includes the same signal number. This is because, when the operating system finds more than one candidate thread for signal delivery or acceptance, it is undefined which candidate it will actually choose, thus leading to an unpredictable behavior.

Two, more powerful variants of `sigwait` are available. The first one returns to the caller additional information about the signal it accepted, besides the signal number:

```
int sigwaitinfo(const sigset_t *restrict set,
    siginfo_t *restrict info);
```

The function stores the additional information into an object of type `siginfo_t` provided by the caller and referenced by the `info` argument. This is the same information passed to a signal handler when the `SA_SIGINFO` flag is set for a signal number. Table 6.12 summarizes the object contents.

The `sigwaitinfo` function returns the number of the signal it accepted, which is guaranteed to be non-negative, upon successful completion. Otherwise it returns −1 and sets `errno` to an error code. The `sigwaitinfo` function may fail for the following reason:

EINTR The function was interrupted by the delivery of an unblocked, caught signal. The function returns to the caller after the corresponding signal handler has finished executing.

The second variant further extends `sigwaitinfo` to restrict the maximum waiting time:

```
int sigtimedwait(const sigset_t *restrict set,
    siginfo_t *restrict info,
    const struct timespec *restrict timeout);
```

The function `sigtimedwait` behaves like `sigwaitinfo`, but returns to the caller with an error indication when the amount of time indicated by the `timeout` argument elapsed before the function could accept any signals. Accordingly, it may fail for the following reason:

EAGAIN No signals were accepted before the time limit established by the `timeout` argument expired.

The `timeout` argument specifies a *relative* time interval, with respect to the time of the call. The interval shall be measured with the `CLOCK_MONOTONIC` clock—to be discussed in Section 6.6—if the system implements it. Otherwise, the function shall make use of `CLOCK_REALTIME`.

As a special case, if the time interval referenced by `timeout` is zero, the function polls pending signals and fails immediately if none of the signals indicated by `set` are pending at the moment. The function's behavior is instead unspecified when the `timeout` pointer is `NULL`.

In summary, the most important aspects to be taken into account from the programming point of view when using `sigwait` or one of its variants are:

1. All these function must be used to wait for a *blocked* signal, that is, a signal that belongs to the calling thread's signal mask.
2. The return values of the three functions differ in subtle ways. More specifically, `sigwait` returns either 0 or −1. On the contrary, `sigwaitinfo` and `sigtimedwait` return either the non-negative number of the signal they accepted or −1.

3. The `sigtimedwait` uses the error code `EAGAIN` to report a timeout. The POSIX standard also defines the more specific error code `ETIMEDOUT` to the same purpose and other functions use it instead.

The last group of functions, listed at the bottom of Table 6.10, enable the caller to initialize and manipulate a signal mask:

```
int sigemptyset(sigset_t *set);
int sigfillset(sigset_t *set);
int sigaddset(sigset_t *set, int signo);
int sigdelset(sigset_t *set, int signo);
int sigismember(const sigset_t *set, int signo);
```

They all take a pointer to a signal set as first argument and some of them also take a signal number as second argument. All functions return zero upon successful completion, otherwise they shall return −1 after setting `errno` to an error code. However, the standard defines no errors for `sigemptyset` and `sigfillset`. Instead, `sigaddset`, `sigdelset`, and `sigismember` may fail because:

EINVAL The signal number `signo` is invalid or unsupported.

Overall, the return value of `sigismember` is the most complex in the group because it can assume three different values:

- Zero, if `signo` is not a member of the signal set referenced by `set`.
- One, if `signo` is a member of the signal set referenced by `set`.
- Minus one, if the function failed because `signo` was invalid or unsupported.

Either `sigemptyset` or `sigfillset` must always be invoked to initialize a signal set before using it in any other ways. Afterwards, its contents can be manipulated with `sigaddset` and `sigdelset`, and queried by means of `sigismember`.

6.6 TIMEKEEPING

In POSIX, the passage of time is expressed and measured by means of one or more time bases, or *clocks*. Each clock has a known resolution and a value, which evolves with time and can be read upon request. Moreover, the value of some clocks can be changed, provided the calling thread has sufficient privileges. Even though the time measured by a clock is often related to the real wall-clock time, this must not necessarily be the case. Specialized clocks can also measure the apparent passage of time related to some system events. For instance, as we will see in the following, they may measure how much processor time a thread has spent executing.

The standard explicitly defines several clocks, which are guaranteed to behave consistently across all implementations, and leaves implementations free to introduce additional, non-portable ones. RTEMS implements all clocks defined by the standard and listed in Table 6.13:

TABLE 6.13

RTEMS Clocks, POSIX API

Clock name	Description
CLOCK_REALTIME	System-wide clock, measures real time
CLOCK_MONOTONIC	Like CLOCK_REALTIME but without backward jumps
CLOCK_PROCESS_CPUTIME_ID	Per-process clock, measures the CPU time given to a process
CLOCK_THREAD_CPUTIME_ID	Per-thread clock, measures the CPU time given to a thread

- The CLOCK_REALTIME clock is a system-wide clock, that is, it is the same for all processes and threads in the system. It measures real time and its value represents the system notion of wall-clock time expressed as the amount of time elapsed since a well-defined point in the past, called the *Epoch*. In turn, the Epoch is defined (somewhat arbitrarily) as the time 00:00:00 on January 1st, 1970 Coordinated Universal Time (UTC).

  In order to accurately keep track of real time, the system may automatically adjust this clock to compensate drifts. As a consequence, threads may observe that its value is *jumping*—that is, suddenly changing its value—forwards or backwards.

- The CLOCK_MONOTONIC is similar to CLOCK_REALTIME but it is guaranteed to never jump backwards, even when the system adjusts it. Backward adjustments are often implemented by slowing down the monotonic clock rate, in this case, so that the clock reaches the desired value gradually.

  Backward clock jumps are undesirable because they easily disrupt the logic of time-based applications. When such a jump occurs, some events that were already in the past of the application timeline are projected into the future again. The consequences of a forward jump are usually less severe because it simply makes the application realize it fell behind its timeline.

  As a side effect of being monotonic, the CLOCK_MONOTONIC measures real time from an unspecified point in the past, and not necessarily from the Epoch. Moreover, threads may not set its value.

- Unlike the previous ones, the CLOCK_PROCESS_CPUTIME_ID is a *per-process* clock, that is, there is one instance of this clock for each process in the system, although all threads still use the same identifier to refer to it. When a thread gets access to this clock, it implicitly gets access to the clock instance pertaining to the process it belongs to.

  The value of the clock is the amount of processor time that the process has spent executing so far.

- The CLOCK_THREAD_CPUTIME_ID also measures processor time, but at the thread granularity. The value of the clock instance associated to a thread is therefore the amount of processor time that the thread has spent executing so far. As before, when a thread gets access to this clock, it implicitly gets access to its own clock instance.

Of these, only CLOCK_REALTIME must be present on all POSIX implementations, the presence of the others depends on which POSIX options the implementation supports. Several timed POSIX services, to be described in Chapters 7 and 9, implicitly make use of CLOCK_REALTIME as their time base.

The absolute time value of a clock is encoded as a struct timespec. The structure has the following two fields:

time_t tv_sec represents the number of seconds, and
time_t tv_nsec represents the number of nanoseconds

since the clock reference time point, for instance, the Epoch for the CLOCK_REALTIME clock. The same data structure is also used to store *relative* time values, expressed as a number of seconds and nanoseconds.

For instance, relative time values can be passed as argument to some POSIX functions. They are also encoded in a struct timespec and they still use a clock as a reference to measure time, but the contents of the structure are interpreted with respect to the value of the reference clock at the time of the call.

An important characteristic of a clock is its *resolution*, which is the smallest time interval it can measure and is also the time distance between two distinct, consecutive values of the clock. The resolution of a POSIX clock cannot be changed at runtime, also because it often depends on the characteristics of the underlying hardware that implements the clock itself.

However, the standard specifies that the resolution of the CLOCK_REALTIME clock (the only one whose implementation is mandatory in all cases) must be at worst _POSIX_CLOCKRES_MIN nanoseconds. In the current edition of the standard, the value of this macro is 20000000, which corresponds to 20 ms.

This value has been chosen to allow clocks driven at the AC power line frequency (50 Hz in several parts of the world, higher in others) to be compliant with the standard. The actual resolution of a clock can be queried at runtime by means of the functions described in the following.

The functions that operate on a clock are very simple and are listed at the top of Table 6.14. In particular, the function:

```
int clock_gettime(clockid_t clock_id, struct timespec *tp);
```

stores into the location pointed by tp the current value of the clock clock_id. clock_id must be one of the identifiers listed in Table 6.13 or a non-portable, implementation-defined identifer.

The function clock_gettime returns zero to the caller upon successful completion. Otherwise, it returns -1 and sets errno to one of the following values:

EINVAL The clock identifier clock_id is invalid or the corresponding clock does not exist in the system.

Similarly, the function:

TABLE 6.14

RTEMS Timekeeping Primitives, POSIX API

Function	Purpose
clock_gettime	Retrieve the value of a clock
clock_getres	Retrieve the resolution of a clock
clock_settime	Set the value of a clock (if possible)
timer_create	Create a timer (initially inactive)
timer_delete	Delete a timer
timer_settime	Activate/deactivate a timer, and set its value and period
timer_gettime	Get the value and period of a timer
timer_getoverrun	Retrieve the overrun counter of a timer
clock_nanosleep	Block the caller until a certain time has elapsed

```
int clock_getres(clockid_t clock_id, struct timespec *res);
```

stores in the location referenced by `res` the resolution of clock `clock_id`. The return value and the possible error codes are the same as for `clock_gettime`.

Finally, the function:

```
int clock_settime(clockid_t clock_id,
    const struct timespec *tp);
```

allows a thread with sufficient privileges to change the value of a clock. More specifically, it sets the value of the clock `clock_id` to the time indicated in the `struct timespec` pointed by `tp`.

Like the other functions just discussed it returns zero to the caller if the value of the clock has been changed successfully. Otherwise, it returns -1 and sets `errno` to one of the following codes:

EINVAL This code shall be used to report a variety of errors:
- The clock identifier `clock_id` is invalid.
- The corresponding clock does not exist in the system.
- The clock exists but its value cannot be changed (as is the case of the `CLOCK_MONOTONIC` clock).
- The given clock value is invalid by itself (for instance, its `tv_nsec` field is less than zero) or is outside the valid range of values allowed for the clock `clock_id`.

Optionally, implementations may also check whether the calling thread has sufficient privileges to perform the requested operation and, if this is not the case, report:

EPERM The calling thread does not have the permission to set the clock.

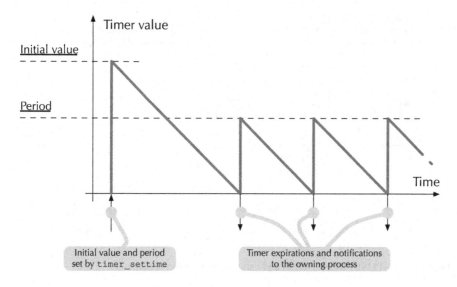

FIGURE 6.5 Principle of POSIX timers update.

Besides providing time values upon request, a subset of the clocks available in the system can also act as a timing reference for *timers*. This subset must always include CLOCK_REALTIME and, if the implementation supports CLOCK_MONOTONIC, it must be included as well.

A timer is a per-process object that is able to send timing signals to the process. It is characterized by its current *value* and, optionally, by a *period* also called *reload value*.

Figure 6.5 shows, in abstract terms, how timers are set up and updated. The figure shows how the value of a timer (on the vertical axis) evolves with time (on the horizontal axis). The main aspects of interest are:

- The initial value and period of a timer are set by means of the function timer_settime, depicted as an upward arrow on the left of the figure.
- The system decrements each timer with a non-zero value using its clock as a reference. The timer and its reference clock are associated upon timer creation, which is performed by calling the function timer_create and not shown in the figure.
- When the value of a timer reaches zero, it *expires* and the system notifies the owning process by means of one of the notification mechanisms specified by the standard, to be discussed in the following. The ultimate result of a notification is that one of the threads of the process that owns the timer executes some *timer handling* code.
- If the timer period is zero, the value of the timer stays at zero after expiration until an explicit action is taken to set it up and start it again with a new

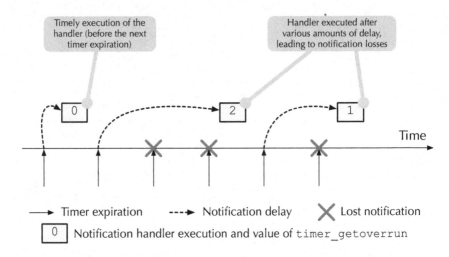

Timely execution of the handler (before the next timer expiration)

Handler executed after various amounts of delay, leading to notification losses

0

2

1

Time

⟶ Timer expiration   ---▶ Notification delay   ✕ Lost notification

0 Notification handler execution and value of `timer_getoverrun`

FIGURE 6.6 POSIX timer overrun and value of the overrun counter.

call to `timer_settime`. A timer with a period of zero is sometimes called a *one-shot* or *aperiodic* timer because it generates exactly one notification.

- If the timer period is not zero (the case shown in Figure 6.5), the timer value is reloaded from its period whenever the timer expires. A timer with a non-zero period is also called *repetitive* or *periodic* timer because, as shown in the figure, it generates a continuous stream of notifications until it is turned off.

- The notifications generated by a repetitive timer are all equally spaced and their distance is equal to the timer period. Instead, the delay between timer setup and the first notification is given by the initial value of the timer. It may or may not coincide with the period.

Each timer cannot have more than one pending notification, because otherwise the amount of memory needed to keep track of them would be potentially unbounded in the case of a repetitive timer. As a consequence, if a timer expires again while the previous notification it generated is still waiting to be delivered, the newly coming notification is simply *lost*.

However, repetitive timers are often used to give to a process the notion of elapsing time, and this kind of behavior worsens the accuracy of this notion. In other words, lost notifications make the affected process "fall behind" the timer and believe that less time has elapsed with respect to reality.

To counteract the inaccuracy, the POSIX standard specifies that each timer has an *overrun counter*, which can be read by means of the `timer_getoverrun` function. The value of the overrun counter of a timer while one of its notifications is being handled is equal to the number of additional notifications from the same timer that have been discarded since the notification was generated. Figure 6.6 better illustrates this point with the help of a couple of examples.

In the figure, upward arrows indicate the stream of notifications coming from a repetitive timer as time goes by. Dashed arrows denote the delay that occurs between a notification and its handling. The execution of the timer handling code is represented by a gray rectangle. The number inside the rectangle shows the value of the timer overrun counter as it would be returned by a `timer_getoverrun` call performed by the handler itself. Crossed notifications are the ones discarded by the system.

- The leftmost part of the picture shows the normal case, in which a notification can be delivered and handled before the timer generates the next. In this case, the value of the overrun counter is zero because no further notifications were generated during the delay between the leftmost notification and its delivery.
- In the next example, two additional notifications were generated and discarded before a notification could be delivered and handled. In this case, the value of the overrun counter is two. The third example illustrates a similar scenario, in which one additional notification was generated.
- Overall, the picture shows that the timer handling code can use the overrun counter it retrieves at the beginning of its execution to compensate for lost notifications. If we imagine, for instance, that the timer handler would like to keep track of how many notifications have been generated, it should just increment a counter by one plus the value of the overrun counter every time it is executed. Looking back at the figure, at the end the handler would obtain a total count of 6 (3 from handler executions, plus $0 + 2 + 1 = 3$ from the overrun counter), which exactly corresponds to the total number of notifications generated by the timer although some of them have not actually been delivered.

The APIs that operate on a timer are summarized at the bottom of Table 6.14. More specifically, the function:

```
int timer_create(clockid_t clockid,
    struct sigevent *restrict evp,
    timer_t *restrict timerid);
```

creates a new timer that uses `clockid` as reference clock and notifies the calling process upon expiration as specified in the data structure referenced by `evp`. Upon successful completion, the function stores into the location pointed by `timerid` the descriptor of the new timer and returns zero. Timers cannot be shared among processes, but can be used by all threads belonging to the process that created them. Immediately after creation, the new timer is inactive and must be activated by means of the function `timer_settime`.

When `timer_create` fails, it returns -1 and sets `errno` to an appropriate status code. The function shall fail for the following reasons:

EINVAL The clock identifier `clock_id` is invalid or the corresponding clock does not exist in the system.

TABLE 6.15

Fields of a struct sigevent, POSIX API

Field (prefixed by sigev_)	Purpose
notify	Type of notification: SIGEV_NONE, SIGEV_SIGNAL, or SIGEV_THREAD
signo	Number of the signal to be generated (for SIGEV_SIGNAL)
value	Argument to the passed to the signal handler (for SIGEV_SIGNAL)
notify_function	Function to be executed in a new thread (for SIGEV_THREAD)
notify_attributes	Thread attributes to be used to create the new thread (for SIGEV_THREAD)

ENOTSUP The clock identified by clock_id cannot be a reference clock for a timer.

EAGAIN The system lacks the resources needed to create the timer or set up its notification mechanism.

As said previously, the data structure of type struct sigevent pointed by evp indicates how the timer shall generate notifications and provides additional information to carry them out. Its fields are listed in Table 6.15. The standard specifies three possible notification methods. The specific method to be used is determined by the sigev_notify field of the structure.

A value of SIGEV_NONE indicates that no notifications should be generated, a feature seldom used with timers. A value of SIGEV_THREAD means that the timer handling code consists of a function, referenced by the sigev_notify_function field of the structure. The function is executed every time a notification is delivered to the process that owns the timer, and execution takes place in its own thread. The field sigev_notify_attributes points to a thread attribute data structure the system shall use to create this thread, as described in Section 6.2. The SIGEV_THREAD notification method will not be further discussed in the following because RTEMS currently does not support it.

The third notification method is SIGEV_SIGNAL, which is supported by RTEMS if its configuration option RTEMS_SCORE_COREMSG_ENABLE_NOTIFICATION is set. In this case, a notifications triggers the delivery of a signal to the process that owns the timer and any timer handling code must be placed in the corresponding signal handler. The field sigev_signo specifies the signal number to be generated and sigev_value, of type union sigval, is the argument to be passed to the signal handling function. Upon delivery, the signal is handled in the standard way discussed in Section 6.5.

An existing timer can be deleted, regardless of whether it is currently active or not, by calling the function timer_delete. Its only argument is the timer descriptor timerid:

```
int timer_delete(timer_t timerid);
```

The return value of the function is zero upon successful completion. The standard recommends the function returns -1 and sets `errno` to an appropriate status code upon detecting the following conditions:

EINVAL The timer descriptor `timerid` is invalid.

The function:

```
int timer_settime(timer_t timerid, int flags,
    const struct itimerspec *restrict value,
    struct itimerspec *restrict ovalue);
```

sets the value and the period of the timer identified by `timerid` as specified in the data structure referenced by `value`. Atomically, it also stores the previous value and period into the structure referenced by `ovalue`, unless `ovalue` is a `NULL` pointer. Both data structures are of type `struct itimerspec`, which shall not be confused with `struct timespec`:

- As described previously, a `struct timespec` holds an absolute or relative time value, expressed as an integral number of seconds and a fraction of a second, in nanoseconds.
- A `struct itimerspec` has two fields and each of them is a `struct timespec`. Namely, the field `it_value` holds the value of a timer and the field `it_interval` holds its period.

Setting the value of a timer to zero deactivates it, regardless of the period. Setting the value of a timer to a non-zero value activates the timer. The value of an active timer then evolves as depicted in Figure 6.5.

The argument `flags` determines whether the value of the timer specified in the `it_value` field of the structure referenced by `value` is absolute or relative. In particular:

- If the flag `TIMER_ABSTIME` is *not* set in `flags`, `it_value` is interpreted in *relative* terms with respect to the time of the call to `timer_settime`. That is, it represents the amount of time that shall elapse from the time of the call to the first expiration of the timer.
- If the flag `TIMER_ABSTIME` is set in `flags`, `it_value` is interpreted in *absolute* terms and represents the absolute time at which the first expiration of the timer shall take place. In other words, the time that shall elapse between the time of the call to `timer_settime` and the first expiration of the timer shall be equal to the difference between `it_value` and the value of the clock associated to the timer at the time of the call.

The function `timer_settime` returns zero when it completed successfully, otherwise it returns -1 and sets `errno` to an error code. It shall detect the following error condition:

EINVAL The contents of the data structure pointed by `value` are invalid.

Moreover, the standard recommends to check and report the following additional error condition:

EINVAL The timer descriptor `timerid` is invalid.

If a thread only wants to retrieve the value and period of a timer without changing them, it can invoke the function:

```
int timer_gettime(timer_t timerid,
    struct itimerspec *value);
```

whose arguments are `timerid`, a timer descriptor, and `value`, a pointer to a `struct itimerspec` where the function will store the timer value and period. As `timer_settime` does, this function may detect that `timerid` is invalid and fail with the `INVAL` error code.

Last, the function:

```
int timer_getoverrun(timer_t timerid);
```

returns the current value of the overrun counter of the timer described by `timerid`. Calling this function with an invalid `timerid` leads to undefined results.

Sometimes the only reason for using a timer is to implement a periodic task, that is, a task that performs certain actions at fixed, predefined time intervals. A periodic timer, which generates a stream of uniformly spaced notifications, would certainly be adequate to this purpose. However, a more straightforward alternative not based on signals is also available. The function:

```
int clock_nanosleep(clockid_t clock_id, int flags,
    const struct timespec *rqtp,
    struct timespec *rmtp);
```

blocks the calling thread until, according to the clock `clock_id`, the time specified in the data structure referenced by `rqtp` has elapsed. The argument `flags` is analogous to the argument of `timer_settime` with the same name, and determines whether the contents of this data structure are interpreted in relative or absolute terms, namely:

- If the flag `TIMER_ABSTIME` is *not* set in `flags`, the contents of the data structure pointed by `rqtp` are interpreted in *relative* terms with respect to the time of the call to `clock_nanosleep`. That is, it represents the amount of time that shall elapse from the time of the call until `clock_nanosleep` returns to the caller.
- If the flag `TIMER_ABSTIME` is set in `flags`, the contents of the data structure pointed by `rqtp` are interpreted in *absolute* terms and represents the absolute time at which `clock_nanosleep` shall return, as measured by the clock `clock_id`.

It should be noted that, after a successful `clock_nanosleep`, the calling thread may actually resume execution *later than* what was specified in `rqtp`. There are two distinct reasons for this:

1. The amount of time specified by `rqtp` is necessarily rounded up to a multiple of the reference clock resolution.
2. Moreover, `clock_nanosleep` only moves the calling thread back to the *ready* state when the time specified by `rqtp` elapses. Whether or not the calling thread also moves to the *running* state and resumes execution immediately depends on processor scheduling decisions. On a single-processor system, the transition to the *running* state may be delayed, for instance, when other higher-priority threads are also ready for execution.

The function `clock_nanosleep` may also return to the caller *earlier* than planned, when the wait is interrupted by the arrival of a signal and the execution of the corresponding signal handler. In this case, to enable the calling thread to compensate for the premature return, the function stores into the data structure pointed by `rmtp` the amount of time it did *not* spend waiting, that is, the difference between the requested time pointed by `rqtp` and the actual duration of the wait. This is *not* done in two circumstances:

1. When the `rmtp` pointer is NULL, meaning that the calling thread is not interested in the information.
2. When the `flags` argument asks for an absolute wait, because in this case the calling thread certainly does not need the information.

The function `clock_nanosleep` returns zero when successful. When it fails, it shall return one of these non-zero error codes:

EINVAL is returned when one of the arguments to `clock_nanosleep` is invalid, namely:
  - The given `clock_id` is unknown to the system.
  - The contents of the data structure pointed by `rqtp` are invalid or, for an absolute wait, are out of range for the `clock_id`.
  - The `clock_id` specifies the CPU time clock of the calling thread.
ENOTSUP The clock `clock_id` does not support `clock_nanosleep`. Among the clocks explicitly defined by the standard, only CLOCK_REALTIME and CLOCK_MONOTONIC do, whereas CPU time clocks do not.
EINTR The function returned prematurely because it was interrupted by a signal. The return takes place after the execution of the corresponding signal handler.

The function `nanosleep` can be used as an even simpler shortcut to ask for a relative wait using the CLOCK_REALTIME clock:

```
int nanosleep(const struct timespec *rqtp,
    struct timespec *rmtp);
```

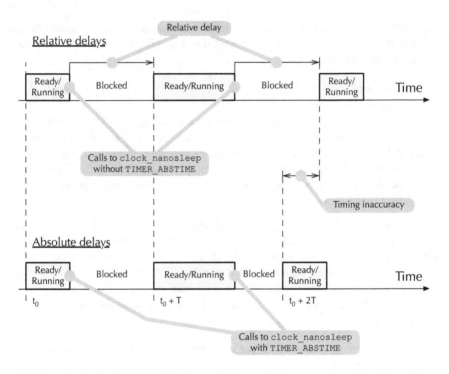

FIGURE 6.7 Timing accuracy of relative versus absolute delays.

This function is equivalent to `clock_nanosleep` with the argument `clock_id` set to `CLOCK_REALTIME` and `flags` set to 0. The arguments `rqtp` and `rmtp` have the same meaning as for `clock_nanosleep`.

Figure 6.7 depicts the effect of relative and absolute delays on the timing accuracy of a periodic task implemented as an outer shell and a body. The outer shell consists of an infinite loop that comprises a passive wait implemented by means of a call to `clock_nanosleep` followed by the body. The body is the code to be executed periodically, with period T.

The upper time diagram shows the timings of a task that makes use of relative delays. Namely, after executing the body starting at t_0, it calls `clock_nanosleep` with a relative time value, in order to achieve a constant execution period T and start executing its body again at $t_0 + T$. After this second execution, the task calls `clock_nanosleep` again to block until $t_0 + 2T$, and so on.

However, relative time delays must be known at the time `clock_nanosleep` is called. Using a fixed delay of $T - R$, where R is an a priori estimate of the response time of the body, makes the task prone to timing inaccuracies, since the actual response time may vary from one execution to another. For instance, as shown in the figure, if the response time of the second execution of the body is longer than expected, the third execution will start late, that is, its timings will be inaccurate.

As shown in the lower time diagram of Figure 6.7, passing an absolute time to clock_nanosleep while keeping the same task organization solves this issue. Namely, passing to clock_nanosleep the absolute times $t_0 + T, t_0 + 2T, \ldots$ guarantees that the task is always unblocked at the intended instants regardless of how the response time of its body changes from one execution to another.

Even this approach falls shorts from ensuring accurate timings if $R > T$, that is, if the response time exceeds the period, but this issue cannot be solved at the task activation and scheduling level. Instead, this is within the domain of the scheduling analysis techniques presented in Section 4.1.

Similarly, it must be remarked once more that unblocking a task at a certain instant does not guarantee that it will start executing immediately thereafter. Also in this case, an appropriate priority assignment and scheduling analysis can assist in limiting the worst-case amount of interference the task may suffer from higher-priority tasks in the system and calculate an upper bound on it.

Overall, choosing the best kind of delay to be used with clock_nanosleep and other timekeeping functions also depends on the purpose of the delay. A relative delay may be useful, for instance, if an I/O device must be given a certain amount of time to perform an operation and report back on its status. In this scenario, it is sensible to measure the delay from when the command has actually been sent to the device, and a relative delay makes sense.

On the other hand, as we have just seen, an absolute delay works better to implement a periodic task because it guarantees that the period will stay constant although the response time of its body varies from one execution to another.

6.7 SUMMARY

This chapter illustrated the main applications programming interfaces for task management and timekeeping specified by the POSIX standard and provided by RTEMS. The functions discussed in this chapter have the primary function of populating a system with new tasks. The scheduling priority and other characteristics of a task can be set at creation time by means of a generic mechanism based on attribute objects.

Another set of functions can be used to terminate a task, in either a voluntary or a forcible way. In the second case, suitable safeguards can be put in place to ensure that task termination takes place in a controlled way. The ordinary execution flow of a task can also be altered upon the occurrence of an asynchronous event by means of a signal.

Finally, a set of timekeeping functions, based on the underlying concepts of clocks and timers, allows programmers to tie task execution to time-related events. In RTEMS, this is primarily accomplished by raising a signal directed to a certain process whenever a timer expires, or by blocking a thread for a specified amount of time.

Part III

Inter-Task Synchronization and Communication

7 Inter-Task Synchronization and Communication (IPC) Based on Shared Memory

CONTENTS

The main focus of this chapter is on classic inter-task synchronization and communication methods that rely on shared memory for data exchange among tasks, semaphores, and monitors in particular. Two more specialized synchronization objects, barriers and events, are also included in the discussion because, although they are less interesting from the theoretical point of view, they are still very useful in practice.

Before delving into this, the chapter introduces the fundamental concepts of race condition, critical region, and lock-based mutual exclusion. Two extremely important issues, priority inversion and deadlock, which may affect all lock-based inter-task synchronization and communication mechanisms, are also presented in this chapter.

To help readers get started faster on the topics they are most interested in, in this and the following chapter, we also steer the general theoretical discussion towards real-time execution concepts and their practical implementation by means of the RTEMS Classic and POSIX API.

7.1 RACE CONDITIONS AND MUTUAL EXCLUSION

7.1.1 AN EXAMPLE OF RACE CONDITION

If a set of tasks must cooperate to solve a certain problem, they will need to exchange some information in virtually all cases. Using a set of *shared variables* to this purpose—that is, variables that multiple tasks can concurrently read and write—is a rather straightforward extension of what is often done in sequential programming.

Sequential programs written in a high-level language are frequently organized as a set of functions and procedures, each with a well-defined purpose. Together, they implement the functionality that the program must provide. These functions may conveniently exchange data by means of a set of global variables defined in the program. All functions have access to them, within the limits set forth by the scoping rules of the programming language, and they can get and set their value as required by the specific algorithm they implement.

Something similar also takes place during a function call, in which the caller calculates the function arguments and stores them into an area of memory whose structure is well-known to both the caller and the callee, often allocated on the stack. The callee then reads its arguments from there and uses them as needed. The return value of the function is handled in an equivalent way. Although the compiler may optimize the exchange by using some processor registers instead of memory, provided the arguments and the return value fit into them, the general idea is still the same.

Unfortunately, trying to apply the same idea to a system in which the actors are concurrently executing tasks immediately gives rise to subtle but deep issues, even in seemingly trivial scenarios. Let us imagine, for instance, that we want to count how many events of a certain kind happen within a sequential program.

As illustrated in Figure 7.1, probably the most straightforward and intuitive solution is to define a global variable k, initialized to zero, and a function that increments it by one. When using the C programming language, the function could be called void inc(void) and would contain a single statement, k = k+1.

Actually, as also shown in the figure, no real-world processor is actually able to increment k in a single, indivisible step, at least when the code is compiled into ordinary assembly instructions. Instead, a typical computer based on the *von Neumann* architecture [52, 122] will perform a sequence of three distinct operations:

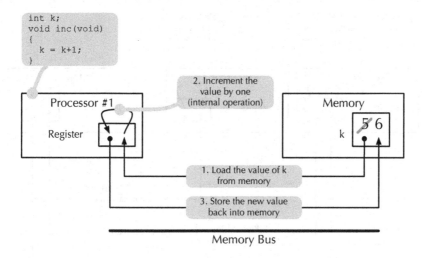

```
int k;
void inc(void)
{
  k = k+1;
}
```

FIGURE 7.1   Increment of a variable on a single-processor system.

1. Load the value of k from memory into an internal processor register. On a simple processor, this register would likely be the accumulator. From the processor's point of view, this is an external operation because it also involves memory besides the processor itself. These two units communicate through a memory bus to which other units, like another processor, may possibly be connected. The load operation does not alter memory contents, that is, k retains its current value after it has been performed.
2. Increment the register that contains the value of k just loaded from memory by one. Unlike the previous one, this operation is completely internal to the processor. It cannot even be observed by memory because it does not require any memory bus cycle in order to be performed. On a simple processor, the result is stored back into the accumulator.
3. Store the register, which now contains the updated value of k, into memory. This is again an external operation involving a memory bus transaction like the first one. Only at this point, the new value of k becomes observable from outside the processor. In other words, if we look at memory contents, k retains its original value until this final step has been completed.

Even though real-world processor architectures are nowadays much more sophisticated than the one shown in Figure 7.1 and their actual behavior while they are accessing memory is way more complex for performance-related reasons, the basic concept is still the same. Most operations that are thought to be indivisible when looking at them from a high-level programming language perspective—even simple, short statements like the one we considered previously—actually correspond to a sequence of low-level elementary steps when examined at the instruction execution level.

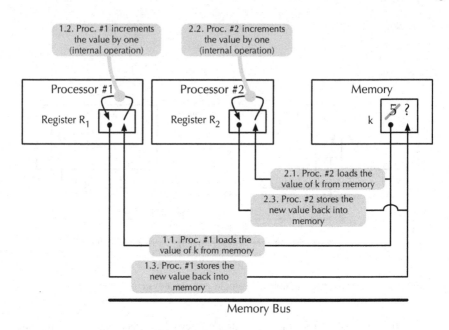

FIGURE 7.2 Race condition when two processors (or cores, or tasks) increment a variable.

This aspect is often overlooked, also because it has no practical consequences as long as the code being considered is executed in a strictly sequential fashion on a single processor. On the contrary, it becomes extremely important when multiple tasks, cores, or processors, execute the same code *concurrently*.

To continue the example, let us imagine a situation in which two tasks want to increment k concurrently because they are both counting the same kind of events. They both call the function inc() to perform the increment. As shown in Figure 7.2, we assume that each process is running on its own physical processor and the two processors are connected to a single-port memory by means of a common memory bus. However, the argument would not change even if we were considering two cores, or even two tasks sharing the same physical processor by way of the multiprogramming techniques presented in Chapter 3.

If we let the initial value of k be 5, the elementary steps *might* be executed in the following sequence:

1.1. Processor #1 starts executing inc() first and loads the value of k from memory and stores it into one of its registers, R_1. Since the current value of k is 5, R_1 will also contain 5.

1.2. Processor #1 keeps executing and increments register R_1 by one, bringing its value to 6.

2.1. Now processor #2 starts executing inc(), too. It loads the value of k from memory and stores it into one of its registers, R_2. Since processor #1 has not stored the updated value of R_1 back to memory yet, processor #2 still loads the value 5 from k and R_2 will also contain 5.

2.2. Processor #2 increments its register R_2 and the value of R_2 becomes 6.

2.3. Processor #2 stores the contents of R_2 back to memory, that is, it stores 6 into k.

1.3. Processor #1 does the same, that is, it also stores 6 into k.

From this example, it is easy to conclude that even a very simple and obviously correct function cannot be directly transferred from a sequential to a concurrent execution environment without adverse side effects. More specifically, two distinct issues became evident:

- The final value of k is incorrect because, even though its initial value was 5 and it has been incremented by one twice, its final value is 6 instead of 7.
- What is even worse from the debugging point of view is that the result is incorrect only *sometimes*. For instance, executing the elementary steps in the sequence 1.1, 1.2, 1.3, 2.1, 2.2, 2.3 would have led to a correct result.

In other words, the correctness of the result depends on how the elementary steps performed by one processor *interleave* with the steps performed by the other. In turn, this depends on the precise timing relationship between the processors, down to the instruction execution level. This is not only hard to determine, it may also easily change from one execution to another, or if the same tasks are executed on a different machine.

Even if we restrict our attention only to single-processor systems, task preemption and interleaving are managed by the operating system scheduling algorithm and depend on a complex set of circumstances, most of which are not under the control of the tasks involved. Also in this case, the scheduling decision taken by the operating system may vary widely from one execution to another.

Unfortunately, these kinds of errors may be hard to reproduce and correct because they are time-dependent and only some low-probability interleavings trigger them. In turn, they may occur only when the computer system is working in the field and disappear during bench testing because the small, but hard to avoid, differences between actual operation and testing slightly disturbed system timings. For the same reasons, even the addition of software-based instrumentation or debugging code to a concurrent application may make a time-dependent error disappear.

All these observations lead to the general definition of a pathological condition, known as *race condition*, which may affect a concurrent system. There is a race condition whenever a set of tasks makes use of some *shared objects* to carry out a computation and the result of this computation depends on the exact way the tasks interleaved.

In the previous statement, the term "shared object" has a rather broad meaning. In the simplest cases, it may refer to a shared variable, like the variable k mentioned in the example. In other cases, it may indicate other, more complex kinds of objects, such as files and devices. Since race conditions undermine the correctness of a concurrent system, one of the main goals of any properly formulated IPC system must be to eliminate them, possibly with programmers' assistance.

Another related concept is the race condition *zone* or *window*. It is defined as the time frame in which, due to the way tasks are scheduled, a race condition may occur. In our example, the decision of preempting one of the two tasks while the increment is in progress on a single-processor system starts a race condition window in which a race condition may occur if the other task is executed and also increments the same variable. The window ends when the scheduler goes back to execute the first task.

When using simple forms of scheduling, like the cyclic executive [18] or any non-preemptive task-based scheduling algorithm, race conditions can be kept under control more easily because the scheduler can switch from one task to another, thus possibly starting a race condition window, only at specific locations within a task. For instance, a non-preemptive scheduling algorithm only switches from one task to another when the running task executes a blocking synchronization primitive or voluntarily relinquishes the processor in some other ways.

These points are therefore well known to programmers, and they can organize their code so as to avoid spreading the statements that make access to shared variables across them. The only exception are interrupt handlers, which can interrupt ordinary tasks at arbitrary points by definition. However, this is a more confined issue that can be tackled with special-purpose methods, like the lock-free and wait-free synchronization and communication methods to be discussed in Chapter 13.

On the contrary, the adoption of a preemptive task-based scheduling algorithm, like the ones analyzed in Chapter 3, makes the extent and location of race condition zones virtually impossible to predict, especially when the number of tasks grows. This is because the task switching points are now chosen autonomously by the operating system scheduler instead of being hard-coded in the code.

Due to this fact, the switching points will also likely change from one task activation to another. Therefore, what is needed is a method to avoid race conditions in a way that is independent from the particular scheduling algorithm in use. This is indeed the main goal of the IPC mechanisms to be discussed in this and the next chapter. In this book, the description of race conditions and how to avoid them is kept at a practical level as much as possible, mainly focusing on their implications from the concurrent programming point of view. Readers interested in a more formal and theoretical discussion can refer, for instance, to Lamport's works [80, 81].

7.1.2 CRITICAL REGIONS

Even in relatively small software systems, having to examine the tasks code as a whole when looking for possible race conditions quickly becomes a daunting proposition. Fortunately, the following considerations, originally due to Hoare [59] and Brinch Hansen [25], allow us to focus the effort only on a much smaller portion of the task's code.

- A task spends part of its execution doing internal operations, that is, executing instructions that do not require or make access to any shared data.

By definition, all these operations cannot lead to any race condition, and the corresponding pieces of code can be safely ignored when we look at the code from the concurrent programming point of view.

- In other cases, a task executes a region of code that makes access to a certain shared object, like the `inc()` function introduced in the previous example. Those are the regions of code that must be identified and scrutinized because they can indeed lead to a race condition. For this reason, they are called *critical regions* or *critical sections* associated with the object.

Moreover, it is possible to identify two general *necessary* conditions for a race condition to occur.

- Two or more tasks must be executed concurrently, in a way that leaves open the possibility of a context switch occurring among them. In other words, they must be within a race condition zone, as defined in Section 7.1.
- These tasks must also be actively working on the same shared object when the context switch occurs, that is, they must be within a critical region.

The conditions outlined above are not yet *sufficient* to cause a race condition because a context switch must typically occur at very specific locations within a critical region to actually corrupt the shared object. It is the combination of all these conditions that makes object corruption a low-probability event that may be hard to reproduce, and makes race conditions difficult to detect, analyze, and fix.

Considering the definition of critical region given previously, one sensible way to avoid race conditions when accessing a certain shared object is to allow only one task at a time to be in a critical region pertaining to that object, within a race condition zone. Since in preemptive, task-based scheduling it may be hard to determine in advance where race condition zones could be—especially across a range of operating systems because this may depend, as explained previously, on specific details of the scheduling algorithm—a more general solution consists of enforcing the *mutual exclusion* among all critical regions associated with the same shared object *at any time*, without considering race condition zones at all.

Going back to the example of race condition shown in Figure 7.2, we can see that both processes have a critical region associated with shared variable k, the body of function `inc()`, because it increments k. Even if the critical region code is correct when executed by one task at a time, the race condition stemmed from the fact that two distinct tasks were allowed to be in their critical region simultaneously.

7.1.3 LOCK-BASED MUTUAL EXCLUSION

The most straightforward implementation of mutual exclusion among critical regions makes use of a *lock-based* synchronization protocol. With this approach, a task that wants to access a shared object, by means of a certain critical region associated with it, must, first of all, acquire some sort of *lock* also associated with the shared object and possibly wait if it is not immediately available.

After acquiring the lock, a task is allowed to use the shared object freely. A context switch may still be allowed to occur at this time, but it will not cause a race condition because any other task trying to enter a critical region pertaining to the same shared object will also try to acquire the lock beforehand, and will be blocked in the attempt. When the original task has completed its operation on the shared object and it is leaving its critical region, it must release the lock. In this way, other tasks can acquire it and be able to access the shared object in the future. The lock release mechanism is also responsible for unblocking one of the tasks already waiting to acquire the lock, if any.

In simple cases, mutual exclusion can be ensured by resorting to special machine instructions that implement the lock-based protocol just discussed. For example, on the Intel 64 and IA-32 architecture [67], the INC instruction increments a memory-resident integer variable by one. When executed, the instruction loads the operand from memory, increments it internally to the processor, and finally stores back the result. It is therefore subject to the same race condition depicted in Figure 7.2. However, when the instruction is preceded by the LOCK prefix, the processor executes the whole sequence *atomically*—that is, indivisibly—even in a multiprocessor or multicore system. More specifically, the processor does not accept interrupts while executing the instruction, and the cache coherency and memory bus access protocols ensure that the operation is carried out atomically also with respect to memory.

These ad-hoc low-level solutions are extremely useful as building blocks of more complex task synchronization mechanisms. However, they are not general enough to be directly applied in all circumstances. For instance, the LOCK prefix cannot force more than one machine instruction to be executed atomically on the Intel architecture, whereas a critical region that updates a shared object frequently contains a longer sequence of instructions.

In the general case, critical regions must be bracketed by two auxiliary pieces of code, usually called the critical region *entry* and *exit* code, which take care of acquiring and releasing the lock, respectively. For some kinds of task synchronization techniques, like the ones described in this chapter, the entry and exit code must be invoked explicitly by the task itself, and hence, the overall structure of the code strongly resembles the one just described. In other cases, for instance, when using the message passing primitives to be discussed in Chapter 9, critical regions and their entry/exit code may be hidden within the inter-task communication primitives, so that they are invisible to the programmer, but the concept is still the same.

The critical region entry and exit code may be copied directly immediately before and after the critical region, when it is relatively short. Otherwise, it may be executed indirectly, by means of appropriate function calls, with the same effect. This also highlights the fact that critical region contents are defined by the dynamic concept of code *execution*, not only by the presence of some code between the critical region entry and exit code. For example, if there is a function call in a critical region, the whole body of the called function must also be considered part of the critical region itself.

Although lock-based mutual exclusion currently is by far the most common approach to avoid race conditions, other, radically different methods also exist. For instance, it is indeed possible to implement meaningful inter-task communication and synchronization *without* using any locks, by means of *lock-free* or *wait-free* techniques. These methods will be discussed in more detail in Chapter 13, since they are especially relevant to multicore systems.

7.1.4 CORRECTNESS CONDITIONS

Going back to lock-based mutual exclusion, theoretical analysis shows that four conditions must be satisfied in order to have an acceptable solution [119]:

1. It must really work as intended, that is, it must allow only one task at a time to execute code within any of the critical regions associated with the same shared object. On the other hand, for efficiency reasons, it is undesirable to "overdo" mutual exclusion. For example, the solution should not prevent a task from entering a critical region while another task is already within a critical region, if these two regions are associated with unrelated shared objects.
2. Internal operations shall not matter. In other words, any task that is performing some internal operations—and hence, is not executing within a critical region at the moment—shall not prevent any other tasks from entering their critical regions, if they so decide.
3. If a task wants to enter a critical region, it must not have to wait forever to do so. This condition guarantees that the task will eventually make progress in its execution.
4. The solution must work regardless of any low-level details of the hardware or software architecture, which may be unknown to programmers and will likely change over time. For instance, the correctness of the solution must not depend on the number of tasks in the system, the number of physical processors or cores, or their relative speed.

In the following, we will introduce several lock-based inter-task communication methods, making use of these conditions to assess their correctness.

7.2 SEMAPHORES

7.2.1 DEFINITION AND PROPERTIES

The first definition of a *semaphore* as a general inter-task synchronization mechanism is due to Dijkstra [42]. Although the original proposal was based on active wait, nowadays most operating systems replaced it with passive wait without changing semaphore's semantics in any way.

Although, strictly speaking, semaphores are not powerful enough to solve every concurrent programming problem that can be conceived [76], they have successfully been used to address many problems of practical significance in diverse concurrent programming domains. Another reason for their popularity is that they are easy to implement in an efficient way, to the point that virtually all operating systems offer semaphores as an inter-task synchronization method.

According to its abstract definition, a semaphore s is an object that contains two items of information:

- a nonnegative integer *value* s.v,
- a *queue* of tasks s.q.

For some kinds of semaphore the programmer chooses the initial semaphore value when creating it. Other kinds of semaphore have a predetermined, fixed value. For a newly initialized semaphore, the queue is always empty. Its purpose is to hold the tasks that are waiting on the semaphore, as will be explained in the following. After initialization, the only way to interact with a semaphore, and possibly modify its value and queue, is to invoke the primitives defined on it. Some implementations make the current value of a semaphore available but, in any case, the value and the queue cannot be directly manipulated after initialization.

An important assumption about semaphore primitives is that they are executed *atomically*, that is, as indivisible units. It is up to the implementation to guarantee that the assumption is true, by means of lower-level mechanisms. For instance, in a single-core system, atomic execution can be ensured by disabling interrupts as described in Chapter 3. In multicore systems, a more sophisticated approach is necessary to achieve atomicity across all cores, as discussed in Chapter 13, possibly using hardware-assisted locks like the one described previously as building block.

The two semaphore primitives, called P() and V() according to the original Dijkstra's nomenclature [42], work as follows.

1. The primitive P(s) has a semaphore s as argument. Its behavior depends on whether or not the value of s is strictly greater than zero.
   - If s.v is strictly greater than zero, P(s) decrements s.v by one and returns without blocking the calling task.
   - Otherwise, it inserts a reference to the calling task into s.q and moves the task into the *Blocked* state of the task state diagram discussed in Section 3.2.2.
   In the second case, the primitive returns to the caller only when the task is moved into the *Running* state again, at a later time.
2. The behavior of the primitive V(s), which also has a semaphore s as argument, depends instead on whether or not the queue of s is empty.
   - If s.q is empty, V(s) increments s.v by one. By construction, s.v will be strictly positive after the increment.
   - Otherwise, s.v is certainly zero. In this case, V(s) picks one of the tasks referenced by s.q, removes the reference from the queue, and unblocks the task by moving it back to the *Ready* state of the task state diagram, so as to make it eligible for execution again.
   By itself, the invocation of V(s) never blocks the calling task. However, the calling task may still stop executing because, for instance, V(s) unblocks another task, and the operating scheduler opts for a preemption because the newly unblocked task has a higher priority.

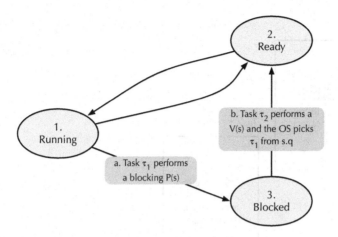

FIGURE 7.3 Task state diagram transitions of τ_1 triggered by semaphore operations.

It is also worth noting that, while the abstract definition of semaphore stipulates that V(s) picks exactly *one* task in s.q, if s.q is not empty, it does not specify which one. The choice does not affect the correctness of the synchronization primitives, but does affect their timing behavior, as will be described in Section 8.1.

The same semaphore primitives are also known by other names, depending on the API and, in some cases, on the specific kind of semaphore being considered. For instance, in the RTEMS Classic API the primitive P() is called *obtain*, while V() is called *release*. In the POSIX API, the scenario is more complex because the names of the primitives depend on the kind of semaphore. The primitives that operate on a general-purpose semaphore are called *wait* and *post*, respectively, while the ones for mutual exclusion semaphores—a special-purpose kind of semaphore to be discussed in the following—are called *lock* and *unlock*.

The execution of a semaphore primitive is related to the transition of some tasks from one task state diagram state to another. We can therefore specialize the abstract task state diagram defined in Chapter 3 to be more explicit about the role of semaphore primitives. Figure 7.3 is a refinement of the abstract task state diagram shown in Figure 3.4 that specifically highlights the transitions involved in semaphore operations, namely, the transition from the *Running* to the *Blocked* state, and from *Blocked* to *Ready*.

- The transition of a task τ_1 from *Running* to *Blocked* takes place when the task itself voluntarily calls the semaphore primitive P(s) on a semaphore s whose value s.v is zero.
- The transition of τ_1 out of the *Blocked* state is necessarily involuntary and must be triggered by another task. This is because a blocked task cannot execute any instruction by definition, including synchronization primitives. Namely, it takes place when another task, for instance, τ_2 executes a V(s)

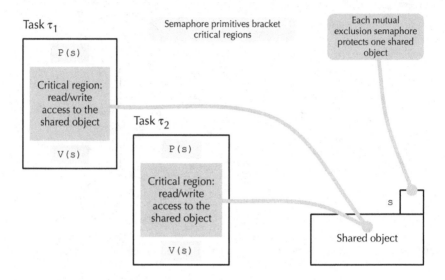

FIGURE 7.4 Typical usage of a mutual exclusion semaphore.

on semaphore s, the operating system finds a reference to τ_1 in s.q and decides to unblock it.

Referring again to Figure 7.3, it is important to note that when task τ_1 is unblocked, it goes into the *Ready* state, rather than *Running*. Being now in the *Ready* state, τ_1 becomes eligible for execution again, but this does not imply it will resume execution immediately. This is a useful design choice because it enforces a clear separation of concerns between two distinct entities in an operating system, namely:

- The synchronization mechanism, in particular the queuing policy of s.q in this case, determines which task may proceed in its execution as a consequence of a V(s). The primary goal of this choice is to ensure that the synchronization itself is correct, for instance, by avoiding race conditions when a semaphore is used to enforce mutual exclusion.
- The scheduling algorithm decides which tasks, among those eligible for execution, shall actually execute at any given time. As discussed in Section 3.2, this decision takes place according to other criteria, for instance, task priorities.

7.2.2 MUTUAL EXCLUSION SEMAPHORES

Semaphores provide a simple and convenient way of enforcing mutual exclusion among an arbitrary number of tasks that want to have access to a certain shared object. As shown in Figure 7.4, in this case the semaphore is used in a standardized way, according to the following steps:

- A distinct semaphore (s in the figure) is associated with each shared object to be protected. Only one semaphore is needed, regardless of how complex the object is internally. The initial value of a mutual exclusion semaphore is invariably 1.
- All critical regions that make access to a certain shared object are surrounded by the primitives P(s) and V(s) on the corresponding semaphore s. In other words, these primitives are used as brackets around each critical region and constitute its entry and exit code, respectively.

In the figure, we assume that s has been properly initialized somehow. In practice, one or more explicit initialization steps are often required, depending on the API being used. Section 7.4 provides more information about this aspect.

Although a formal proof is beyond the scope of this book, we can observe that, by intuition, the approach just described certainly fulfills the first correctness condition presented in Section 7.1.4. When one task τ_1 wants to enter its critical region, it must first pass through the critical region entry code, that is, P(s). Depending on the value of s, this operation may have two different outcomes:

1. If s.v is 1, the effect of P(s) is to decrement s.v to 0 without blocking τ_1, thus allowing the task to continue into the critical region. This is correct because s.v being 1 implies that no other tasks are currently within a critical region.
2. If s.v is 0, τ_1 is blocked in the critical region entry code, that is, at the critical region boundary. This is because, if s.v is 0, another task was allowed to enter its critical region and it did not exit from it yet.

As a consequence, if two tasks τ_1 and τ_2 try to enter their critical regions, both controlled by the same semaphore s, only one of them—for example τ_1—is allowed to proceed immediately because it finds s at 1. Instead, the other task finds s.v at 0, is enqueued on s.q, and is blocked. At the same time, this does not prevent other tasks from entering their critical regions, provided they pertain to a different shared object, and hence, are controlled by *another* semaphore. This property fulfills the second part of the first correctness condition, which disallows "useless" waits.

When a task, τ_1 in our example, reaches the critical region exit code and invokes V(s) two distinct things may happen:

1. If there is at least one task blocked on s, s.v does not change (it stays at 0) and exactly *one* of the blocked tasks, τ_2 in our example, is unblocked. As a consequence, it is now allowed to enter its critical region as soon as the scheduler decides to execute it. Mutual exclusion is still guaranteed because no other tasks can enter any critical region controlled by s until the task just unblocked exits from its critical region and executes V(s). Moreover, if yet another task tried to enter a critical region associated with the same semaphore s, it would be blocked as well, because s.v is still 0.
2. When a task exits from the critical region by executing V(s) and there are no other tasks waiting on s, like it happens to τ_2 in our example, s.v goes back to

1 (its initial value) so that exactly one process will be allowed to enter into the critical region immediately, without being blocked, in the future.

No race conditions during the execution of P(s) and V(s) are possible because, as stated in Section 7.2.1, the implementation of these primitives must necessarily guarantee their atomicity.

For what concerns the second correctness condition, it can easily be observed that the only case in which the mutual exclusion semaphore prevents a task from entering a critical region takes place when another task is already within a critical region controlled by the same semaphore. By construction, tasks doing internal operations shall not execute any primitive on any mutual exclusion semaphore, and hence, they cannot prevent other tasks from entering their critical regions.

Whether or not a mutual exclusion semaphore satisfies the third correctness condition depends on its implementation and, in particular, the queuing policy of its queue. The condition can easily be fulfilled if the policy is first-in, first-out (FIFO), so that the V() primitive always unblocks the task that has been waiting on the semaphore for the longest time.

However, in Section 8.1 we will see that this is often not adequate in a real-time system because it makes the system prone to the unbounded priority inversion issue and hinders schedulability analysis, as hinted in Section 4.1.3. On the other hand though, when using a different queuing policy, some processes may in principle be subject to an indefinite wait and this possibility must be ruled out by other means. For instance, since a successful schedulability analysis implies that all tasks complete without violating their deadlines, it also implies that none of them can ever be subject to an indefinite, unbounded wait.

To informally check that the fourth, and last, correctness condition is also satisfied, it is sufficient to observe that the definition of a semaphore and its primitives completely abstracts away from any architectural details about the system and does not contain any reference to any task or processor characteristics. Similarly, semaphore implementations are also required not to introduce such dependencies to be considered adequate for use.

7.2.3 SYNCHRONIZATION SEMAPHORES

Besides mutual exclusion, a semaphore is also useful to implement synchronization among tasks, for instance, when we want to enforce a precedence constraint and block a task τ_2 until a certain event, generated by another task τ_1, occurs. In this case, the semaphore is often called synchronization semaphore.

As shown in Figure 7.5, when task τ_1 makes use of a shared object to transfer data to τ_2, we must make sure that τ_2 reads from the shared object only after τ_1 has updated it completely. In its simplest form, this precedence constraint can be enforced by means of a semaphore s initialized to zero. Then, task τ_2 performs a P(s) before reading from the shared object, while τ_1 calls V(s) after updating it. In this way, assuming that τ_2 executes first:

- Task τ_2 blocks in P(s) because s.v is zero when it invokes the primitive.

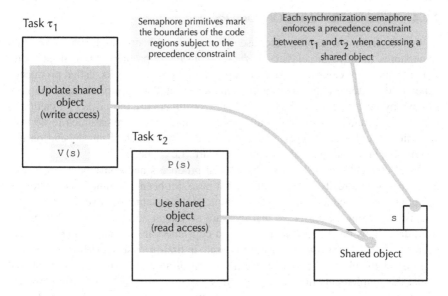

FIGURE 7.5 Using a synchronization semaphore to enforce a precedence constraint.

- When τ_1 is done with the update, it executes V(s), thus unblocking τ_2.

In this way, τ_2 has a consistent view of the shared object, which contains all the new data that τ_1 wrote into it, when it eventually executes. Moreover, s.v is still at zero after both P() and V() have been completed, and hence, the semaphore is ready for the next synchronization round.

The example also emphasizes the important role the value of a semaphore plays in memorizing and keeping track of past events. This becomes evident if we analyze what happens if τ_1 runs *before* τ_2 has had the opportunity of blocking:

- When τ_1 executes V(s), it finds that s.q is empty. Therefore, it increments s.v to 1.
- When eventually τ_2 invokes P(s) it finds s at 1, meaning that the synchronization condition it would like to wait for has already been fulfilled in the past. Accordingly, it continues immediately, without blocking, after bringing s.v back to zero.

It is also worth noting that this is a one-way synchronization. In other words, it prevents τ_2 from consuming data that τ_1 has not produced yet, but it does not prevent τ_1 from overrunning and overwriting its own data if it runs more than once before τ_2 had the opportunity to consume them. As will become clearer in the next section, we need one semaphore for each unidirectional synchronization condition that the concurrent program must respect.

7.2.4 PRODUCERS AND CONSUMERS

In this section, we explore the use of mutual exclusion and synchronization semaphores to solve the classic *producers–consumers* problem outlined, in a preliminary form, by Courtois [38]. In this problem a group of tasks, called producers, generate data items and make them available to another group of tasks, the consumers, by means of the `prod()` function. Consumers use the `cons()` function to retrieve data items. To simplify the discussion, data items are assumed to be integer values, held in `int`-typed variables, but the solution shall not depend on this assumption and work with data items of any arbitrary type.

An N-element buffer, interposed between producers and consumers, holds data items that have already been produced, but have not been consumed yet. Data consumption is assumed to be destructive, that is, each consumer removes the data items it consumes from the buffer, so that they are no longer accessible to others. Moreover, data items must be made available to consumers in the same order they have been produced, that is, the buffer must operate in first-in, first-out (FIFO) order.

In order to solve this problem, we must first of all define an appropriate shared object that implements the buffer. The most natural and straightforward approach, which also satisfies the FIFO access order requirement, is to use a circular buffer. As shown in Figure 7.6, such a buffer is composed of three elements:

1. A shared array of N integer elements, `int buf[N]`. Without loss of generality, we assume $N = 8$ in the figure.
2. A shared index `int in`, which points to a free element in the buffer, that is, the element to be filled next.
3. A shared index `int out`, which points to the oldest full element in the buffer.

Both `in` and `out` start at zero and are incremented upon insertion and removal of an element from the buffer, respectively. To make the buffer circular, they are incremented modulus N, that is, they go back to zero when they are incremented beyond $N - 1$.

As an example, Figure 7.6 shows the state of the data structure when 4 data items have been inserted and 2 have been removed after initialization. The value of `in` is 4, meaning that `buf[4]` must be filled when inserting the next element into the circular buffer, and `out` is 2, which indicates that `buf[2]` contains the oldest element in the buffer. Full buffer elements are denoted by black-filled circles and empty elements by empty circles.

Unless the circular buffer is never filled completely, in order to leave a so-called *guard* element always empty, there is an inherent ambiguity regarding its state. Namely, the condition `in == out` could indicate that the buffer is completely empty (as it is initially) or completely full. In the solution we are describing here, this ambiguity is resolved at the task synchronization instead of the data structure level, and hence, no guard element is needed.

According to the problem statement and the definition of circular buffer just given, in order to insert and remove an element from the circular buffer we need the following fragments of code in the `prod()` and `cons()` functions, respectively:

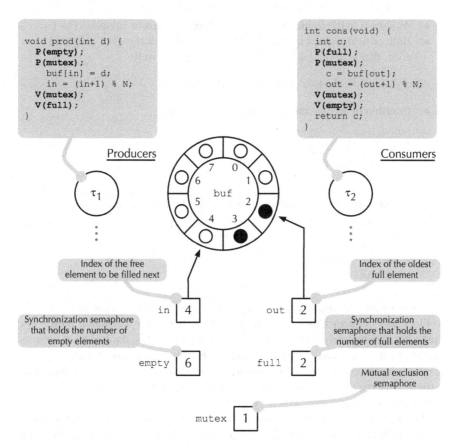

```
void prod(int d) {
  P(empty);
  P(mutex);
    buf[in] = d;
    in = (in+1) % N;
  V(mutex);
  V(full);
}
```

```
int cons(void) {
  int c;
  P(full);
  P(mutex);
    c = buf[out];
    out = (out+1) % N;
  V(mutex);
  V(empty);
  return c;
}
```

Producers   Consumers

τ_1 τ_2

Index of the free element to be filled next   Index of the oldest full element

Synchronization semaphore that holds the number of empty elements   in | 4   out | 2   Synchronization semaphore that holds the number of full elements

empty | 6   full | 2

Mutual exclusion semaphore

mutex | 1

FIGURE 7.6   Synchronization and mutual exclusion semaphores to coordinate access to a circular buffer.

- If d is the data item to be inserted and we assume for the time being that the buffer is not full, the insertion can be performed by means of:

```
buf[in] = d;
in = (in+1) % N;
```

- Symmetrically, in order to remove a data item from the buffer and store it into c, assuming that the buffer is not full, we need to execute:

```
c = buf[out];
out = (out+1) % N;
```

These fragments of code are critical regions because they both make access to a shared object, the circular buffer. According to Section 7.2.2, they must be associated to, and protected by, the same mutual exclusion semaphore, called mutex. More specifically:

- As shown in the figure, the `mutex` semaphore must be initialized to 1.
- Both critical regions must be bracketed by the critical region entry and exit code, that is, `P(mutex)` and `V(mutex)`, respectively.

After doing this, the two critical regions become:

- Insertion of an element in the `prod()` function:

```
P(mutex);
  buf[in] = d;
  in = (in+1) % N;
V(mutex);
```

- Removal of an element in the `cons()` function:

```
P(mutex);
  c = buf[out];
  out = (out+1) % N;
V(mutex);
```

It is now time to turn our attention to task synchronization. First of all, we need to block any consumer that wants to get a data item from the shared buffer when the buffer is completely empty. A blocked consumer must be unblocked as soon as a producer puts a new data item into the buffer. However, this is not the only synchronization condition of interest. Symmetrically, we also need to block a producer when it tries to insert a data item into a buffer that is already completely full, so as to avoid overwriting existing data.

As described in Section 7.2.3, we need one semaphore for each synchronization condition that the concurrent program must respect. In this case, we have two conditions and hence we need two semaphores:

1. The semaphore `full` counts how many full elements there are in the buffer. Its initial value is 0 because at the very beginning the circular buffer is empty. Consumers perform `P(full)` before removing a data item from the buffer, and possibly block if there are no data in the buffer. On the other hand producers perform a `V(full)` after storing an additional data item into the buffer. In this way, they either unblock a waiting producer or increment the count of full elements.
2. The semaphore `empty` counts how many empty elements there are in the buffer. Its initial value is N because the buffer is completely empty at the beginning, so all its N elements are indeed empty. Producers perform a `P(empty)` before putting more data into the buffer to either update the count of empty elements (if the buffer is not completely full) or block (otherwise). After removing one data item from the buffer, consumers perform a `V(empty)` to either unblock one waiting producer or increment the count of empty elements.

As illustrated in Figure 7.5, the `P()` and `V()` primitives must then be placed at the boundaries of the code fragments that must satisfy each execution precedence constraint. More specifically:

- Since a consumer must be forced to wait until a producer has produced a data item if the buffer is empty, P(full) must precede the consumers' critical region and V(full) must follow the producers' critical region.
- Symmetrically, a producer must be forced to wait until a consumer has consumed a data item if the buffer is completely full. Therefore, P(empty) must precede the producers' critical region and V(empty) must follow the consumers' critical region.

Since the initial value of empty is *N* the first *N* producers are granted access to their critical region even without intervening consumers. This gives producers the ability to completely fill the circular buffer without blocking. The full code, written using a C-like syntax, is shown at the top of Figure 7.6 with semaphore primitives highlighted.

The placement of the return c statement within the cons() function is critical and deserves special attention. At first sight, it may seem that it could be put anywhere after the statements that extracted the data item from the circular buffer. However, it must necessarily follow both V(mutex) and V(empty) for the following reasons:

- If the function returns to the caller before executing V(mutex), the critical region guarded by mutex stays open. As a consequence, any other tasks trying to enter their critical region in the future will block on the critical region entry code P(mutex).
- If the function returns to the caller after closing the critical region but before executing V(empty), the state of the semaphore is no longer synchronized with the state of the circular buffer. More specifically, although one data item has been extracted from the buffer, the state of the empty semaphore does not reflect this. Therefore, consumers trying to deposit more data items into the buffer in the future will be blocked on P(empty) even though the buffer is not really full.

These are two simple examples of a pathological condition that may affect a concurrent system and causes tasks to wait indefinitely. It is known as *deadlock* and will be discussed in Section 8.2. Unfortunately, to complicate matters further, neither of these mistakes can reliably be detected by the C compiler because it sees P() and V() as ordinary function calls and is unaware of their complex semantics.

7.3   MONITORS

7.3.1   DEFINITION AND PROPERTIES

As seen in the previous section, semaphores can be defined easily and their behavior can be fully described in a relatively short space. On one hand, it turns out that they are also easy to implement, so virtually all operating systems support them. On the other hand, semaphores are also a very low-level task synchronization mechanism and, for this reason, they are difficult to use correctly.

We already saw that even a small mistake in the placement of a `return` statement may disrupt a semaphore-based concurrent program without necessarily being noticed by the compiler, but the same is true also when semaphore primitives are misplaced ever so slightly. For instance, we may consider the following, alternative definition of the function `prod()` in the producers–consumers problem:

```
void prod(int d)  {
  P(mutex);
  P(empty);
    buf[in] = d;
    in = (in+1) % N;
  V(mutex);
  V(full);
}
```

With respect to the correct solution shown in Figure 7.6, the only difference is that we swapped the two `P()` primitives shown in boldface. After all, the code still makes sense by intuition, because we could justify the placement of the primitives in the following way:

- Before storing a new data item into the circular buffer a producer must make sure that it has exclusive access to the circular buffer itself. Hence, it must execute a `P(mutex)` and possibly block if other tasks are operating on the buffer at the moment.
- Moreover, the producer must also ensure that there is some free space in the buffer before writing into it. To do this, it must execute a `P(empty)` to update the count of free buffer elements held in `empty.v` and, if necessary, block until at least one free element becomes available for use.
- The producer can actually operate on the buffer only after both prerequisites (mutual exclusion and free space availability) have been satisfied.

Unfortunately, this kind of reasoning is incorrect because it can sometimes lead to a deadlock. More specifically, when a producer tries to store an element into a buffer that is completely full, the following sequence of events may occur:

- The producer succeeds in gaining exclusive access to the shared buffer by executing a non-blocking `P(mutex)`. From this point on, the value of the semaphore `mutex` becomes zero.
- Since the buffer is full, `empty.v` is zero because this semaphore counts the number of free elements. Therefore, `P(empty)` blocks the producer *within* its critical region because it did not release the mutual exclusion semaphore `mutex` before blocking.

After this, the only way of unblocking the producer would be to execute a `V(empty)` from another task. The only tasks that could possibly execute this primitive are consumers but, in order to do this, they must first go through a critical region controlled by `mutex`. This is impossible because the current value of `mutex` is zero.

Therefore, the whole system of producers and consumers soon grinds to a halt because:

- Any incoming producer blocks on P(mutex) because mutex.v is zero.
- The first N incoming consumers execute a non-blocking P(full) and eventually bring full.v to zero but cannot retrieve any data from the buffer because they block on P(mutex).
- If further consumers arrive, they do not even reach P(mutex) because they block on P(full).

This example also provides the opportunity to remark once more that a deadlock, like many other issues in concurrent programming, often occurs with low probability and may easily go unnoticed. In this case, it is easy to see that a deadlock never occurs if producers and consumers interleave so as to never fill the shared buffer completely.

To address these shortcomings, Brinch Hansen [26] and Hoare [60] proposed a more structured and higher-level IPC mechanism, called *monitor*. A third variant of the same mechanism was also defined and pioneered by the Mesa programming language [51], which directly influenced POSIX monitors. A monitor is a composite object, which contains a *shared object* and a set of *methods* that operate on it. By definition, a monitor must exhibit the following two main properties:

- *Information hiding.* The shared object within a monitor can be accessed only by means of the methods belonging to that monitor. There is no way to access the shared object in other ways. At the same time, monitor methods are not allowed to access any shared object belonging to other monitors. Only the monitor methods are public and can be freely invoked from outside the monitor.
- *Mutual exclusion.* The monitor implementation must guarantee that only one task will be actively executing within any of the monitor methods at any given instant. This property is similar to what a mutual exclusion semaphore does with respect to the critical regions it controls, but more relaxed because it allows multiple tasks to be within monitor methods, as long as at most one of them is executing at any given time.

Both properties are relatively easy to implement in practice if the monitor is a built-in construct of the programming language, as originally stipulated by its proponents. In this case, the language compiler knows exactly which methods are associated to a certain monitor and can automatically add the necessary mutual exclusion code to them. Similarly, the language compiler also has all the information it needs to enforce the information-hiding rule just discussed and flag errors appropriately while parsing the source code.

In other cases, like in the C language, the concept of monitor is unknown to the compiler. Although the POSIX international standard [68] provides all the monitor's building blocks, programming discipline is needed in some key areas to properly realize them because the compiler cannot be of much help in this respect. More specifically:

- The programmer is responsible for defining a mutual exclusion semaphore for each monitor and invoking `P()` and `V()` on it when entering and exiting all monitor methods. On the same lines, when dealing with multiple monitors, the programmer must also correctly associate each method to the monitor it belongs to, and hence, invoke these primitives on the right semaphore.
- Similarly, the programmer must implement information hiding using one of the underlying mechanisms provided by the language, also accepting all their limitations and shortcomings. For instance, defining a `static` shared object in C makes it accessible only within a compilation unit rather than globally, and hence, partially hides it.

7.3.2 CONDITION VARIABLES

The two properties discussed in the previous section avoid race conditions in accessing the shared object belonging to a monitor from its methods. In particular, the mutual exclusion property bears striking similarities with mutual exclusion semaphores, and it is no surprise that a mutual exclusion semaphore indeed can be used to enforce it, like it happens in POSIX.

However, in Section 7.2.3 we also discussed synchronization semaphores and no counterpart for them has been introduced within the monitor framework so far. This counterpart does exist and is called *condition variable*. Condition variables are the third kind of component that may belong to a monitor, besides the shared object and methods.

They are also subject to the information hiding rule and, accordingly, they cannot be referenced in any way, except from methods belonging to the same monitor. As for the shared object, if the programming language is unaware of monitors, it is the programmer's responsibility to appropriately hide condition variables so that they cannot be misused. The following two primitives are defined on a condition variable `c`:

- `wait(c)` blocks the invoking task and releases the monitor, thus allowing other tasks to enter it and execute its methods. These two steps are executed atomically, that is, in a single indivisible action.
- `signal(c)` wakes up one of the tasks blocked on `c`, if any. If no tasks are blocked on `c`, it has no effect.

Informally speaking, if a task discovers it cannot conclude its work immediately while executing one of the monitor methods, it invokes `wait` on a condition variable. In this way, it blocks and gives to other tasks the opportunity to enter the monitor and perform their duty. When one of these tasks discovers that the first task can now continue, it invokes `signal` on the condition variable.

Since tasks may have many distinct reasons for blocking within a monitor, we can introduce a distinct condition variable for each reason. In this way, blocked tasks can be divided into groups, or categories, depending on which condition variable

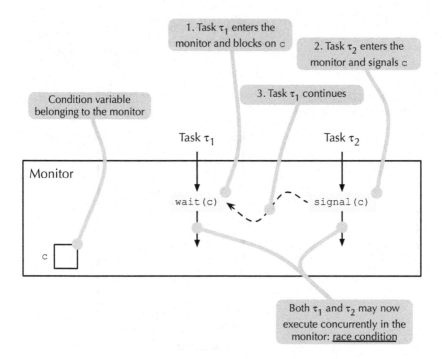

FIGURE 7.7   Potential race condition after wait/signal in a monitor.

they are blocked on, and then unblocked selectively by signaling the appropriate
condition variable.

Although the definition of `wait` and `signal` may already seem satisfactory by
intuition, it is still incomplete. Namely, we must introduce additional constraints on
task execution to guarantee that the synchronization mechanism condition variables
introduce does not hinder mutual exclusion, which monitors must invariably guarantee. Otherwise, we fall into a race condition after the following wait/signal sequence
on condition variable `c`, depicted in Figure 7.7 and involving two tasks:

1. Assuming the monitor is initially free, task τ_1 enters the monitor, starts executing
   one of its methods, and then blocks by means of `wait(c)`.
2. The wait primitive releases the monitor, and hence, a second task τ_2 is allowed to
   enter the monitor and execute one of its methods. There is no race condition up to
   this point because τ_1 is blocked and no additional tasks can enter the monitor due
   to the mutual exclusion constraint at the monitor boundary.
3. While executing in the monitor, task τ_2 invokes `signal(c)`, thus unblocking τ_1.

After this sequence of actions, *both τ_1 and τ_2* may concurrently execute within
the monitor. As a consequence, they are both allowed to manipulate its shared object
in an uncontrolled way, a scenario that clearly leads to a race condition.

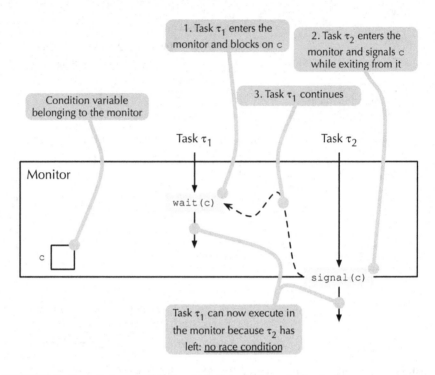

FIGURE 7.8 Forced placement of signal in a Brinch Hansen's monitor.

The simplest way to address this issue was proposed by Brinch Hansen and is depicted in Figure 7.8.

It is based on constraining the placement of `signal` primitives within the monitor methods. In particular, if a task ever invokes `signal` within a monitor method, it must be its very last action within the monitor and implicitly causes the task to exit from it. In other words, as shown in the figure, all invocations of `signal` must always be at the boundary between the monitor and the outside world.

In this way, as also shown in Figure 7.8, only task τ_1 can execute within the monitor after a `wait`/`signal` sequence. Task τ_2 actually keeps running concurrently with τ_1, but this is of no concern because τ_2 is now outside of the monitor and the information hiding rule implicitly prevents any race condition. Although our example only involves two tasks for simplicity, it can be proved [26] that this approach solves the problem in general.

An advantage of following this path is that it does not add any additional complexity or runtime overhead to the `wait` and `signal` primitives with respect to their original, intuition-based definition. Moreover, if monitors are a programming language construct, the compiler can easily detect any violation of the constraint while parsing the source code. However, the burden of designing monitor methods so that `signal` only appears in the right places is left to programmers.

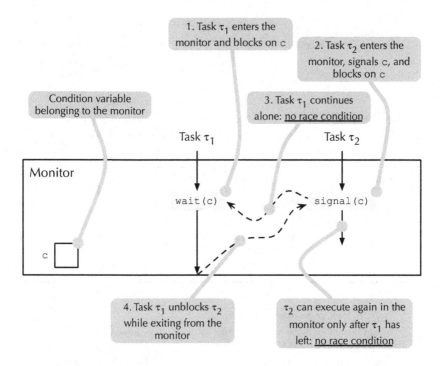

FIGURE 7.9 Blocking signal in a Hoare's monitor.

On the contrary, Hoare's approach [60] is more flexible and gives programmers more freedom because it imposes no constraints at all on the placement of `signal`, which can therefore be invoked anywhere in monitor methods. The price to be paid is a greater complexity of `wait`, `signal`, and of the monitor exit code. In addition, the semantics of `signal` become less intuitive and clear because it may now *block the caller*.

The general mechanism of condition variables in a Hoare's monitor is illustrated in Figure 7.9. It revolves around the following rules:

- A task that invokes `signal(c)`, like τ_2 in the figure, blocks if it successfully unblocks another task, like τ_1, which has been waiting on condition variable c. In this way, τ_1 continues its execution within the monitor, but no race condition occurs because τ_2 is now blocked and no other tasks are allowed to enter the monitor due to the mutual exclusion mechanism at the monitor boundary.
- When task τ_1 exits from the monitor or waits again, by means of another `wait`, one of the tasks blocked on a `signal`, like τ_2 in the figure, is allowed to continue past it. Otherwise, one of the tasks waiting to enter the monitor is allowed in. Finally, if no tasks are waiting to enter the monitor, the monitor is released. As a consequence, tasks waiting to enter the monitor anew will be let in, one at a time, only when the task actively executing

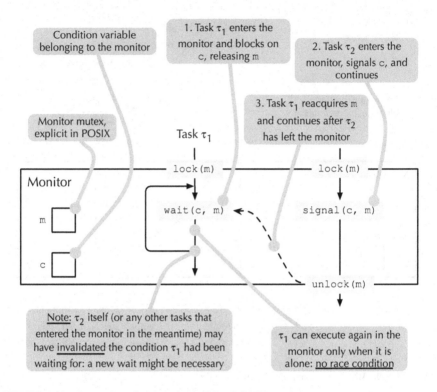

FIGURE 7.10 Semantics of wait and signal in a POSIX monitor.

in the monitor leaves or blocks, and no tasks are blocked in a `signal`. Again, there are no race conditions because τ_2 may continue within the monitor only after τ_1 has left or has blocked again.

As with Brinch Hansen's proposal, it can be proven that Hoare's method works with an arbitrary number of tasks and any number of condition variables, even though the example we depicted in Figure 7.9 is very simple.

The approach to avoid race conditions in a monitor according to the POSIX standard [68] differs from the previous two in a number of significant aspects. Programmers should keep these differences in mind because writing code for a certain flavor of monitor, and then executing it on another, may clearly lead to incorrect results.

1. As already outlined previously, monitors are not part of the C language. Therefore, programmers are responsible for defining a mutual exclusion semaphore for each monitor—called m in Figure 7.10—and initialize it appropriately. As shown in the figure, they are also in charge of correctly bracketing all monitor methods with the P (m) and V (m) primitives, called lock (m) and unlock (m) in POSIX. Moreover, they should make m visible and accessible only from monitor methods.

2. Another important consequence of lacking language support for monitors is that the `wait` primitive in POSIX has two arguments instead of one. Since the compiler has no way to associate condition variables with the monitor they belong to, it also may not autonomously infer what is the mutual exclusion semaphore that `wait` is supposed to release. Therefore, the programmers must pass the right semaphore as an additional argument.

3. The underlying reasoning behind the POSIX approach is that a task that has just been unblocked after invoking a `wait`, like τ_1 in the figure, must reacquire the mutual exclusion semaphore before proceeding further. This is done implicitly, within the `wait` primitive. Instead, the signaling task, τ_2 in the figure, continues immediately. In other words, the POSIX approach avoids race conditions in the opposite way as Hoare's, given that it postpones the *signaled* task instead of the *signaling* task, thus following the same approach as the Mesa programming language [51].

4. When the task currently executing within the monitor exits from it or blocks in a `wait`, it releases the mutual exclusion semaphore it holds, by either invoking `unlock` explicitly (if it is exiting from the monitor), or implicitly (as part of its `wait`). As a result, it unblocks one of the tasks waiting to resume or start its execution in the monitor, if any. More specifically, the task can be:

   • waiting to reacquire the mutual exclusion semaphore after having been unblocked, like τ_1 in the figure, or

   • waiting to enter the monitor from the outside, being blocked on `lock(m)`,

   and the decision depends on the queuing policy of the mutual exclusion semaphore m. If no tasks are waiting, the value of m goes back to one, to remember that the monitor has become free and no tasks are currently executing within it.

To summarize, if we assume that no other tasks besides τ_1 and τ_2 try to operate on the monitor, we can describe the behavior of the simple POSIX wait/signal sequence of Figure 7.10 as follows:

• Task τ_1 acquires the mutual exclusion semaphore m with `lock(m)` and enters the monitor without blocking.

• While executing in the monitor, τ_1 waits on condition variable c by means of `wait(c, m)`. As a result, it also releases the monitor by performing an implicit `unlock(m)`. Both operations are performed as an atomic unit according to the definition of `wait`.

• Another task, τ_2 in the example, who has been blocked in the `lock(m)` at the monitor boundary, is allowed to continue.

• Task τ_2 performs `signal(c)` and continues, since this is a non-blocking primitive. Task τ_1 cannot continue immediately because it must first reacquire m.

• When τ_2 leaves the monitor, it performs a `unlock(m)`, thus allowing τ_1 to continue after waiting.

• Task τ_1 can now continue its execution within the monitor.

The most important side effect of this approach from the practical standpoint is that, as also highlighted in the example, when task τ_1 blocks on a condition variable because it has to wait until a condition has been fulfilled, it cannot be completely certain that the condition it has been waiting for will still be satisfied when it will eventually continue its execution in the monitor.

This is because other tasks, like the signaling task τ_2 itself or any other tasks allowed in the monitor in the meantime, may have invalidated the condition. The first scenario can be avoided by making sure that tasks do not invalidate the condition they have been signaling before they exit the monitor.

Instead, there is no general way to avoid the second scenario. As mentioned before, the choice of which task executes in the monitor after τ_2 exits depends on the queuing policy of the mutual exclusion semaphore m. It may be τ_1, but the choice may also fall on other tasks waiting to enter the monitor. Therefore, when using a POSIX monitor, the `wait` primitive should always be enclosed in a loop that checks whether the condition of interest is indeed satisfied after the primitive returns and invokes `wait` again if this is not the case. As a side effect, it may happen that a task has to invoke `wait` more than once before it can wait for a condition successfully and then proceed, thus introducing some unwanted execution time variability.

To conclude this section, Figure 7.11 shows how the producers–consumers problem can be solved using the simplest kind of monitor considered so far, that is, the Brinch Hansen's monitor. Section 7.4 contains instead an example of use of POSIX monitors. Since the C language does not support monitors, the code shown in the figure is written by means of the following, hypothetical C language extensions:

- The `monitor` keyword introduces a monitor. Like in a `struct` definition, a pair of braces syntactically groups together the shared object, condition variables, and methods that belong to the monitor.
- In a monitor, the `condition` keyword defines a condition variable, with a syntax equivalent to a C variable definition in which `condition` replaces the data type specifier.

With respect to the semaphore-based solution shown in Figure 7.6 there are several important differences. More specifically:

- The mutual exclusion semaphore `mutex` is no longer needed because the Brinch Hansen's monitor construct enforces the mutual exclusion among monitor methods by itself.
- In the semaphore-based solution, the value of the synchronization semaphores was used to keep track of how many empty and full elements there were in the circular buffer. Condition variables have no value, and hence, the monitor-based solution has to keep the count in the shared variable `count`. The variable starts at zero and counts the number of full elements in the buffer.
- The two synchronization semaphores `empty` and `full` have been replaced by two condition variables called `not_full` and `not_empty`, respectively. There is a strong analogy between the role of a synchronization

```
#define N 8
monitor ProducersConsumers
{
    int buf[N];
    int in = 0, out = 0;
    condition not_full, not_empty;
    int count = 0;

    void prod(int v)
    {
        if(count == N)   wait(not_full);
        buf[in] = v;
        in = (in + 1) % N;
        count = count + 1;
        if(count == 1)   signal(not_empty);
    }

    int cons(void)
    {
        int v;
        if(count == 0)   wait(not_empty);
        v = buf[out];
        out = (out + 1) % N;
        count = count - 1;
        if(count == N-1)   signal(not_full);
        return v;
    }
};
```

FIGURE 7.11 Producers–consumers problem solved by means of a Brinch Hansen's monitor.

semaphore and that of the corresponding condition variable. For instance, the `empty` semaphore gives producers the ability to block until the buffer has at least one empty element by means of a `P(empty)`. Symmetrically, producers can block until the buffer is not completely full by invoking `wait(not_full)`.

- Both `wait` and `signal` are executed conditionally in the monitor-based solution, after checking the condition that tasks would like to wait upon, or signal, respectively. For instance, a producer executes `wait(not_full)` only after detecting that `count == N`, and a consumer calls `signal(not_full)` only after removing one element from a buffer that was completely full, thus making `count == N-1`. On the contrary, semaphore primitives are invoked unconditionally in the semaphore-based solution.

TABLE 7.1

RTEMS Semaphore Primitives, Classic API

Function	Purpose
rtems_semaphore_create	Create a semaphore
rtems_semaphore_ident	Find the identifier of a semaphore given its name
rtems_semaphore_delete	Delete a semaphore
rtems_semaphore_obtain	Perform a P() on a semaphore
rtems_semaphore_release	Perform a V() on a semaphore
rtems_semaphore_flush	Unblock all threads waiting on a semaphore
rtems_semaphore_set_priority	Set the per-scheduler ceiling priority of a semaphore

7.4   RTEMS API FOR SHARED-MEMORY IPC

7.4.1   CLASSIC API

Semaphore creation and semaphore attributes

The RTEMS operating system offers semaphores as one of its main synchronization devices, through its *Semaphore Manager*. As summarized in Table 7.1, a semaphore is created by means of the function:

```
rtems_status_code rtems_semaphore_create(
    rtems_name name,
    uint32_t count,
    rtems_attribute attribute_set,
    rtems_task_priority priority_ceiling,
    rtems_id *id);
```

This function creates a semaphore with the given name and, if successful, stores its unique identifier into the location pointed by id. This identifier must then be used to refer to the semaphore with all the other semaphore primitives except rtems_semaphore_ident. Like all the other RTEMS object names, the name argument is a 32-bit value. Users can synthesize it with the help of the function rtems_build_name, which takes 4 ASCII characters as arguments, when they wish to make it more human-readable.

The most important argument of the function rtems_semaphore_create is attribute_set. It determines the specific variety of semaphore to be created among the ones that RTEMS supports. It is the bitwise or of the constants listed in Table 7.2. Each variety represents a different trade-off point between implementation efficiency and expressive power of the semaphore.

The first two attributes, RTEMS_LOCAL and RTEMS_GLOBAL, determine the visibility and accessibility of the semaphore to be created on a multi-node system. Namely, the semaphore is always created on the local node where the calling thread

TABLE 7.2

RTEMS Semaphore Attributes, Classic API

Attribute	Meaning
RTEMS_LOCAL	Local semaphore
RTEMS_GLOBAL	Globally accessible semaphore
RTEMS_COUNTING_SEMAPHORE	Ordinary, counting semaphore (as in Section 7.2)
RTEMS_BINARY_SEMAPHORE	Binary semaphore that supports recursive locks
RTEMS_SIMPLE_BINARY_SEMAPHORE	Binary semaphore that does *not* support recursive locks
RTEMS_FIFO	Use a first-in, first-out (FIFO) queuing policy
RTEMS_PRIORITY	Use a priority-based queuing policy
RTEMS_NO_INHERIT_PRIORITY	Do not use the priority inheritance protocol
RTEMS_NO_PRIORITY_CEILING	Do not use the immediate priority ceiling protocol
RTEMS_NO_MULTIPROCESSOR _RESOURCE_SHARING	Do not use the multiprocessor resource sharing protocol
RTEMS_INHERIT_PRIORITY	Use the priority inheritance protocol
RTEMS_PRIORITY_CEILING	Use the immediate priority ceiling protocol
RTEMS_MULTIPROCESSOR _RESOURCE_SHARING	Use the multiprocessor resource sharing protocol

resides. Then, semaphores with the RTEMS_LOCAL attribute can be used only by threads executing on the same node. Instead, information about a RTEMS_GLOBAL semaphore is broadcast to all other nodes in the system, so that remote threads can access it.

Considering that the creation of a global semaphore implies more overhead than the creation of a local semaphore, and remote semaphore operations are less efficient because they are carried out with the help of a proxy agent on the local node, semaphores should be kept local unless doing otherwise is strictly necessary.

The second group of attributes listed in Table 7.2 has to do with the expressive power of the semaphore. More specifically:

- A RTEMS_COUNTING_SEMAPHORE is the most powerful kind of semaphore that RTEMS supports and conforms to the theoretical definition of semaphore given in Section 7.2. The argument count is the initial value of the semaphore and can be any non-negative integer that can be represented in a uint32_t data type.
- A RTEMS_BINARY_SEMAPHORE is a semaphore whose values can be only 1 (unlocked) and 0 (locked), making it especially useful to implement mutual exclusion. Moreover, this kind of semaphore supports the *recursive* lock feature, which enables a thread to perform multiple, nested P() on the same semaphore without self-deadlocking. The operating system keeps a count of the nested P() performed by the thread and releases the

semaphore only when the number of `V()` executed by the thread matches the number of nested `P()` performed by the same thread.

- A `RTEMS_SIMPLE_BINARY_SEMAPHORE` is even simpler and does not support recursive locks. If a thread performs a `P()` on a semaphore it already owns, it self-deadlocks. In return, this kind of semaphore is more efficient and can be deleted while it is locked.

In addition, RTEMS semaphores support two queuing policies:

- The `RTEMS_FIFO` policy enqueues threads blocked on a semaphore in first-in, first-out order.
- The `RTEMS_PRIORITY` policy enqueues threads based on their priorities instead.

Selecting the `RTEMS_PRIORITY` policy is a prerequisite for enabling the priority inheritance or the immediate priority ceiling protocol (both to be discussed in Section 8.1) on a semaphore. This is done through another set of attributes:

- The `RTEMS_INHERIT_PRIORITY` attribute specifies that the semaphore must use the priority inheritance protocol.
- The `RTEMS_PRIORITY_CEILING` attribute specifies that the semaphore must use the immediate priority ceiling protocol. The ceiling of the semaphore is specified by means of the argument `priority_ceiling` of `rtems_semaphore_create`.

The complementary attributes `RTEMS_NO_INHERIT_PRIORITY` and `RTEMS_NO_PRIORITY_CEILING`, respectively, indicate that these protocols shall not be used. The priority inheritance and priority ceiling protocols cannot be enabled together and are supported only for local, binary semaphores whose queuing policy is priority-based.

In addition, RTEMS implements two protocols specifically designed for symmetric multiprocessor (SMP) and multicore systems: the multiprocessor resource sharing protocol (MrsP) [28] and the $O(M)$ independence-preserving protocol (OMIP) [23]. In a nutshell, MrsP is an extension of the priority ceiling protocol that can be enabled for local, binary semaphores with priority-based queuing. Like priority ceiling, it is based on assigning to each semaphore ceiling priorities, but instead of a single priority there is one for each scheduler instance in the system. Ceiling priorities must be calculated and assigned by the user.

Similarly, OMIP works with the same kind of semaphore and is an extension of the priority inheritance protocol. Like its ancestor, it does not require any user-specified semaphore configuration. Although the peculiarities and limitations of both protocols will be better discussed in Chapter 13 of the book, some information about how they are used by means of the Classic and POSIX API will still be given in this chapter, to give readers one single place where these APIs are fully described.

The MrsP protocol must be selected explicitly during semaphore creation by specifying the `RTEMS_MULTIPROCESSOR_RESOURCE_SHARING` attribute. Instead,

OMIP automatically and implicitly replaces the priority inheritance protocol on multicore systems, when the RTEMS_INHERIT_PRIORITY attribute is specified. In both cases, the RTEMS_BINARY_SEMAPHORE and RTEMS_PRIORITY attributes must also be set upon semaphore creation.

The function rtems_semaphore_create, like many other RTEMS functions, returns a status code to report the outcome of the operation. The RTEMS_SUCCESSFUL status code indicates that the semaphore has successfully been created. Upon failure, the function returns one of the following status codes:

RTEMS_INVALID_NAME The name of the semaphore passed in the name argument is invalid.

RTEMS_INVALID_ADDRESS The id argument is a NULL pointer, and hence, it does not point to a valid location where the newly created semaphore identifier could be stored.

RTEMS_NOT_DEFINED The argument attribute_set contains an undefined, or otherwise invalid, attribute.

RTEMS_INVALID_NUMBER The initial value of the semaphore given in the count argument is invalid for the kind of semaphore being created. For instance, binary semaphores that use the MrsP protocol cannot be created in a locked state. Hence, for this kind of semaphore count must necessarily be one.

RTEMS_TOO_MANY The new semaphore could not be created because the maximum number of local or global semaphores in the system (depending on the kind of semaphore the attribute set asked for) has been reached. These limits are set in the RTEMS application compile-time configuration outlined in Section 2.4.2 and fully described in [105].

RTEMS_MP_NOT_CONFIGURED The attribute set given as argument asked for a global semaphore, but multi-node support has not been enabled in the system configuration.

After a successful call to rtems_semaphore_create, the semaphore identifier obtained from it can directly be used to operate on the semaphore. The semaphore identifier can also be obtained from the semaphore name by calling the function:

```
rtems_status_code rtems_semaphore_ident(
    rtems_name name,
    uint32_t node,
    rtems_id *id);
```

This function looks for a semaphore whose name is given by the name argument on node node and stores its identifier into the location pointed by id. If node is the special value RTEMS_SEARCH_ALL_NODES, the function searches all nodes in the system, starting from the local node, that is, the node where the calling thread resides.

If multiple semaphores have the same name, the function is guaranteed to return the identifier of a semaphore with the given name, but exactly which one is left unspecified. Upon failure, the function returns one of the following status codes:

RTEMS_INVALID_NAME The function did not find any semaphore with the requested `name` within the scope of its search.

RTEMS_INVALID_NODE The `node` identifier is invalid.

The function:

```
rtems_status_code rtems_semaphore_delete(
    rtems_id id);
```

deletes the semaphore whose identifier `id` has been passed as argument. After a successful call to this function, the identifier shall no longer be used. The semaphore may have been created by another thread, but must reside on the same node as the calling thread. The return value of `rtems_semaphore_delete` is RTEMS_SUCCESSFUL upon successful completion or one of the following status codes:

RTEMS_INVALID_ID The semaphore identifier is invalid.

RTEMS_ILLEGAL_ON_REMOTE_OBJECT The semaphore could not be deleted because it does not reside on the same node as the calling thread.

RTEMS_RESOURCE_IN_USE The semaphore is actually a mutex and it could not be deleted because it is currently locked.

Semaphore operations

Referring back to Table 7.1, the functions `rtems_semaphore_obtain` and `rtems_semaphore_release` are the RTEMS counterparts of the abstract P() and V() operations discussed in Section 7.2. More specifically, the function:

```
rtems_status_code rtems_semaphore_obtain(
    rtems_id id,
    rtems_option option_set,
    rtems_interval timeout);
```

performs a P() on the semaphore identified by `id`. The additional arguments `option_set` and `timeout` determine whether the calling thread may block—and for how long—to complete the semaphore operation, thus leading to three possibilities:

- If `option_set` is set to RTEMS_NO_WAIT, the `timeout` argument is ignored and the function returns to the caller the status code RTEMS_UNSATISFIED if the semaphore operation cannot be completed immediately, that is, without blocking.
- If `option_set` is set to RTEMS_WAIT and `timeout` is set to the special value RTEMS_NO_TIMEOUT, the function may potentially block the caller forever, until the semaphore operation can be completed or an error occurs.
- If `option_set` is set to RTEMS_WAIT and `timeout` is set to a timeout value expressed in ticks (see Chapter 5) the function may block the caller for up to `timeout` ticks. If the timeout expires before the

semaphore operation could be completed, the function returns the status code `RTEMS_TIMEOUT`.

Additional reasons for failure are reflected in the following status codes that `rtems_semaphore_obtain` may also return:

RTEMS_INVALID_ID The semaphore identifier is invalid.

RTEMS_OBJECT_WAS_DELETED The semaphore was deleted (by another thread) while the calling thread was blocked on it.

RTEMS_UNSATISFIED The semaphore is an MrsP semaphore and a deadlock condition was detected. Moreover, this status code is also returned when trying to obtain an MrsP semaphore more than once from the same task without releasing it in between, because MrsP semaphores do not support recursive locking.

The counterpart of `V()` is the function:

```
rtems_status_code rtems_semaphore_release(
    rtems_id id);
```

It returns either `RTEMS_SUCCESSFUL`, if the semaphore was released successfully, or one of the following status codes upon failures:

RTEMS_INVALID_ID The semaphore identifier is invalid.

RTEMS_NOT_OWNER_OF_RESOURCE The semaphore could not be released because it is a mutex and the calling thread did not obtain it beforehand.

RTEMS_INCORRECT_STATE The operation failed because the semaphore is an MrsP semaphore and the calling task did not respect the expected acquisition and release order. Semaphores using this protocol must be released in the opposite order with respect to the order in which they were obtained.

The last semaphore operation supported by the RTEMS classic API has not direct theoretical counterpart. The function:

```
rtems_status_code rtems_semaphore_flush(
    rtems_id id);
```

unblocks *all tasks* waiting on the semaphore, thus emptying its waiting queue, without changing the semaphore value. Each task will receive a `RTEMS_UNSATISFIED` status code from the `rtems_semaphore_obtain` it had been engaged in, to highlight that the semaphore has not been correctly obtained, and hence, some properties normally guaranteed by the execution of the `P()` operation (like mutual exclusion in the case of a mutex) may not be true.

Besides `RTEMS_SUCCESSFUL`, the `rtems_semaphore_flush` function may also return the following status codes:

RTEMS_INVALID_ID The semaphore identifier is invalid.

RTEMS_ILLEGAL_ON_REMOTE_OBJECT The semaphore is globally accessible and resides on a remote node. The flush operation is not supported in this case.

RTEMS_NOT_DEFINED The flush operation is not supported for MrsP semaphores.

Per-scheduler ceiling

The MrsP protocol requires a per-scheduler ceiling priority to be associated to each mutex. When creating a mutex that uses this protocol, all these priorities are initially set to the same value, that is, the one specified in the `priority_ceiling` argument of `rtems_semaphore_create`. This is likely not adequate, hence these priorities can be individually retrieved and set after semaphore creation by means of the function:

```
rtems_status_code rtems_semaphore_set_priority(
    rtems_id semaphore_id,
    rtems_id scheduler_id,
    rtems_task_priority new_priority,
    rtems_task_priority *old_priority);
```

This function sets the priority of semaphore `semaphore_id` with respect to the scheduler `scheduler_id` to the value `new_priority`, also storing the previous priority value into the location pointed by `old_priority`. The caller must pass a valid pointer even though it is not interested in the old priority value. If the new priority value is `RTEMS_CURRENT_PRIORITY` the function only retrieves the current per-scheduler priority without changing it. In both cases, the function operates only on semaphores residing on the same node as the calling thread.

The function `rtems_semaphore_set_priority` returns `RTEMS_SUCCESSFUL` when it succeeds or one of the following status codes when it fails:

RTEMS_INVALID_ID Either the semaphore identifier, or the scheduler identifier, or both, are invalid.

RTEMS_INVALID_ADDRESS The `old_priority` pointer is `NULL`.

RTEMS_INVALID_PRIORITY The `new_priority` value is not valid for the scheduler `scheduler_id`.

RTEMS_NOT_DEFINED The semaphore does not make use of the MrsP protocol and per-scheduler ceiling priorities are defined only for this kind of semaphore. For an ordinary priority ceiling semaphore, the ceiling priority is specified directly as an argument of `rtems_semaphore_create`.

RTEMS_ILLEGAL_ON_REMOTE_OBJECT The semaphore does not reside on the same node as the calling thread.

Binary semaphores

RTEMS also provides a specific, separate implementation of binary semaphores. Unlike the full-fledged semaphores provided by the Semaphore Manager, which are allocated by the kernel and referred to by means of an `rtems_id`, the storage for this kind of binary semaphore must be provided by the user and references to the semaphore are made by means of a pointer to the semaphore itself. A copy of a semaphore is not a valid semaphore itself and shall not be used for synchronization. Moreover, they cannot be accessed from remote nodes of a multi-node system.

TABLE 7.3

RTEMS Binary Semaphore Primitives, Classic API

Function	Purpose
rtems_binary_semaphore_init	Initialize a user-allocated binary semaphore
rtems_binary_semaphore_destroy	Destroy a binary semaphore
rtems_binary_semaphore_wait	Perform a P() on a binary semaphore
rtems_binary_semaphore_try_wait	Non-blocking variant of P()
rtems_binary_semaphore _wait_timed_tick	Timed variant of P()
rtems_binary_semaphore_post	Perform a V() on a binary semaphore
rtems_binary_semaphore_set_name	Set the human-readable name of a binary semaphore
rtems_binary_semaphore_get_name	Retrieve the name of a binary semaphore

Table 7.3 summarizes the primitives available for binary semaphores. With respect to full-fledged semaphores, they offer a streamlined interface that may make user code simpler and more efficient. The function:

```
void rtems_binary_semaphore_init(
    rtems_binary_semaphore *binary_semaphore,
    const char *name);
```

initializes the semaphore pointed by `binary_semaphore` and assigns the given `name` to it. The `name` may be `NULL` but, if it is not, it is the user's responsibility to ensure that the string it refers to remains valid through the whole lifetime of the semaphore, or until the name of the semaphore is changed. The user is also in charge of allocating the object of type `rtems_binary_semaphore` pointed by `binary_semaphore` before calling this function. Binary semaphores can also be statically initialized by means of the macro `RTEMS_BINARY_SEMAPHORE_INITIALIZER`. More specifically, the statement:

```
rtems_binary_semaphore binary_semaphore
    = RTEMS_BINARY_SEMAPHORE_INITIALIZER(name);
```

is equivalent to a call to `rtems_binary_semaphore_init`. It is worth noting that, in both cases, the initial value of the semaphore is *zero* and not one. As a consequence, if the binary semaphore is used for mutual exclusion, it is mandatory to perform an isolated V() on the semaphore right after initialization.

When no longer in use, a semaphore can (and should) be destroyed by means of the following function:

```
void rtems_binary_semaphore_destroy(
    rtems_binary_semaphore *binary_semaphore);
```

After destroying a semaphore, the corresponding `rtems_binary_semaphore` object shall no longer be used, unless it is first re-initialized. Destroying a semaphore while it is in use leads to unpredictable results.

The following function performs a P() operation on a binary semaphore referenced by `binary_semaphore`, thus possibly blocking the caller. Threads are inserted in the semaphore wait queue in priority order.

```
void rtems_binary_semaphore_wait(
    rtems_binary_semaphore *binary_semaphore);
```

As for full-fledged semaphores, a non-blocking and a timed variant of this function are available:

```
int rtems_binary_semaphore_try_wait(
    rtems_binary_semaphore *binary_semaphore);

int rtems_binary_semaphore_wait_timed_ticks(
    rtems_binary_semaphore *binary_semaphore,
    uint32_t ticks);
```

Unlike `rtems_binary_semaphore_wait`, which cannot fail, both these functions return an `int` whose value indicates whether or not they concluded successfully. More specifically, a return value of zero indicates success, whereas one of the following non-zero values denote failure:

EAGAIN The function `rtems_binary_semaphore_try_wait` was unable to acquire the semaphore immediately.

ETIMEDOUT The function `rtems_binary_semaphore_wait_timed_ticks` was unable to acquire acquire the semaphore before the specified number of `ticks` elapsed.

Since it never blocks the caller, `rtems_binary_semaphore_try_wait` is the only function in this group that may be called from an interrupt context.

The function:

```
void rtems_binary_semaphore_post(
    rtems_binary_semaphore *binary_semaphore);
```

performs a V() on the binary semaphore referenced by `binary_semaphore`. It can be invoked either from a thread or an interrupt context.

The last two functions in Table 7.3 enable the caller to retrieve and modify the human-readable name associated to a binary semaphore, used for debugging purposes:

```
const char *rtems_binary_semaphore_get_name(
    const rtems_binary_semaphore *binary_semaphore);

void rtems_binary_semaphore_set_name(
    rtems_binary_semaphore *binary_semaphore,
    const char *name);
```

TABLE 7.4
RTEMS Mutex Primitives, Classic API

Function	Purpose
mutex_init	Initialize a standalone mutex [1]
mutex_destroy	Destroy a standalone mutex
mutex_set_name	Set the human-readable name of a mutex
mutex_get_name	Retrieve the human-readable name of a mutex
mutex_lock	Perform a P() on a mutex
mutex_unlock	Perform a V() on a mutex
mutex_recursive_init	Initialize a standalone recursive mutex [2]
mutex_recursive_destroy	Destroy a standalone mutex
mutex_recursive_set_name	Set the human-readable name of a recursive mutex
mutex_recursive_get_name	Retrieve the human-readable name of a recursive mutex
mutex_recursive_lock	Perform a P() on a recursive mutex
mutex_recursive_unlock	Perform a V() on a recursive mutex

[1] Static initializer RTEMS_MUTEX_INITIALIZER also available.
[2] Static initializer RTEMS_RECURSIVE_MUTEX_INITIALIZER also available.

When setting the semaphore name, the caller must ensure that the storage pointed by name remains valid through the whole lifetime of the semaphore, or until the semaphore name is changed again. The name argument may also be a NULL pointer.

Mutual exclusion and recursive mutual exclusion devices

Besides general-purpose and binary semaphores, RTEMS also provides even more specialized and more efficient synchronization devices, specifically designed for mutual exclusion. They automatically use the priority inheritance (on single-core systems) or the OMIP protocol (on multicores). Like binary semaphores, they are standalone object, that is, the storage space they need is completely allocated and managed by the user. There are two distinct flavors of mutual exclusion device, or mutex:

- A mutex that does not support recursive locks is an object of type rtems_mutex and is managed by means of the functions listed in the top half of Table 7.4. Any attempt to recursively lock a non-recursive mutex, that is, any attempt by a task to lock a mutex that it already owns, leads to undefined results.
- A mutex that does support recursive locks is an object of type rtems_recursive_mutex and is managed by means of the functions listed in the bottom half of the table. In this case, recursive lock attempts succeed and lead to the expected result.

The functions:

```
void rtems_mutex_init(
    rtems_mutex *mutex,
    const char *name);

void rtems_recursive_mutex_init(
    rtems_recursive_mutex *mutex,
    const char *name);
```

initialize a mutex and a recursive mutex, respectively. They take as arguments a pointer to the object to be initialized and a pointer to a string of characters that will become the human-readable name of the mutex, used for debugging purposes. If no name is needed, a NULL pointer can be passed as name. If name is not NULL, the storage it points to must persist for the whole lifetime of the mutex. A newly created mutex or recursive mutex is initially unlocked.

Calling the initialization function is mandatory for dynamically allocated mutex objects. Statically allocated objects can also be initialized statically, by means of the following initialization macros:

```
rtems_mutex mutex = RTEMS_MUTEX_INITIALIZER(name);

rtems_recursive_mutex mutex =
    RTEMS_RECURSIVE_MUTEX_INITIALIZER(name);
```

When a mutex or recursive mutex is no longer needed, it must be destroyed by means of the functions:

```
void rtems_mutex_destroy(rtems_mutex *mutex);

void rtems_recursive_mutex_destroy(
    rtems_recursive_mutex *mutex);
```

Both functions take as argument a pointer to the object to be destroyed. A mutex can no longer be used after destruction, unless it is initialized anew. If a mutex is destroyed while there are tasks waiting on it, the result is undefined.

The human-readable name of a mutex can be changed after creation by means of the functions:

```
void rtems_mutex_set_name(
    rtems_mutex *mutex, const char *name);

void rtems_recursive_mutex_set_name(
    rtems_recursive_mutex *mutex, const char *name);
```

Symmetrically, the name can also be retrieved, given a pointer to the mutex object, by means of the functions:

```
const char *rtems_mutex_get_name(
    const rtems_mutex *mutex);
```

```
const char *rtems_recursive_mutex_get_name(
    const rtems_recursive_mutex *mutex);
```

The functions:

```
void rtems_mutex_lock(rtems_mutex *mutex);
```

```
void rtems_recursive_mutex_lock(
    rtems_recursive_mutex *mutex);
```

lock the mutex or recursive mutex pointed by `mutex`, respectively, blocking the caller if necessary. For this reason, these functions must be invoked from a task context with interrupts enabled. When blocked on the mutex, tasks wait and are awakened in priority order. The function `rtems_recursive_mutex_lock` operates on a recursive mutex and is able to detect if the calling task already owns the mutex, that is, it locked the mutex one or more times without unlocking it. If this is the case, the function returns to the caller immediately.

A locked mutex can be unlocked by invoking the functions:

```
void rtems_mutex_unlock(rtems_mutex *mutex);
```

```
void rtems_recursive_mutex_unlock(
    rtems_recursive_mutex *mutex);
```

Like the mutex lock functions, also these functions must be called from a task context with interrupts enabled. If the calling task does not own the mutex at the time of the call, the result is unpredictable. Similarly, attempting to unlock again an unlocked mutex is also illegal and leads to an unpredictable result. The function `rtems_recursive_mutex_unlock` actually unlocks a recursive mutex only if the current unlock operation matches the outermost `rtems_recursive_mutex_lock` performed by the same task. Otherwise, it returns to the caller immediately.

7.4.2   POSIX API

The POSIX API offers three main inter-task synchronization objects: general-purpose semaphores, mutual exclusion semaphores (called *mutex*), and condition variables. General-purpose semaphores adhere to the abstract definition given in Section 7.2 quite closely. In addition, all blocking synchronization primitives have non-blocking and timed counterparts. Both are useful to ensure that a real-time task, after engaging in a synchronization, always has a way to bail out within a well-defined amount of time, regardless of whether or not other software components are functioning correctly.

Mutual exclusion semaphores are specialized semaphores whose value can only be 1 or 0, to indicate an unlocked and locked semaphore, respectively. In addition, $P(m)$ and $V(m)$ on a mutual exclusion semaphore m are restricted to appear in this order and as pairs within each task that makes use of m, to bracket critical regions

TABLE 7.5

RTEMS Semaphore Primitives, POSIX API

Function	Purpose
sem_init	Create an unnamed semaphore
sem_destroy	Destroy an unnamed semaphore
sem_open	Create/open a named semaphore
sem_close	Close a named semaphore
sem_unlink	Mark a named semaphore for destruction
sem_wait	Perform a P() on a semaphore
sem_post	Perform a V() on a semaphore
sem_trywait	Non-blocking P()
sem_timedwait	Variant of P() with timeout
sem_getvalue	Get current value of a semaphore

according to the standard mutual exclusion semaphore usage paradigm (see Section 7.2.2).

In exchange for these limitations, mutual exclusion semaphores may be implemented more efficiently than general-purpose semaphores and provide additional features. For instance, they incorporate some of the protocols to address unbounded priority inversion (see Section 8.1). Moreover, the POSIX *recursive* lock feature enables a task to lock the same mutual exclusion semaphore multiple times without self-deadlocking. Although not strictly required from the theoretical point of view, this feature may be very convenient to simplify software implementation. Mutual exclusion semaphores, unlike general-purpose semaphores, can also be used in concert with condition variables to implement monitors (see Section 7.3).

Last, but not the least, all inter-task synchronization objects can be declared to be private, that is, accessible only by threads belonging to a single process, or be shared among multiple processes. Even though this declaration does not affect object semantics, it offers the POSIX implementation more optimization opportunities, which may make private objects more efficient than shared ones. In the following, we will only describe private objects in detail. Since RTEMS implements a multi-task execution environment, but not multiple processes, private objects are the direct counterpart of Classic API objects.

General-purpose semaphores

Table 7.5 summarizes the main POSIX primitives for general-purpose semaphores. These semaphores are available in two flavors: *unnamed* and *named*, which differ in the way they are created and destroyed, but are otherwise used in a uniform way.

As a general remark the use of any POSIX functions, like the ones listed in the table, requires the inclusion of one or more header files. Since the goal of this book is not to be a reference manual of the POSIX standard, in the following we will only categorize and summarize POSIX functions without going into these fine details. Readers should refer to the standard itself [68] for authoritative information.

Unnamed semaphores are created by means of the `sem_init` function, defined as:

```
int sem_init(sem_t *sem, int pshared, unsigned value);
```

The function initializes the data structure of type `sem_t` pointed by `sem`, which will represent the semaphore and must have been allocated in advance by the caller. The semaphore has the specified non-negative initial `value`. The `pshared` argument, if not zero, indicates that the semaphore may be shared among multiple processes. Otherwise, the semaphore can be used only by threads belonging to the same process as the calling thread. Since RTEMS implements the single-user, single-process (SUSP) execution environment outlined in Chapter 6, it ignores this argument.

As is commonly done in POSIX, the return value of the `sem_init` function is zero if the function succeeded, or -1 if the function failed. In the latter case, the variable `errno` variable is set to a status code that provides more information on the reason for the failure:

EINVAL indicates that `value` is invalid, that is, it is either negative or greater than `SEM_VALUE_MAX`. The value of `SEM_VALUE_MAX` is guaranteed to be at least 32767 across all POSIX implementations.

ENOSPC means that the system lacks resources to create the new semaphore, often because the system-wide limit on the maximum number of semaphores, `SEM_NSEMS_MAX` has been reached.

EPERM indicates that the caller does not have sufficient privileges to create the semaphore.

The maximum value of a semaphore, like other runtime invariant values in POSIX, must be defined as a symbolic constant in the header `limits.h` when it is fixed and known at compile time. If the value is indeterminate at compile time (for instance, because it is configurable or depends on the amount of system resources available at runtime) the definition is omitted and process can retrieve the value by means of the `sysconf` function.

After creating a semaphore, a pointer to its `sem_t` data structure must be used as an argument of all the other functions that operate on it, including `sem_destroy`:

```
int sem_destroy(sem_t *sem);
```

The effect of `sem_destroy` is to destroy the semaphore pointed by `sem`. It is responsibility of the caller to ensure that no threads are blocked on the semaphore being destroyed because this would lead to an undefined behavior in general, although implementation may choose to return the `EBUSY` error indication. Similarly, the semaphore must no longer be used after destroying it, unless the semaphore is re-initialized beforehand, by means of another call to `sem_init`. The possible status codes are:

EINVAL The `sem` argument does not represent a valid semaphore.
EBUSY There are threads currently locked on the semaphore.

Named semaphores exist as global objects in the system. A unique name (a character string) is used to get access to them. As it happens with other POSIX objects, like files, the same function `sem_open` can be used both to create a semaphore and to connect the calling process to an existing semaphore:

```
sem_t *sem_open(const char *name, int flags, ...);
```

Its arguments are the `name` of the semaphore the function should operate upon and a set of `flags`, which affects some aspects of the function's behavior. Generally speaking, the `name` must conform to the rules for a pathname in order to be valid. Moreover, to guarantee that two processes refer to the same semaphore when they use the same `name`, it must start with a slash (`/`). Depending on the `flags`, the function may need two additional arguments, indicated by `...` in the prototype above.

After a successful call to this function, the calling process gets a pointer to the `sem_t` structure that represents the semaphore and all threads belonging to the calling process can use it. If an error occurs, the function returns the special value `SEM_FAILED` and sets `errno` to a status code.

The `flags` argument is the bitwise or combination of the following two flags:

O_CREAT When set, this flags allows `sem_open` to create the semaphore if it does not exist already. In this case, `sem_open` takes two additional arguments:
 - `mode_t mode` determines the access permissions of the newly created semaphore, and
 - `unsigned value` is its initial value.
O_EXCL This flag shall only be set together with `O_CREAT`. When set, it causes `sem_open` to fail when trying to create a semaphore that already exists. The existence check and the creation are carried out in an atomic step with respect to other tasks doing the same.

The most common status codes that `sem_open` may report through `errno` are:

ENOENT The semaphore does not exist, but `flags` did not include `O_CREAT`.
EEXIST The semaphore already exists, but both `O_CREAT` and `O_EXCL` were set in `flags`.
EINVAL The given `name` is not a valid name for a semaphore, or the initial `value` of the semaphore specified together with `O_CREAT` was greater than `SEM_VALUE_MAX`.
EACCES The semaphore could not be accessed or created due to permission issues.

Other codes, like `EMFILE`, `ENFILE`, `ENOMEM`, and `ENOSPC` indicate that the system lacks various kinds of resources it needs to create the semaphore.

The counterpart of `sem_destroy` for named semaphores are `sem_close` and `sem_unlink`. The function `sem_close`, invoked as:

```
int sem_close(sem_t *sem);
```

indicates that the calling process no longer intends to use semaphore `sem`. As for `sem_destroy`, no further operations on `sem` are allowed after a successful `sem_close`. The only possible status code is:

EINVAL The argument `sem` does not point to valid semaphore.

A successful execution of `sem_close` does not destroy the semaphore unless it has been previously marked by means of `sem_unlink`:

`int sem_unlink(**const char** *name);`

The execution of `sem_unlink` has two possible outcomes:

1. If no processes have the semaphore open, it destroys the semaphore immediately.
2. Otherwise, it marks the semaphore for later destruction and returns.

In the second case, the semaphore is destroyed as soon as the number of processes that have it open drops to zero. In the meantime, any further attempt to open the semaphore fails, as if the semaphore did not exist, and the creation of a semaphore with the same name creates a new semaphore distinct from the previous one. The fact that `sem_unlink` takes as argument the symbolic *name* of the semaphore, rather than a *pointer* to its descriptor, enables processes to destroy or mark a semaphore for destruction without having opened it first.

The function `sem_unlink` returns zero if it succeeds. Otherwise, it returns -1 and sets `errno` to a status code. Possible reasons for failure include:

ENOENT No semaphores named `name` exist in the system.
EACCES The caller lacks the permission to destroy the semaphore.

Although unnamed and named semaphores can be used interchangeably with most other primitives, their creation and destruction functions must not be mismatched. For instance, a named semaphore shall not be destroyed with `sem_destroy`, because this would lead to undefined results.

For both kinds of semaphore, a group of functions implements several variants of the abstract P() and V() primitives. More specifically, the function:

`int sem_wait(sem_t *sem);`

performs a P() on semaphore `sem`, blocking the caller if necessary. Unlike its abstract counterpart, `sem_wait` may fail. In this case, the most common status codes are:

EINVAL The argument `sem` does not point to valid semaphore.
EDEADLK The operation was not performed because it would have resulted in a deadlock.

The function `sem_trywait` is the non-blocking variant of `sem_wait` and has the same prototype:

```
int sem_trywait(sem_t *sem);
```

Accordingly, it shall fail if the P() could not be completed immediately, reporting the following status code:

EAGAIN The value of the semaphore sem was zero, so the operation could not be completed immediately.

A timed-wait variant is also available, with a slightly more complex interface:

```
int sem_timedwait(sem_t *restrict sem,
  const struct timespec *restrict abstime);
```

The additional argument abstime indicates the absolute instant in time when any wait initiated by the function must end and the function must return to the caller, even though the P() could not be completed successfully. In this case, the function reports the following status code:

ETIMEDOUT The timeout specified by abstime expired before the P() could be concluded.

Both sem_trywait and sem_timedwait may also report the same status codes as sem_wait. In the case of sem_timedwait the meaning of EINVAL has been extended to also indicate that the abstime argument was invalid.

The timeout is expressed as a structure of type struct timespec that contains a time value specification and has already been introduced in Section 6.6. In the case of sem_timedwait the reference clock is CLOCK_REALTIME, the system-wide realtime clock that measures the time elapsed since the Epoch.

The V() operation is performed by the sem_post POSIX function:

```
int sem_post(sem_t *sem);
```

There are no variants of this function because V() is, by itself, a non-blocking operation. The only argument of sem_post is the semaphore sem it shall operate upon. Passing an argument that does not refer to a valid semaphore may cause the function to fail with the following status code in errno:

EINVAL The argument sem does not point to valid semaphore.

The last semaphore-related POSIX function to be discussed here is sem_getvalue:

```
int sem_getvalue(sem_t *restrict sem, int *restrict val);
```

As it happens for the other semaphore-related functions, passing to sem_getvalue an argument that does not refer to a valid semaphore may cause the function to fail with the following status code in errno:

EINVAL The argument sem does not point to valid semaphore.

This function stores into the variable pointed by `val` the current value of semaphore `sem`. However, it is not as useful as it may seem because it cannot be executed atomically with respect to any other synchronization primitive. For instance, trying to use a combination of `sem_getvalue` and `sem_wait` as follows is not a good idea:

```
sem_t sem;
int sval, st;
...
st = sem_getvalue(&sem, &sval);
if(st == 0 && sval > 0)   st = sem_wait(&sem);
...
```

This fragment of code is *not* equivalent to `sem_trywait` because nothing prevents another task from changing the value of the semaphore between the invocation of `sem_getvalue` and `sem_wait`, thus possibly causing `sem_wait` to block the caller even though `sval` was greater than zero.

Mutual exclusion semaphores

Mutual exclusion semaphores, called *mutex* in the POSIX standard, are a specialized kind of semaphore. They are optimized, as their name itself suggests, for mutual exclusion. Accordingly, there are several important differences with respect to the general-purpose semaphores described previously:

- Mutual exclusion semaphores can only have either 1 (unlocked) or 0 (locked) as a value.
- The `P()` and `V()` primitives that operate on a mutual exclusion semaphore may not be placed arbitrarily in the code. They must always appear as pairs that bracket the critical regions associated with the semaphore.
- In return for these restrictions, mutual exclusion semaphores offer operating system designers more opportunities for optimization. Therefore, they are likely to be more efficient than general-purpose semaphores.
- Specialization also gives another benefit. For mutual exclusion semaphores it becomes possible to address the unbounded priority inversion problem described in Section 8.1.
- Last, but not the least, mutual exclusion semaphores, unlike general-purpose semaphores, can be used as a building block to implement monitors.

As for other POSIX objects, each mutex is characterized by a set of attributes, stored in a mutex attribute object of type `pthread_mutexattr_t` that must be mentioned upon mutex creation. Table 7.6 lists the POSIX function related to mutex attributes and attribute objects that RTEMS supports. As usual, the mutex attribute object must be initialized before use by means of the function:

```
int pthread_mutexattr_init(pthread_mutexattr_t *attr);
```

TABLE 7.6

RTEMS Mutex Attributes, POSIX API

Function	Purpose
pthread_mutexattr_init	Initialize a mutex attribute object
pthread_mutexattr_destroy	Destroy a mutex attribute object
pthread_mutexattr_getpshared	Get the pshared mutex attribute [1]
pthread_mutexattr_gettype	Get the type mutex attribute
pthread_mutexattr_getprotocol	Get the protocol mutex attribute
pthread_mutexattr_getprioceiling	Get the prioceiling mutex attribute
pthread_mutexattr_setpshared	Set the pshared mutex attribute [1]
pthread_mutexattr_settype	Set the type mutex attribute
pthread_mutexattr_setprotocol	Set the protocol mutex attribute
pthread_mutexattr_setprioceiling	Set the ceiling mutex attribute

[1] Unused in RTEMS.

TABLE 7.7

Default Values of Mutex Attributes, POSIX API

Attribute	Value
pshared	PTHREAD_PROCESS_PRIVATE
type	PTHREAD_MUTEX_DEFAULT
protocol	PTHREAD_PRIO_NONE
prioceiling	Maximum priority supported by the scheduler

This function initializes the mutex attribute object pointed by attr and sets all the attributes it contains to their default values, listed in Table 7.7. The meaning of its return value and the status codes it returns are the same as for the pthread_attr_init function described in Chapter 5.

Symmetrically, the function:

```
int pthread_mutexattr_destroy(pthread_mutexattr_t *attr);
```

destroys the mutex attribute object pointed by attr and releases any dynamic memory possibly allocated for it by pthread_mutexattr_init. Also in this case, the returned status codes are the same as for pthread_attr_destroy.

The pshared attribute can be set to either PTHREAD_PROCESS_PRIVATE (the default) or PTHREAD_PROCESS_SHARED by means of the functions:

```
int pthread_mutexattr_getpshared(
    const pthread_mutexattr_t *restrict attr,
    int *restrict pshared);
```

TABLE 7.8

Possible Values of the `type` Mutex Attribute

Value	Meaning
PTHREAD_MUTEX_DEFAULT	Undefined behavior on recursive lock and errors
PTHREAD_MUTEX_NORMAL	Self-deadlock on recursive lock, undefined behavior on errors
PTHREAD_MUTEX_RECURSIVE	Enable lock count for recursive lock and unlock
PTHREAD_MUTEX_ERRORCHECK	Enable additional error checks, also w.r.t. condition variables

```
int pthread_mutexattr_setpshared(
    pthread_mutexattr_t *attr, int pshared);
```

Both functions take a pointer `attr` to an initialized mutex attribute object as first argument. For `getpshared` the second argument is a pointer to an `int` variable where the function will store the value of the attribute fetched from the attribute object. Instead the second argument of `setpshared` is an `int` that represents the value the attribute must be set to.

As it happens with most of the functions that get and set an attribute to be discussed in the following, both `getpshared` and `setpshared` may optionally return a non-zero status code when they detect that `attr` is invalid:

EINVAL The `attr` argument does not point to an initialized attribute object.

In addition, `setpshared` may also fail if its argument `pshared` is not one of the legal values of the `pshared` attribute:

EINVAL The `pshared` argument is neither PTHREAD_PROCESS_PRIVATE nor PTHREAD_PROCESS_SHARED.

When the `pshared` attribute of a mutex is set to PTHREAD_PROCESS_PRIVATE, the mutex shall be used only by threads belonging to the same process as the thread that created the mutex, otherwise the result is undefined. Instead, if the attribute is set to PTHREAD_PROCESS_SHARED, any thread that has access to the area of memory where the mutex object is stored may operate on it.

In general, the added flexibility may come together with a performance penalty, so it is important to use PTHREAD_PROCESS_SHARED only when needed. In the specific case of RTEMS, getting and setting this attribute is supported but has no effect because this operating system supports a multithreaded, single-process execution model.

As detailed in Table 7.8, the `type` attribute of a mutex lets programmer choose among different levels of error checks and, most notably, enable the recursive lock feature previously mentioned. It can be retrieved from, and stored into, an attribute object by means of the functions:

TABLE 7.9

Possible Values of the `protocol` Mutex Attribute

Value	Meaning
PTHREAD_PRIO_NONE	Disable all protocols against priority inversion
PTHREAD_PRIO_INHERIT	Priority inheritance protocol
PTHREAD_PRIO_PROTECT	Immediate priority ceiling protocol, see the `prioceiling` attribute

```
int pthread_mutexattr_gettype(
    const pthread_mutexattr_t *restrict attr,
    int *restrict type);

int pthread_mutexattr_settype(
    pthread_mutexattr_t *attr, int type);
```

A related attribute, whose possible values are listed in Table 7.9, is `protocol`. It can be accessed by means of the functions:

```
int pthread_mutexattr_getprotocol(
    const pthread_mutexattr_t *restrict attr,
    int *restrict protocol);

int pthread_mutexattr_setprotocol(
    pthread_mutexattr_t *attr, int protocol);
```

and allows the caller to enable either the priority inheritance protocol or the immediate priority ceiling protocol. Both algorithms have been discussed in Section 8.1. For what concerns protocols suitable for multicores, MrsP cannot be selected through the POSIX interface because this interface does not provide a way to set the required per-scheduler ceiling priorities. Instead, RTEMS automatically uses OMIP instead of priority inheritance on multicore systems.

When choosing the immediate priority ceiling protocol, the `prioceiling` attribute must also be set to the desired ceiling. This attribute can be set and retrieved by means of the functions:

```
int pthread_mutexattr_getprioceiling(
    const pthread_mutexattr_t *restrict attr,
    int *restrict prioceiling);

int pthread_mutexattr_setprioceiling(
    pthread_mutexattr_t *attr, int prioceiling);
```

TABLE 7.10

RTEMS Mutex Primitives, POSIX API

Function	Purpose
pthread_mutex_init	Initialize a mutex
pthread_mutex_destroy	Destroy a mutex
pthread_mutex_lock	Lock (perform a P() operation on) a mutex
pthread_mutex_unlock	Unlock (perform a V() operation on) a mutex
pthread_mutex_trylock	Non-blocking variant of lock
pthread_mutex_timedlock	Timed variant of lock
pthread_mutex_getprioceiling	Get the current ceiling of a mutex
pthread_mutex_setprioceiling	Set the ceiling of a mutex after creation

In POSIX, mutex are represented by an object with data type pthread_mutex_t. Table 7.10 summarizes the main functions that operate on them. The function:

```
int pthread_mutex_init(
    pthread_mutex_t *restrict mutex,
    const pthread_mutexattr_t *restrict attr);
```

initializes the mutex pointed by the mutex argument with the attributes pointed by the attr argument. If default mutex attributes are appropriate, attr can be left NULL or, even more simply, the mutex can also be statically allocated and initialized with the help of the macro PTHREAD_MUTEX_INITIALIZER:

```
pthread_mutex_t a_mutex = PTHREAD_MUTEX_INITIALIZER;
```

In both cases, only the pthread_mutex_t object initialized in one of these ways may be used as a mutex. In other words, a copy of a mutex is not itself a valid mutex. If the mutex is to be shared among multiple threads or processes, they all must have access to the area of memory where the mutex is stored. This requirement is easily met by threads belonging to the same process, because they implicitly share the same address space. Instead, when sharing a mutex among multiple processes, it must be stored in an explicitly shared memory segment.

The pthread_mutex_init function returns zero upon successful completion. It shall fail and return a non-zero status code for the following reasons:

ENOMEM The mutex could not be initialized due to lack of memory.
EAGAIN The mutex could not be initialized due to lack of other resources.
EPERM The caller does not have the permission to create a mutex.

Care must be taken not to initialize the same mutex twice without destroying it first because this leads to undefined behavior, although implementations may detect this error and make pthread_mutex_init return:

EBUSY The mutex has already been initialized.

In addition, implementations may check whether or not `attr` refers to an initialized mutex attribute object and return the following status code if the check fails:

EINVAL The `attr` argument is neither `NULL` nor a reference to an initialized mutex attribute object.

The function:

```
int pthread_mutex_destroy(pthread_mutex_t *mutex);
```

destroys the mutex referenced by the `mutex` argument. The mutex may no longer be used after destroying it, unless it is initialized again. Moreover, attempting to destroy a mutex that is currently locked or referenced in some other ways (for instance, through a condition variable, to be discussed next) leads to undefined behavior. Optionally, `pthread_mutex_destroy` may detect this condition and fail, returning:

EBUSY The argument refers to a locked mutex or a mutex referenced by another thread.

Both `pthread_mutex_init` and `pthread_mutex_destroy` may also optionally return:

EINVAL The `mutex` pointer is invalid.

All the other mutex-related functions take a reference to a mutex object as an argument. The function:

```
int pthread_mutex_lock(pthread_mutex_t *mutex);
```

performs a P() on the given `mutex`, blocking the caller if necessary. A non-blocking variant of the same function is also available, which returns immediately if the `mutex` is already locked:

```
int pthread_mutex_trylock(pthread_mutex_t *mutex);
```

Both functions fail and return a non-zero status code when they detect one of the following error conditions:

EINVAL The priority ceiling protocol is enabled for the mutex and the calling task has a priority higher than the mutex ceiling.
EAGAIN The mutex is recursive and it could not be acquired without exceeding the maximum lock count.

Moreover, `pthread_mutex_lock` shall also fail if:

EDEADLK Additional error checks are enabled for the mutex (it is of type `PTHREAD_MUTEX_ERRORCHECK`) and the operation would lead to a self-deadlock because the calling task already owns the mutex.

The same status code may also be returned if `pthread_mutex_lock` was able to detect that the operation would lead to other kinds of deadlock. As outlined previously, being non-blocking, `pthread_mutex_trylock` also fails if:

EBUSY The mutex is already locked.

A timed variant of `pthread_mutex_lock` also exists, which enables the caller to exercise an even finer control on the maximum amount of time that could be spent waiting:

```
int pthread_mutex_timedlock(
    pthread_mutex_t*restrict mutex,
    const struct timespec *restrict abstime);
```

As for `sem_timedwait`, the additional argument `abstime` points to a `struct timespec` and indicates the absolute instant in time when any wait initiated by the function must end and the function must return to the caller. Besides the status codes specified for `pthread_mutex_lock`, this function may also return:

EINVAL The structure pointed by `abstime` is invalid.
ETIMEDOUT The timeout specified by `abstime` expired before the P() could be concluded.

A mutex that is currently locked by a thread can be released only by the same thread, by means of the function:

```
int pthread_mutex_unlock(pthread_mutex_t *mutex);
```

If there are threads blocked on the mutex then one of them, chosen according to the scheduling policy, acquires the mutex and is allowed to continue, otherwise the mutex becomes unlocked. For a recursive mutex, this is done only when the number of outstanding lock operations performed on the mutex by the calling thread goes down to zero, otherwise this function has no other effect besides decrementing the lock counter itself. As for `pthread_mutex_lock` and `_trylock`, it is recommended that the `_unlock` function returns:

EINVAL The `mutex` pointer is invalid.

when it detects that the object pointed by `mutex` is not a valid, initialized mutex.

The last two functions listed in Table 7.10 pertain only to a mutex for which the immediate priority ceiling protocol is enabled, that is, for a mutex whose `protocol` attribute has been set to `PTHREAD_PRIO_PROTECT`. They allow the caller to retrieve and set the ceiling of a mutex after creation and have the following prototype:

```
int pthread_mutex_getprioceiling(
    const pthread_mutex_t *restrict mutex,
    int *restrict prioceiling);
```

TABLE 7.11

RTEMS Condition Variable Attributes, POSIX API

Function	Purpose
pthread_condattr_init	Initialize a condition variable attribute object
pthread_condattr_destroy	Destroy a condition variable attribute object
pthread_condattr_getpshared	Retrieve the pshared attribute [1]
pthread_condattr_setpshared	Set the pshared attribute [1]

[1] Unused in RTEMS.

```
int pthread_mutex_setprioceiling(
    pthread_mutex_t *restrict mutex,
    int prioceiling, int *restrict old_ceiling);
```

The function pthread_mutex_getprioceiling stores in the location pointed by prioceiling the current priority ceiling of mutex.

Instead, the function pthread_mutex_setprioceiling must first lock the mutex, as pthread_mutex_lock would do, but without necessarily adhering to the immediate priority ceiling protocol. Then, it must store the current ceiling of the mutex into the location pointed by old_prioceiling and change it to the new value prioceiling. Last, it must unlock the mutex and return to the caller.

Both functions shall return the following non-zero status code when they detect the corresponding error condition:

EINVAL The protocol attribute of the given mutex is PTHREAD_PRIO_NONE.
EPERM The caller does not have sufficient privileges to perform the requested operation.

Moreover, pthread_mutex_setprioceiling shall also fail when the internal pthread_mutex_lock failed, and return the same status codes. It should be noted that, according to the POSIX standard, implementations are free to follow or not follow the immediate priority ceiling protocol when locking the mutex as part of pthread_mutex_setprioceiling. In the first case, programmers must be careful not to incur in the following failure reason:

EINVAL The mutex uses the PTHREAD_PRIO_PROTECT (immediate priority ceiling) protocol, the implementation adheres to this protocol when locking the mutex within pthread_mutex_setprioceiling, and the priority of the calling thread was higher than the current ceiling of the mutex.

TABLE 7.12

RTEMS Condition Variable Primitives, POSIX API

Function	Purpose
pthread_cond_init	Initialize a condition variable
pthread_cond_destroy	Destroy a condition variable
pthread_cond_wait	Wait on a condition variable
pthread_cond_timedwait	Like pthread_cond_wait, but with a timeout
pthread_cond_signal	Signal a condition variable
pthread_cond_broadcast	Unblock all threads blocked on a condition variable

Condition variables

A POSIX condition variable is an object of type pthread_cond_t. They are used in concert with mutexes to implement monitors, as described in Section 7.3. Condition variables can be configured in the usual way, by means of attributes stored in an attribute object of type pthread_condattr_t. As shown in Table 7.11, this kind of attribute object is initialized and destroyed by means of the pthread_condattr_init and pthread_condattr_destroy functions, respectively. They are analogous to their counterpart that work on mutex attribute objects:

```
int pthread_condattr_init(pthread_condattr_t *attr);
int pthread_condattr_destroy(pthread_condattr_t *attr);
```

The only condition variable attribute supported (but ignored) by RTEMS at the time of this writing is pshared, which has the same meaning as the attribute with the same name defined for mutex attribute objects and already discussed previously.

Table 7.12 lists the primitives related to condition variables. The function:

```
int pthread_cond_init(
    pthread_cond_t *restrict cond,
    const pthread_condattr_t *restrict attr);
```

initializes the condition variable referenced by cond, setting its attributes as specified by attr. As for a mutex, if default attributes are appropriate for a condition variable, attr can be left NULL, and static allocation and initialization becomes possible as well:

```
pthread_cond_t cond = PTHREAD_COND_INITIALIZER;
```

As for a mutex, a copy of a condition variable is not a valid condition variable and shall not be used. The function pthread_cond_init returns zero upon successful completion or a non-zero status code if it detected an error. The function shall fail if:

ENOMEM The condition variable could not be initialized due to lack of memory.

EAGAIN The condition variable could not be initialized due to lack of other resources.

Initializing the same condition variable twice without destroying it first leads to undefined behavior, although implementations may detect this fact and make `pthread_cond_init` return the following status code:

EBUSY The condition variable has already been initialized.

Similarly, implementations may check whether or not `attr` refers to an initialized condition variable attribute object and return the following status code if the check fails:

EINVAL The `attr` argument is neither `NULL` nor a reference to an initialized condition variable attribute object.

The function:

```
int pthread_cond_destroy(pthread_cond_t *cond);
```

destroys the condition variable referenced by `cond`. After the condition variable has been destroyed, any reference to it becomes invalid until the condition variable is initialized again, by means of a new call to `pthread_cond_init`.

Also in this case, destroying a condition variable that has not been initialized, or destroying a condition variable while some threads are waiting on it leads to undefined behavior, although some implementations of `pthread_cond_destroy` may carry out more thorough checks and fail with the following status codes:

EINVAL The condition variable has not been initialized.
EBUSY There are threads currently waiting on the condition variable.

The function:

```
int pthread_cond_wait(
    pthread_cond_t *restrict cond,
    pthread_mutex_t *restrict mutex);
```

blocks the calling thread on the condition variable referenced by `cond` and releases the mutex referened by `mutex` in an atomic operation, thus performing the POSIX counterpart of the abstract `wait` operation described in Section 7.3.2.

The calling thread must own the mutex referenced by `mutex`, which is assumed to protect the boundary of the monitor to which `cond` belongs. Otherwise, undefined behavior results unless `mutex` is of type `PTHREAD_MUTEX_ERRORCHECK`, in which case the function shall fail and report a non-zero status code as detailed in the following.

When the calling thread is eventually unblocked successfully by some other thread, the `pthread_cond_wait` function must reacquire the `mutex`, blocking again if necessary, before returning. As usual, the function shall return zero to indicate that it completed successfully. The implementation shall also detect the following error and return the corresponding non-zero status code:

EPERM The mutex is of type `PTHREAD_MUTEX_ERRORCHECK` and the calling thread does not own the mutex.

Moreover, it is recommended that the function also checks the validity of the objects referenced by `cond` and `mutex` and report an error if the check fails:

EINVAL The `cond` argument does not refer to a valid and initialized condition variable object, or the `mutex` argument does not refer to a valid and initialized mutex object.

The timed variant of the primitive has an additional argument `abstime`:

```
int pthread_cond_timedwait(
    pthread_cond_t *restrict cond,
    pthread_mutex_t *restrict mutex,
    const struct timespec *restrict abstime);
```

Unlike `pthread_cond_wait`, this function shall fail and return to the caller when the absolute time specified by `abstime` is reached before the calling thread is unblocked because the condition variable has been signaled or broadcast. However, the actual return may take place later than expected because the function must nonetheless acquire the mutex again before returning, and this may take extra time.

In addition to the reasons previously listed for `pthread_cond_wait`, this function shall also fail because:

ETIMEDOUT The timeout specified by `abstime` expired before the wait could be successfully concluded.

EINVAL The structure pointed by `abstime` is invalid.

The function:

```
int pthread_cond_signal(pthread_cond_t *cond);
```

performs the POSIX counterpart of the abstract `signal` operation on a condition variable. The argument `cond` is a reference to the condition variable to be signaled.

Besides the differences between the abstract operation and its concrete POSIX counterpart already described in Section 7.3, the POSIX standard explicitly states that *spurious wakeups* might also occur. In other words, a single `pthread_cond_signal` might wake up *more than one thread* blocked on the condition variable.

The standard leaves this possibility open to allow a more efficient implementation of condition variables on multiprocessor and multicore systems. At the same time, it does not hinder the efficiency of the upper layers of software built upon condition variables. This is because spurious wakeups occur with low probability and, for the reasons described in Section 7.3.2, a POSIX thread that continues after a `pthread_cond_wait` has to check that the condition it was waiting for is indeed satisfied in any case.

The standard does not define any failure reasons for the function `pthread_cond_signal`. However, it recommends that the function checks the validity of `cond` and returns the following status code if the check fails:

EINVAL The argument `cond` does not refer to a valid, initialized condition variable.

If the implementation opts not to perform the recommended check, the result of operating on an invalid condition variable is undefined.

A variant of `pthread_cond_signal` unblocks all threads currently waiting on a condition variable referenced by `cond`:

int `pthread_cond_broadcast(pthread_cond_t *cond);`

The possible status code the function may return are the same as for `pthread_cond_signal`.

Although the abstract definition of monitor previously given does not comprise a function like this and it could be implemented in terms of `pthread_cond_signal`, it is nonetheless more efficient and quite convenient from the practical point of view whenever a thread knows that it has made possible for multiple threads currently blocked on the same condition variable to continue their work. Mutual exclusion among them is still guaranteed because they will have re-acquire the mutex that protects the monitor before returning from their `pthread_cond_wait` or `_timedwait` and starting to operate on monitor data.

7.5 BARRIERS

7.5.1 GENERAL DEFINITION

A barrier is a synchronization object that enables a number of tasks to wait until all of them have reached a programmer-defined *milestone* in their execution, and then continue concurrently. This is useful in a lot of different circumstances, for instance:

- A set of cooperating tasks may need to go through an initialization phase that they perform independently from each other before starting normal operation. A barrier can be used to ensure that all these tasks have completed their initialization—that is, their milestone—before any of them enters normal operational mode.
- Especially on multicore systems, it may be fruitful to split a time-consuming, computation-intensive job into chunks, to be performed concurrently by N tasks running on different cores. A barrier can guarantee that all tasks have completed their share of the job before they continue with further processing.

As shown in Figure 7.12, a barrier must be created before use, like it happens with all other synchronization objects. Since barriers use up system resources, it is also advisable to destroy them when they are no longer needed. In order to synchronize a set of N tasks, appropriate calls to the *barrier wait* function must be inserted in each

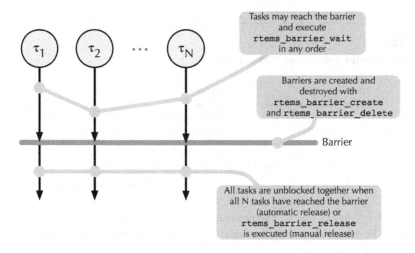

FIGURE 7.12 Task synchronization by means of a barrier (Classic API).

of them, immediately after they reached their milestone and before they perform any subsequent operation.

If the number N of tasks to be synchronized is fixed and known in advance, upon barrier creation, an *automatic-release* barrier can automatically unblock the $N - 1$ tasks already blocked on the barrier when the last one (that is, the N-th) reaches the synchronization point. In other words, with this kind of barrier the first $N - 1$ tasks that reach the barrier get blocked. When the N-th and last task eventually calls the barrier wait function, it does not block at all and also unblocks all the other tasks.

On the other hand, if the number of tasks to be synchronized may vary from one scenario to another or is unknown when the barrier is created, a *manual-release* barrier can be opened by explicitly calling a *barrier release* function. This unblocks all the tasks currently blocked on the barrier, regardless of how many there are.

7.5.2   CLASSIC API

The classic API functions that operate on a barrier are depicted in Figure 7.12 and listed in Table 7.13. The following function creates a barrier:

```
rtems_status_code rtems_barrier_create(
    rtems_name name,
    rtems_attribute attribute_set,
    uint32_t maximum_waiters,
    rtems_id *id);
```

As most other RTEMS objects, a barrier has a user-defined `name` and a unique identifier, which the function `rtems_barrier_create` stores into the location pointed by the argument `id` upon successful completion. The barrier identifier must be used to refer to the barrier in all subsequent operations on the barrier itself, except

TABLE 7.13
RTEMS Barrier Primitives, Classic API

Function	Purpose
rtems_barrier_create	Create a barrier
rtems_barrier_ident	Find the identifier of a barrier given its name
rtems_barrier_delete	Delete a barrier
rtems_barrier_wait	Wait on a barrier until it opens
rtems_barrier_release	Manually open (release) a barrier

rtems_barrier_ident. The argument attribute_set specifies whether the barrier to be created must be an automatic-release or a manual-release barrier. It may have the following values:

RTEMS_BARRIER_AUTOMATIC_RELEASE The barrier is automatically opened (released) and all tasks waiting on it are unblocked when maximum_waiters have called the barrier wait operation. A manual, premature release is still possible when less than the specified number of tasks are waiting, by means of the corresponding function, to be discussed next.

RTEMS_BARRIER_MANUAL_RELEASE The barrier never opens automatically and the argument maximum_waiters is ignored. The barrier must be opened manually by calling the function rtems_barrier_release.

The rtems_barrier_create function returns RTEMS_SUCCESSFUL when it completes successfully, or one of the following status codes upon failure:

RTEMS_INVALID_NAME The barrier name is invalid.
RTEMS_INVALID_ADDRESS The id pointer is NULL.
RTEMS_TOO_MANY The maximum number of barriers allowed in the system has been reached and no more can be created. The maximum number of barriers is set via the RTEMS_CONFIGURE_MAXIMUM_BARRIERS configuration item, as described in Section 2.4.

The utility function:

```
rtems_status_code rtems_barrier_ident(
    rtems_name name,
    rtems_id *id);
```

looks for a barrier with a given name and stores its identifier into the location pointed by id. Besides RTEMS_SUCCESSFUL, it may return the error codes:

RTEMS_INVALID_NAME No barriers with the given name have been found.
RTEMS_INVALID_NODE Barriers can only be accessed locally, but the barrier name correspond to a barrier object that resides on a remote node.

When a barrier is no longer in use it should be destroyed to free the resources associated with it. This is done by the function:

```
rtems_status_code rtems_barrier_delete(
    rtems_id id);
```

This function may fail for the following reasons:

RTEMS_INVALID_ID The barrier identifier given in the id argument is invalid.

If some threads are blocked on the barrier when it is deleted, they are all unblocked. In this case, their barrier wait operation, rtems_barrier_wait, fails and returns the RTEMS_OBJECT_WAS_DELETED status code.

A thread signals that it has reached its milestone and waits for the others by calling the function:

```
rtems_status_code rtems_barrier_wait(
    rtems_id id,
    rtems_interval timeout);
```

This function may block the calling thread until the barrier is manually opened (for a manual-release barrier) or until the number of threads specified in maximum_waiters upon barrier creation has reached the milestone. As a consequence, for manual-release barriers the calling thread is always blocked, whereas for automatic-release barriers it may or may not be blocked depending on whether or not the desired number of threads that arrived at the barrier has been reached.

The argument timeout specifies the maximum amount of time the function is allowed to block the caller before failing. The special value RTEMS_NO_TIMEOUT indicates that the calling thread is willing to potentially wait forever. The rtems_barrier_wait returns RTEMS_SUCCESSFUL when it succeeds or one of the following status codes when it fails:

RTEMS_INVALID_ID The barrier identifier given in id is invalid.
RTEMS_TIMEOUT The given timeout was not RTEMS_NO_TIMEOUT and it expired before the barrier was opened, either manually or automatically.
RTEMS_OBJECT_WAS_DELETED The barrier was deleted while it was still closed. All tasks blocked on it have been unblocked prematurely.
RTEMS_UNSATISFIED The calling thread could not wait on the barrier because a deadlock was detected in the thread queue enqueue logic.

Both an automatic-release and a manual-release barrier can be opened (released) by calling the function:

```
rtems_status_code rtems_barrier_release(
    rtems_id id,
    uint32_t *released);
```

The argument id is the identifier of the barrier to be opened. The function stores into the location pointed by released the number of threads that

TABLE 7.14

RTEMS Barrier Primitives, POSIX API

Function	Purpose
pthread_barrier_init	Initialize an automatic-release barrier
pthread_barrier_destroy	Destroy a barrier
pthread_barrier_wait	Wait on a barrier until it opens

were waiting on the barrier and have been unblocked. The function returns either RTEMS_SUCCESSFUL when successful, or one of the following status codes upon failure:

RTEMS_INVALID_ID The barrier identifier given in id is invalid.

For manual-release barriers, this is the only way to open a barrier. For automatic-release barriers, the barrier can be opened either manually or automatically. In the second case, the threads waiting on the barrier have no way to tell the two scenarios apart.

7.5.3 POSIX API

The POSIX standard specifies a kind of barrier that is simpler than the ones offered by the Classic API of RTEMS and, basically, corresponds to an automatic-release barrier. The POSIX functions that operate on a barrier are listed in Table 7.14.

The function:

```
int pthread_barrier_init(
    pthread_barrier_t *restrict barrier,
    const pthread_barrierattr_t *restrict attr,
    unsigned count);
```

is the POSIX counterpart of rtems_barrier_init. In POSIX, a barrier is implemented as an object of type pthread_barrier_t, whose storage must be allocated and managed by the user. The function pthread_barrier_init initializes the barrier object referenced by barrier. The barrier will open automatically when count threads perform the barrier wait operation. Only the barrier object referenced by barrier must be used in subsequent barrier operations, namely, the copy of a barrier object is not itself a valid object and shall not be used.

Unlike rtems_barrier_init, which accepts barrier attributes directly as an argument, the argument attr of pthread_barrier_init is instead a pointer to a separate barrier attribute object. The barrier attribute object is of type pthread_barrierattr_t and must be initialized and set up on its own, by means of the functions listed in Table 7.15. If default attributes are desired, attr can be a

TABLE 7.15

RTEMS Barrier Attributes, POSIX API

Function	Purpose
pthread_barrierattr_init	Initialize a barrier attribute object with default attributes
pthread_barrierattr_destroy	Destroy a barrier attribute object
pthread_barrierattr_getpshared	Get the pshared attribute [1]
pthread_barrierattr_setpshared	Set the pshared attribute [1]

[1] Unused in RTEMS.

NULL pointer. The function pthread_barrier_init returns zero upon successful completion. Upon failure, it shall return one of the following non-zero status codes instead:

ENOMEM There is not enough memory to initialize the barrier object.

EAGAIN The system lacks some resources, other than memory, to initialize the barrier object.

EINVAL The value of count is zero. This value is invalid because the barrier would invariably be open.

Trying to initialize an already initialized barrier object leads to undefined behavior, whether or not there are threads blocked on it, although implementations are encouraged to perform more thorough error checks and return one of the following additional status codes if these checks fail:

EINVAL The attr argument is not NULL but it does not point to an initialized barrier attribute object.

EBUSY The barrier argument refers to a barrier object that has already been initialized and there possibly are other threads blocked on it.

Barrier attribute objects must be initialized before use by means of the function:

int pthread_barrierattr_init(pthread_barrierattr_t *attr);

The function initializes the barrier attribute object pointed by attr and fills it with default attribute values. If attr points to a barrier attribute object that has already been initialized, results are undefined. The function returns zero upon successful completion. Upon failure, it shall instead return:

ENOMEM There is not enough memory to initialize the barrier attribute object.

It is advisable to destroy barrier attribute objects when they are no longer in use, to reclaim the resources that the system allocated for them. This is done by calling the following function, which accepts a reference to a barrier attribute object as argument:

```
int pthread_barrierattr_destroy(
    pthread_barrierattr_t *attr);
```

It is not mandatory for `pthread_barrierattr_destroy` to check the validity of the `attr` argument, but implementations are encouraged to do so and return the following status code if the check fails:

EINVAL The `attr` argument does not refer to an initialized barrier attribute object.

Similarly, a barrier object should be destroyed when it is no longer in use, by means of the function:

```
int pthread_barrier_destroy(pthread_barrier_t *barrier);
```

After calling this function, the barrier object referenced by `barrier` must no longer be used. In general, the results of destroying a barrier object that has not been initialized or while there are some threads blocked on it is undefined. However, implementations are encouraged to perform additional checks and possibly return the following status codes:

EINVAL The barrier referenced by `barrier` has not been initialized or has been destroyed before calling this function.

EBUSY The `barrier` argument refers to a barrier object that is currently in use, that is, there are some threads blocked on it.

In the second case, and this is the opposite of the approach taken by the RTEMS Classic API, the barrier object is not destroyed and the threads stay blocked on the barrier.

A thread signals that it reached its milestone and possibly waits until the barrier opens by invoking the function:

```
int pthread_barrier_wait(pthread_barrier_t *barrier);
```

The only argument of the function is `barrier`, a pointer to the barrier object to be used for synchronization. Its semantics are simpler than its Classic API counterpart, `rtems_barrier_wait`, because the POSIX standard only specifies automatic-release barriers. Therefore:

- If the number of threads that invoked `pthread_barrier_wait` on the same barrier, including the calling thread, is still less than the number specified upon barrier creation—in the `count` argument of `pthread_barrier_init`—this function blocks the caller.
- If the specified `count` has been reached, this function does not block the caller and opens the barrier, thus also releasing all the other threads blocked on the barrier itself.

TABLE 7.16
RTEMS Events, Classic API

Function	Purpose
rtems_event_send	Send a set of events to a task
rtems_event_receive	Receive, and possibly wait for, a set of events

Upon successful completion of a barrier synchronization, all threads return from pthread_barrier_wait and continue execution. Exactly one of them, chosen arbitrarily, receives the reserved return value PTHREAD_BARRIER_SERIAL_THREAD, whereas all the others receive zero. Therefore, care should be taken not to consider PTHREAD_BARRIER_SERIAL_THREAD as an error indication.

Strictly speaking, calling pthread_barrier_wait on a barrier that is not initialized leads to undefined behavior, but the standard encourages implementations to detect this error and return the following status code instead:

EINVAL The barrier referenced by barrier has not been initialized or has been destroyed before calling this function.

7.6   EVENTS

With respect to semaphores and monitors, events provide a lower-level, but higher-performance inter-task synchronization mechanism, which also supports the exchange of a limited quantity of information. Events can be sent and received only through the Classic API, by means of the directives listed in Table 7.16.

Each task has 32 event flags associated with it, numbered from 0 to 31, in one-to-one correspondence with events. Tasks can refer to events by means of the macros RTEMS_EVENT_0 ... RTEMS_EVENT_31. When another task or an interrupt handler sends an event to a task, the corresponding event flag is set and the event becomes *pending* for that task.

If an event is sent to a task in which the corresponding event flag is already pending, the new event is ignored. In other words, event flags have no queues associated with them. Therefore, if an event is sent to a task more than once before being received, all events but the first are lost. An event carries no information by itself, except the event number, which can nevertheless be used to convey some information to the target task.

A task can receive events, and possibly block while doing so, by calling the event receive directive, specifying as argument a set of events it is interested in. Depending on the caller's choice, the event receive directive blocks the calling task until either *all* events in the set, or *at least one* of them, become pending. At this point, the directive clears the event flags corresponding to the events it received and returns to the caller. The caller is informed about the set of events that has been received and can then act accordingly.

Additional flags and parameters of the event receive directive allow the caller to ask for a non-blocking operation, which only polls event flags, or set an upper cap on the blocking time. The event send and receive directives are automatically serialized when they access event flags, hence there is no risk of events being lost or duplicated when these directives execute concurrently. Interrupt handlers have no event flags and cannot use the event receive directive, although they can send events to tasks.

The rtems_event_send sends a set of events to a task:

```
rtems_status_code rtems_event_send(
    rtems_id id,
    rtems_event_set event_in
);
```

The target task is identified by means of its unique id, which is returned upon task creation and can also be retrieved at a later time given the task name, as described in Section 5.4. The special value RTEMS_SELF is allowed and enables a task to send an event to itself without explicitly specifying its own task identifier. The set of events to be sent is represented by event_in, an instance of the rtems_event_set data type. An event set is built by a bitwise OR of the event macros discussed previously. For instance, RTEMS_EVENT_0 | RTEMS_EVENT_4 is a set that contains events 0 and 4. The macro RTEMS_ALL_EVENTS is an event set that contains all 32 events.

Sending events to a task may unblock that task, if it was waiting for these events and their arrival fulfills its waiting condition. This may lead to the preemption of the task that called rtems_event_send if the target task is managed by the same scheduler and has a higher scheduling priority. If the target task is not waiting for the events being sent, or their arrival is not yet sufficient to fulfill its waiting condition, these events become pending and will be used to satisfy receive operations in the future. The rtems_event_send directive returns to the caller one of the following status codes:

RTEMS_SUCCESSFUL The events have been sent successfully.
RTEMS_INVALID_ID The events have not been sent because the task identifier id was invalid.

A task receives events by invoking the rtems_event_receive directive:

```
rtems_status_code rtems_event_receive(
    rtems_event_set event_in,
    rtems_option option_set,
    rtems_interval ticks,
    rtems_event_set *event_out
);
```

The event_in argument specifies the set of events the task is interested in, while option_set indicates the kind of receive operation to be performed. The `option_set` argument is built by bitwise OR-ing together the macros that represent the options to be selected. Two mutually exclusive options determine how `event_in` is interpreted:

RTEMS_EVENT_ALL indicates that the receive directive is satisfied when all events in `event_in` have been sent to the calling task.

RTEMS_EVENT_ANY means that the receive directive is satisfied when one or more events in `event_in` have been sent to the calling task.

Two more options determine whether or not the receive directive shall wait until it is satisfied:

RTEMS_NO_WAIT indicates that the receive directive shall return immediately and report an error if it cannot be satisfied immediately. The argument `ticks` is ignored when this option is set.

RTEMS_WAIT means that the receive directive shall wait up to the maximum amount of time specified by the `ticks` argument to be satisfied. If `ticks` is set to the special value `RTEMS_NO_TIMEOUT` the directive may potentially wait forever.

The default options, which can be requested by using `RTEMS_DEFAULT_OPTIONS` as `option_set`, are `RTEMS_EVENT_ALL` and `RTEMS_WAIT`.

When invoked, the `rtems_event_receive` directive examines the event flags of the calling task to check whether the condition specified by `event_in` and `option_set` is already satisfied. In this case, the directive succeeds and returns immediately. Otherwise, if the `option_set` indicates that the directive shall wait, it blocks the calling task until the waiting condition is satisfied by the arrival of some new events or the amount of time specified by the `ticks` argument expires.

When successful, the directive clears the event flags of the events it is reporting to the caller and stores this set of events in the output argument pointed by `event_out`. In all cases, `rtems_event_receive` returns to the caller a status code that reflects its outcome. The possible status codes are:

RTEMS_SUCCESSFUL The directive concluded successfully and stored the events it received in the location pointed by `event_out`.

RTEMS_INVALID_ADDRESS The `event_out` pointer was invalid, the directive failed without receiving any events.

RTEMS_UNSATISFIED The `option_set` specified that the directive should not wait (the `RTEMS_NO_WAIT` option was set) and the set of pending events at the time of the call was insufficient to satisfy the directive.

RTEMS_TIMEOUT The directive waited for the amount of time specified by `ticks`, then failed because the pending events at the time of the call, plus the events sent to the task afterwards, were insufficient to satisfy the directive.

Finally, the special value `RTEMS_PENDING_EVENTS` can be used as `event_in` to indicate that `rtems_event_receive` shall immediately return the set of pending events, without clearing the corresponding event flags. This allows the caller to "peek" event flags without affecting the outcome of future receive operations.

7.7   SUMMARY

This chapter described the inter-task communication and synchronization methods based on shared memory for data transfer. The chapter started with an introduction to race conditions in Section 7.1 and explained how lock-based mutual exclusion can solve this ubiquitous problem that affects virtually all concurrent software systems.

The next two sections, Sections 7.2 and 7.3 provided information about semaphores and monitors, which are probably the most classic inter-task communication and synchronization methods. Then, the chapter discussed the RTEMS APIs related to semaphores and monitors in Section 7.4.

The chapter ended with a description of two more specialized, but still very useful in practice, inter-task communication and synchronization methods, barriers and events, in Sections 7.5 and 7.6. While this chapter focused only on the mechanisms of inter-task communication and synchronization, their impact on task execution and scheduling will be the topic of the next chapter.

8 IPC, Task Execution, and Scheduling

CONTENTS

Any kind of lock-based task interaction in a real-time system, and especially mutual exclusion, must be designed, analyzed, and used with great care because it inevitably introduces timing dependencies among tasks with different priorities.

Accordingly, this chapter tackles two extremely important issues, *priority inversion* and *deadlock*, which may impair the timings of any real-time system if its lock-based inter-task synchronization and communication mechanisms are used improperly. In both cases, we discuss appropriate suitable design-time and implementation-time methods that can solve those issues.

8.1 PRIORITY INVERSION

A very common occurrence in a real-time system is that a high-priority and a low-priority task share some data and have critical regions controlled by the same semaphore. It may then happen that the low-priority task enters its critical region first and forces the high-priority task to wait if it also tries to do the same.

This phenomenon is known as *priority inversion* and, on the one hand, is essential to ensure that mutual exclusion (as any other lock-based synchronization method) works correctly. On the other hand, though, it undermines the task priority assignment because, if the mutual exclusion mechanism were not in effect, the high-priority task would always be preferred over the lower-priority one for execution and would never be blocked by it.

If we stay with lock-based synchronization, a certain amount of priority inversion is unavoidable but, to make things worse, in Section 7.2 we suggested that plain mutual exclusion semaphores with a FIFO queuing policy may lead to an issue known as *unbounded* priority inversion.

Informally speaking, this is a pathological condition in which a task is potentially postponed by lower-priority tasks as it tries to enter a critical region and the postponement may last for an unbounded amount of time. If not adequately addressed, priority inversion can therefore adversely affect system schedulability, to the point of making the response time of some tasks completely unpredictable.

8.1.1 MUTUAL EXCLUSION AND PRIORITY INVERSION

Even though proper software design techniques may alleviate the priority inversion issue, for example, by avoiding useless or redundant critical regions, it is also clear that the problem cannot be completely solved in this way, unless the system is designed banning all forms of mutual exclusion.

Fortunately, it is possible to improve the semaphore-based mutual exclusion mechanism to guarantee that the worst-case blocking time B_i^{TI} endured by each individual task in the system is bounded. The worst-case blocking time can then be calculated and used to refine the response time analysis (RTA) method as discussed in Section 4.1.3, in order to determine worst-case response times.

In order to better understand how unbounded priority inversion may arise, what it really means in practice, and to appreciate how mutual exclusion semaphores can be improved to address this issue, let us discuss a very simple example involving only three tasks. The same example will also be used in the following to illustrate how the countermeasures being described work.

Let us consider three periodic tasks τ_1, τ_2, and τ_3, listed in order of decreasing priority. Their running intervals are depicted in Figure 8.1 as rectangles filled with different shades of gray. A darker color corresponds to a higher priority. For the sake of simplicity, we assume that these tasks are under the control of a fixed-priority scheduler on a single-core processor.

The figure shows how the tasks evolve with time. Each task has its own horizontal timeline. As mentioned previously, while a task is running, there is a rectangle with the appropriate color on its timeline. Time intervals in which a task is ready for execution, but not running, are represented by a solid line. Dotted lines represent time intervals in which a task is waiting, because either its next instance has not been released yet, or it is blocked on a semaphore.

For the sake of the example we will also assume that τ_1 and τ_3 share some information, stored in a certain shared memory object O and protected by a mutual exclusion semaphore s. Accordingly, τ_1 and τ_3 make access to O only within suitable critical regions, properly delimited by P(s) and V(s) as described in Section 7.2.2. Instead, τ_2 does not share any information with τ_1 and τ_3. Its only relationship with

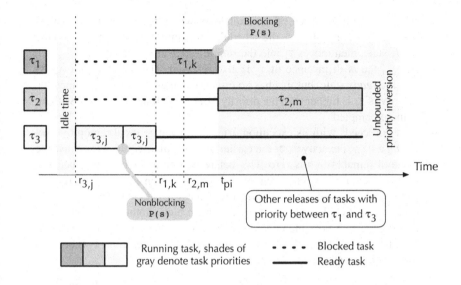

Blocking
P(s)

Nonblocking
P(s)

Other releases of tasks with
priority between τ_1 and τ_3

Running task, shades of
gray denote task priorities

Blocked task
Ready task

FIGURE 8.1 Unbounded priority inversion with a plain mutual exclusion semaphore.

them is that its priority is lower than the priority of τ_1, but higher than the priority of τ_3.

Even in a moderately complex system, this peculiar priority relationship may very well be unknown to the programmers who wrote τ_1, τ_2, and τ_3. When different parts of a system are designed and implemented by different people, it is in fact normal that programmers are unaware of the exact architecture of the parts of the system they are not directly involved with.

Moreover, according to the Rate Monotonic (RM) priority assignment, the priority of a task depends only on its period. When different software subsystems—likely written by distinct groups of programmers, at different times, and made of several tasks each—are put together to build the complete application, the priorities of tasks belonging to different subsystems can easily become intertwined.

Referring back to Figure 8.1, the following sequence of events may occur:

- At the far left of the figure, before $r_{3,j}$, the system is idle and no tasks are running.
- At $r_{3,j}$ the j-th instance of τ_3 (the lowest-priority task) becomes ready for execution, while τ_1 and τ_2 are still waiting for their next instance to be released. Since at this time τ_3 is the only ready task, the scheduler moves it to the *running* state and starts executing it.
- After executing for a while, $\tau_{3,j}$ attempts to enter its critical region by means of a P(s) located at the critical region's boundary. Since no other tasks are accessing the shared object O at the moment, the P(s) is non-blocking. Hence, $\tau_{3,j}$ is allowed to continue immediately and keeps running within the critical region.

- At $r_{1,k}$ the k-th instance of τ_1 is released, while $\tau_{3,j}$ is still in its critical region. As a consequence, the scheduler preempts τ_3, puts it back into the *ready* state, then moves τ_1 into the *running* state and executes it.
- At $r_{2,m}$ the m-th instance of τ_2 is also released. The scheduler moves τ_2 into the *ready* state, but this has no effect on the execution of τ_1. Since the priority of τ_1 is higher than the priority of τ_2, it continues execution without being preempted.
- As τ_1 proceeds with its execution, tries to enter its critical region by executing a P (s), exactly as τ_3 did earlier. At this point, τ_1 blocks because the value of semaphore s is zero. This behavior is completely correct because the purpose of s is to enforce mutual exclusion when accessing shared object O and τ_3 is still within a critical region associated with O.
- Since τ_1 is now blocked, the scheduler must pick another task to execute. Among the two possible candidates, τ_2 and τ_3, the scheduler chooses τ_2 because its priority is higher than the priority of τ_3. Therefore, τ_2 is brought to the *running* state and executed.

Starting from t_{pi} in Figure 8.1, a priority inversion region begins, because τ_3 (the lowest-priority task in the system) blocks τ_1 (the highest-priority one) and the system executes τ_2 (another task with a lower priority than τ_1). This region is highlighted with a light gray background in the figure. Although the existence of the priority inversion region depends on how the mutual exclusion mechanism for shared object O has been designed, it is crucial to determine *for how long* it may last.

First of all we can observe that, somewhat contrary to intuition, the duration of the priority inversion region does not depend on any of the two tasks directly responsible for it, τ_1 and τ_3, because:

- Task τ_1 is blocked and, by definition, it cannot perform any further action on its own, until τ_3 unblocks it by exiting from its critical region and executing V (s).
- Task τ_3 is ready for execution but it cannot proceed unless the scheduler decides to run it. Until then, it may not continue through its critical region and eventually reach the V (s) at its boundary.

The duration of the priority inversion region depends instead on how much time τ_2 and any other tasks whose priority is lower than the priority of τ_1 but higher than the priority of τ_3 spend running. As long as at least one of these tasks is ready for execution, the scheduler will never run τ_3 and, as a consequence, τ_3 will never unblock τ_1.

Bringing this line of reasoning to an extreme, these tasks could take turns entering the *ready* state and being executed so that, even if none of them keeps running for an inordinate amount of time individually, one of them is always running at any given instant. This is the reason why τ_1 might be blocked for an *unbounded* amount of time by tasks like τ_2, although they all have a priority lower than τ_1 itself.

To conclude, the amount of blocking experienced by τ_1 does not depend on the lower-priority task τ_3 that is directly blocking it, but mainly on other lower-priority tasks like τ_2. Unfortunately, as discussed previously, in a complex software system built by integrating multiple components, the programmers who wrote τ_1 and τ_3 may even be unaware that such tasks exist.

It is also important to remark that, as for many other concurrent programming issues, this is not a systematic error. Instead, it is a time-dependent issue that arises only in certain specific, low-probability scenarios like the one depicted in Figure 8.1. Hence, it may easily go undetected when the system is bench tested.

8.1.2 THE PRIORITY INHERITANCE PROTOCOL

Looking back at the unbounded priority inversion example presented in the previous section, we may notice that the issue stemmed from a bad interaction between two key operating system mechanisms, namely, task *synchronization* and *scheduling*. So far, these mechanisms have been defined to be completely independent from each other.

By intuition, we can imagine that the issue may be solved by making the two mechanisms cooperate. This cooperation can be implemented, for instance, by allowing the mutual exclusion mechanism to temporarily *boost* task priorities as needed. This is exactly the way the *priority inheritance* protocol proposed by Sha, Rajkumar, and Lehoczky [110] works.

The general idea of the priority inheritance protocol is to dynamically increase the priority of a task as soon as it is blocking some higher-priority tasks, and run the task at increased priority until it is no longer blocking them. For example, as long as a task (like τ_3 in Figure 8.1) is blocking a higher-priority task (like τ_1 in the same figure) it inherits its priority.

In general, as long as a task is blocking a set of higher-priority tasks, it inherits the highest priority among them. This prevents any mid-priority task from preempting the low-priority task and unduly making the blocking experienced by the high-priority tasks any longer than necessary or, even worse, unbounded.

More formally, the priority inheritance protocol relies on the following set of hypotheses and assumptions about the system being considered:

- The concept of task priority becomes more complex because it is now necessary to distinguish between the *initial*, or *baseline*, priority given to a task by the scheduling algorithm and its *current*, or *active*, priority.
- The baseline priority is used as the initial, default value of the active priority but the priority inheritance protocol may increase the latter when the task is blocking some higher-priority tasks.
- All tasks are under the control of a fixed-priority scheduler and run on a single-core processor. The scheduler works according to active priorities.
- If there are two or more highest-priority tasks ready for execution, the scheduler picks them in first-come, first-served (FCFS) order.
- Semaphore wait queues are also ordered by active priority. In other words, when a task executes a $V(s)$ on a semaphore s and there is at least one

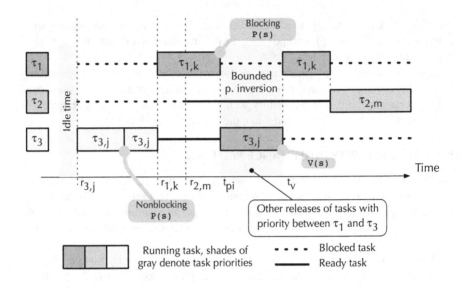

FIGURE 8.2 Bounded priority inversion in the priority inheritance protocol.

task blocked on a P(s), the highest-priority waiting task is unblocked and moved into the *ready* state.

- Semaphore waits due to mutual exclusion are the only source of blocking in the system. Other causes of blocking such as, for example, I/O operations, must be taken into account separately, as discussed in Section 4.1.3.

The priority inheritance protocol itself consists of the following set of rules [126]:

1. When a task τ_1 attempts to enter a critical region that is "busy" because its controlling semaphore has already been taken by another task τ_3, it blocks. At the same time, τ_3 *inherits* the active priority of τ_1 if the active priority of τ_1 is higher than its own.

2. As a consequence, τ_3 executes the rest of its critical region with an active priority at least equal to the priority it just inherited. In general, a task inherits the highest active priority among all tasks it is blocking.

3. When a task τ_3 exits from a critical region and it is no longer blocking any other task, its active priority goes back to the baseline priority.

4. Otherwise, if τ_3 is still blocking some tasks—this may happen when critical regions are nested into each other—it inherits the highest active priority among them.

Although this falls short of being a formal proof of correctness, we can now analyze how the priority inheritance protocol performs on the simple example previously shown in Figure 8.1. The result is illustrated in Figure 8.2 and commented in the following.

- The system behaves as before until t_{pi} because no task is blocking any other task until that point. All tasks are scheduled according to their initial, or baseline, priority.
- The priority inheritance protocol is called into action at t_{pi}, when τ_1 blocks on P(s). At this point, τ_3 inherits the priority of τ_1 because it is blocking τ_1 through semaphore s.
- Therefore, regardless of the presence of other mid-priority tasks, like τ_2, the scheduler resumes the execution of τ_3 because it operates according to the active priority of the ready tasks. Accordingly, this portion of τ_3's execution has been colored with a darker shade of gray in the figure.
- Task τ_3 eventually exits from its critical section at t_v and releases the mutual exclusion semaphore by means of a V(s). The first effect of this primitive is to unblock τ_1, which acquires semaphore s and returns to the *ready* state.
- The second effect pertains to the priority inheritance protocol. Since now τ_3 is no longer blocking any higher-priority task, its active priority goes back to its baseline priority. As a consequence, the scheduler immediately preempts τ_3 to execute τ_1.
- After t_v, τ_1 keeps executing until it terminates or blocks again. Only at this point mid-priority tasks, like τ_2, get a chance of being executed.

The example shows that the length of the priority inversion region is now bounded and limited to the interval between t_{pi} and t_v highlighted in the figure. This is because:

- Any tasks with a priority lower than the baseline priority of τ_1 cannot preempt τ_3, and hence, they cannot increase the amount of time τ_3 spends in its critical region.
- Any tasks with a priority higher than the baseline priority of τ_3, but lower than the baseline priority of τ_1, cannot preempt τ_3 either, because of the priority boost the priority inheritance protocol granted to it.

Tasks with a priority higher than the baseline priority of τ_1 can still preempt τ_3. This may be due to two distinct reasons:

- If their baseline priority is higher than the baseline priority of τ_1, they can interfere with the execution of τ_1 in any case, regardless of what the priority inheritance protocol does, but schedulability analysis already takes this interference into account.
- If their baseline priority was lower than the baseline priority of τ_1, but their active priority is currently higher, this is due to the priority inheritance protocol and, also in this case, it can be proven that the additional time spent by τ_3 in its critical region for this reason is bounded.

If no such tasks exist, the worst-case length of the priority inversion region is equal to the maximum amount of time that τ_3 can possibly spend within its critical region. In the general case, the worst-case blocking is given by Equation (4.7) in Section 4.1.3.

However, the same example also highlights that, within the priority inversion region $[t_{pi}, t_v]$, τ_3 now blocks *both* τ_1 and τ_2, whereas is only blocked τ_1 in the previous example. This fact leads us to observe that the priority inheritance protocol—like any other algorithm dealing with unbounded priority inversion—entails a trade-off between two contrasting goals:

- ensure there is a finite upper bound on the length of priority inversion regions, and
- introduce additional blocking in the system.

More specifically, for the priority inheritance protocol we identify two distinct kinds of blocking:

1. *Direct blocking* occurs when a high-priority task tries to acquire a mutual exclusion semaphore before accessing a shared object, while the semaphore is held by a lower-priority task. This is the kind of blocking affecting τ_1 in this case. Direct blocking is an unavoidable consequence of mutual exclusion and its goal is to ensure the consistency of shared objects.
2. *Push-through blocking* is the additional blocking introduced by priority inheritance and is the kind of blocking experienced by τ_2 in the example. It occurs when an intermediate-priority task (like τ_2) is not executed even though it is ready because a lower-priority task (like τ_3) has inherited a higher priority. This kind of blocking may affect a task even if it does not actually use any shared object, exactly as it happens with τ_2 in the example, but it is necessary to avoid the unbounded priority inversion.

8.1.3 THE PRIORITY CEILING PROTOCOL

Referring back to the example shown in Figure 8.1, we may observe that one underlying cause of the unbounded priority inversion was the preemption of τ_3 by τ_1 at $r_{1,k}$, while τ_3 was in its critical region. If some mechanism could delay the context switch from τ_3 to τ_1 until after the execution of V(s) by τ_3, that is, until the end of its critical region, the issue would not have occurred.

On a single-core system, a very straightforward solution to the unbounded priority inversion problem is to forbid preemption completely while a critical region is being executed. This may be obtained by temporarily disabling the operating system scheduler or, even more drastically, turning interrupts off. As a result, any task that successfully enters a critical region implicitly gains the highest possible priority in the system, so that no other task can preempt it. The task goes back to its regular priority when it exits from the critical region.

The main advantage of this strategy is its simplicity of implementation. Moreover, it is easy to be convinced that it really works. Informally speaking, if no tasks can ever lose the processor while they are holding a mutual exclusion semaphore, they will not directly block any higher-priority tasks by definition.

However, as hinted previously, the method invariably introduces a new kind of blocking, of a different nature and quite pervasive. More specifically, if any other task τ_2 becomes ready while a low-priority task τ_3 is within a critical region, it will not get executed until τ_3 exits from the critical region. This happens regardless of the priority of τ_2 and whether or not it has any relationship with τ_3.

The problem has been solved anyway, because the amount of blocking endured by τ_2 is indeed bounded by the worst-case amount of time τ_3 may actually spend running within its critical region. However, all tasks in the system are affected by this blocking.

For this reason, this way of proceeding is appropriate only for very short critical regions, because it causes pervasive, and often unnecessary, blocking. In addition, it can hardly be applied to multi-core processors, in which globally disabling interrupts is a complex and often time-consuming affair. This does not mean the underlying idea is ineffective, though. Indeed, we shall see that the approach just discussed is merely a strongly simplified version of the priority ceiling emulation protocol, to be described next.

Even if the priority inheritance protocol described in Section 8.1.2 enforces an upper bound on the number and the duration of the blocks that a high-priority task τ_1 can encounter, it has a couple of shortcomings:

- In the worst case, if τ_1 tries to acquire in sequence n mutual exclusion semaphores that have been locked by n lower-priority tasks, it will be blocked for the duration of n critical regions, one for each lower-priority task. This phenomenon is called *chained blocking*.
- The priority inheritance protocol, by itself, does not prevent *deadlocks*. They must therefore be addressed by some other means, for example, by imposing and respecting a total order on the semaphores in the system, as discussed in Section 8.2.3.

All of these issues are addressed by the *priority ceiling protocol* and its *immediate* variant, also proposed by Sha, Rajkumar, and Lehoczky [110]. They both possess the important property that a high-priority task can be blocked by lower-priority tasks *at most once* during its execution, and they also prevent *deadlocks*. Moreover, the worst-case blocking time is still bounded as stated by Equation (4.8) in Section 4.1.3.

The underlying concept of the priority ceiling protocol is to extend the priority inheritance protocol with an additional rule that controls whether or not tasks can immediately acquire a *free* mutual exclusion semaphore. The goal of the additional rule is to ensure that, if a task τ_3 has already acquired a semaphore and it could block a higher-priority task τ_1 for this reason, then no other semaphores that could also block τ_1 can be acquired by any other task except τ_3 itself.

As a consequence a task can be blocked not only because it attempted to acquire a busy semaphore, as it happens for all kinds of mutual exclusion semaphore, but also when acquiring a *free* semaphore could lead a higher-priority task to be blocked more than once. In other words, the priority ceiling protocol trades off some useful properties for yet another form of blocking that did not exist in the priority inheritance protocol. The new kind of blocking that the priority ceiling protocol introduces, in addition to direct and push-through blocking mentioned previously, is called *ceiling blocking*.

The underlying hypotheses of the original priority ceiling protocol are the same as the ones of the priority inheritance protocol, listed in Section 8.1.2. In addition:

- It is assumed that each semaphore has a fixed *ceiling* value associated with it. It can easily be calculated during software design and implementation, by looking at the application code, because it is defined as the maximum initial priority of all tasks that use the semaphore.

As in the priority inheritance protocol, each task has a current (or active) priority that is greater than or equal to its initial (or baseline) priority, depending on whether it is blocking some higher-priority tasks or not. The priority inheritance rules are exactly the same in both cases. Moreover, the priority ceiling protocol operates according to the following additional rule:

1. When a task τ tries to acquire a semaphore, its active priority is checked against the ceiling of all currently busy semaphores, except the ones that τ has already acquired in the past and not released yet.
2. If the active priority of τ is higher than all those ceilings, τ can proceed with the semaphore operation and possibly block if the semaphore is busy.
3. Otherwise, τ is blocked until this condition becomes true, regardless of whether the semaphore it is trying to acquire is busy or free. Afterwards, τ can proceed with the semaphore operation.

The immediate priority ceiling protocol (also called priority ceiling *emulation* protocol) takes a more straightforward approach. Also for this reason, it has been specified together with priority inheritance in the POSIX standard [68] and is available in RTEMS. Namely, the immediate priority ceiling protocol raises the priority of a task to the ceiling associated with a semaphore *as soon as* the task acquires it, rather than only when the task is actually blocking a higher-priority task. More formally, at each instant the active priority of a task is equal to the maximum among its baseline priority and the ceilings of all semaphores it has acquired and not released yet so far.

It can be proven that, when using the immediate priority ceiling protocol on a single-core system, a task can only be blocked at the very beginning of its execution, that is, as soon as it is released. Additional differences with respect to the original priority ceiling protocols are:

- The immediate priority ceiling protocol is easier to implement, as blocking relationships must not be monitored.
- It leads to less context switches, since blocking may only occur prior to the first execution.
- On average, it requires more priority movements, as this happens with all semaphore operations, rather than only if a task is actually blocking another.

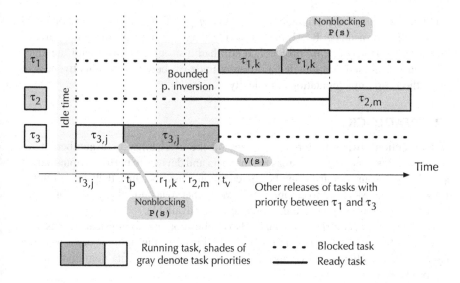

FIGURE 8.3 Bounded priority inversion in the immediate priority ceiling protocol.

To conclude this section, let us refer back to the simple scenario shown in Figure 8.1 and analyze the behavior of the immediate priority ceiling protocol on it. As shown in Figure 8.3:

- Task instance $\tau_{3,j}$ assumes an active priority equal to the baseline priority of τ_1 as soon as it executes its P(s) at t_p. The P(s) is nonblocking as before and $\tau_{3,j}$ keeps executing in the critical region with the increased priority.
- Neither the release of $\tau_{1,k}$ at $r_{1,k}$, nor the release of $\tau_{2,m}$ at $r_{2,m}$ cause the preemption of τ_3 because, even though the baseline priority of both these tasks is higher than τ_3's.
- As a consequence, both $\tau_{1,k}$ and $\tau_{2,m}$ are blocked by $\tau_{3,j}$ until it executes its V(s) at t_v. Note that, as stated previously, these task instances are indeed blocked at the very beginning of their execution, that is, as soon as they are released.
- At t_v task instance $\tau_{3,j}$ returns to its baseline priority, the priority inversion region that started at $r_{1,k}$ ends, and the scheduler picks $\tau_{1,k}$ for execution. From this point on $\tau_{2,m}$ is no longer blocked by $\tau_{3,j}$ either. Instead, it suffers ordinary interference from $\tau_{1,k}$ due to the baseline priority order.
- The P(s) that $\tau_{1,k}$ executes at a later time is nonblocking. This further confirms the fact that, with the immediate priority ceiling protocol, a task can be blocked only once at the beginning of its execution, and $\tau_{1,k}$ already suffered such a block.
- Eventually, $\tau_{2,m}$ runs when $\tau_{1,k}$ completes its execution.

Other variants of the original priority ceiling protocol are or were in practical use as well. For instance, RTAI, one of the Linux real-time extensions, made use of the *adaptive priority ceiling* protocol. It still had a strong similarity with the algorithms just discussed, but entailed a different trade-off between the effectiveness of the method and implementation complexity.

8.2   DEADLOCK

Task execution always requires and relies on the availability of a number of *resources*, for instance, memory areas to store data and devices to perform input–output operations. Resources must often be shared among tasks because they are in limited supply. Hence, tasks have to compete with each other to acquire them, possibly waiting if the resources they need are not immediately available.

The concept of acquiring a resource also applies not only to physical objects, like the ones just mentioned, but also to software entities, like the right to access a shared object in a mutually exclusive way, or the ability to retrieve a data item that another task produced.

Waiting for resources in an uncontrolled way is however dangerous because it may lead a set of tasks to block indefinitely and prevent them from concluding their job. This is what happened with the incorrect solution to the producers–consumers problem described at the end of Section 7.2.4. Even if the probability of this phenomenon, usually called *deadlock*, is generally low, it must still be dealt with adequately, especially in systems that are performing critical activities.

8.2.1   DEFINITION AND EXAMPLES OF DEADLOCK

A deadlock can be defined in the most general way as a condition in which a set of tasks is blocked waiting for an event that can only be generated by another task in the same set. When focusing on resources, there is a deadlock when all tasks in a set are waiting for some resources that have been previously acquired by other tasks in the same set. The consequences of a deadlock are twofold:

- The deadlocked *tasks* will no longer make any progress, that is, they will stay blocked forever.
- All the *resources* allocated to them will never be released, and hence, they will never again be available to other tasks.

Four conditions, originally formulated by Havender [55] and Coffman et al. [37], are *individually necessary* and *collectively sufficient* for a deadlock to occur. These conditions are of both theoretical and practical significance. From the theoretical point of view, they define deadlock in a general way, which abstracts away as much as possible from any irrelevant characteristics of the tasks and resources involved. Practically speaking, they have been used as the basis for a family of deadlock prevention algorithms. These algorithms are all based on the fact that, if an appropriate policy is able to prevent one of the four condition from ever being fulfilled in a

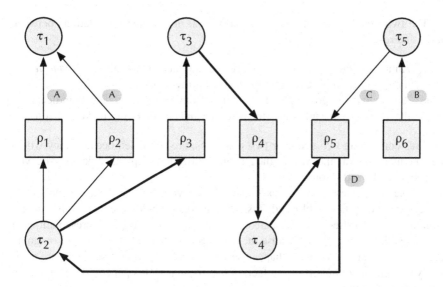

FIGURE 8.4   An example of resource allocation graph.

system, then no deadlock can possibly occur in the system by definition. The four conditions are:

1. *Mutual exclusion*: Each resource can be used by at most one task at a time, so at any given time a resource can only be either free or assigned to one task. If a task tries to acquire a resource currently assigned to another task, it must wait.
2. *Hold and Wait*: The tasks involved in a deadlock must have successfully acquired at least one resource in the past, and have not released it yet as they wait for more resources. In other words, they must both *hold* some resources and *wait* for others.
3. *Non-preemption*: Any resource involved in a deadlock cannot be taken away from the task it has been assigned to, unless the task *voluntarily* releases it.
4. *Circular wait*: The task and resources involved in a deadlock can be arranged to form a circular chain of wait operations. That is, tasks can be ordered so that the first task is waiting for a resource assigned to the second, the second task is waiting for a resource assigned to the third, and so on up to the last task, which is waiting for a resource assigned to the first.

Another very useful tool for reasoning about deadlock is the *resource allocation graph* proposed by Holt [61]. It has the twofold goal of capturing in a rigorous way the state of a system for what concerns resource allocation at a given instant, and assisting in detecting deadlock conditions.

As shown in Figure 8.4, a resource allocation graph is a directed graph with two kinds of nodes and two kinds of arcs. In the most common notation adopted, for example, by Tanenbaum and Woodhull [119] and Silbershatz et al. [112]:

• *circular* nodes represent tasks, and
• *square* nodes represent resources.

The two kinds of arcs convey instead *request* and *ownership* relations between tasks and resources. In particular:

- An arc directed from a task to a resource indicates that the task *requested* the resource and is currently blocked, waiting to acquire it.
- An arc in the opposite direction, that is, from a resource to a task, indicates that the task currently *owns* the resource and can therefore make use of it.

Arcs connecting two tasks, or two resources, are forbidden because they would bear no meaning. In other words, the resource allocation graph must be bipartite with respect to tasks and resources.

Unlike for request arcs, the existence of an ownership arc does not imply that the corresponding task is blocked. Tasks that neither own nor are waiting for any resources, as well as free resources, are often not shown in graphical representations of resource allocation graphs, because these nodes have neither incoming nor outgoing arcs. Moreover, the geometric shapes used to represent tasks and resources are obviously completely unimportant. In fact, in the original Holt's paper [61] the notation was exactly the opposite.

The resource allocation diagram depicted in Figure 8.4 represents a set of 5 tasks and 6 resources. As an example, by looking at the diagram, we can observe that:

- Task τ_1 owns resources ρ_1 and ρ_2 (as indicated by the two arcs labeled A) and is not waiting for any additional resources.
- Instead, task τ_5 owns ρ_6 (arc B) but is blocked, because it also needs ρ_5 to proceed (arc C). Its request cannot be granted at the moment because ρ_5 is owned by τ_2 (arc D).

In general, tasks that have at least one outgoing arc are blocked because they lack some resources they need to proceed. Instead, tasks with no outgoing arcs can proceed from the resource allocation point of view. In the figure, all tasks except τ_1 are blocked. Similarly, resources can have at most one outgoing arc, because they can be assigned to at most one task at a time. The presence of an outgoing arc means the resource is busy, its absence indicates that the resource is free. Incoming arcs denote that some tasks are waiting for the resource and also necessarily imply the presence of one outgoing arc, otherwise tasks would have been blocked improperly.

This data structure can be used by an operating system to keep track of the evolving allocation state of system resources. In this case:

- When a task τ_i requests a resource ρ_j, the operating system checks whether there is an outgoing arc from ρ_j or not. If there is no such an arc, it may assign the resource to τ_i immediately, adding an ownership arc from ρ_j to τ_i. Otherwise, it blocks τ_i and adds a request arc from τ_i to ρ_j to memorize the pending request. Note that deadlock avoidance algorithms, discussed in Section 8.2.4, may compel a task to wait, even if the resource it is requesting is free.

- When a task τ_i releases a resource ρ_j, the operating system deletes the ownership arc from ρ_j to τ_i. The arc must necessarily exist because τ_i must have acquired ρ_j before releasing it. This fact also gives the operating system the ability of checking whether the operation τ_i is trying to perform is legitimate or not.

- If ρ_j has just been released, but has incoming request arcs, the operating system may now pick one of the tasks blocked on ρ_j and reverse the direction of the corresponding request arc to transform it into an ownership arc. Also in this case, deadlock avoidance algorithms may affect which task, if any, the operating system is going to pick.

It has been proved that the presence of a *cycle* in the resource allocation graph is a necessary and sufficient condition for a deadlock to occur. Due to this property, the graph can be used to check whether a certain sequence of resource requests and releases leads to a deadlock. To this purpose, the system must keep track of requests and releases as described previously, and then check if there is a cycle in the graph at each step.

For instance, the resource allocation graph shown in Figure 8.4 indicates a deadlock because $\tau_2 \to \rho_3 \to \tau_3 \to \rho_4 \to \tau_4 \to \rho_5 \to \tau_2$ is a cycle. The related arcs are drawn in bold in the figure. Like for any other directed graph, arc orientation must be taken into account when looking for cycles. For this reason, $\tau_2 \to \rho_1 \to \tau_1 \leftarrow \rho_2 \leftarrow \tau_2$ is *not* a cycle.

Besides generically flagging the presence of a deadlock, a cycle in the resource allocation graph also provides additional information: If there are one or more cycles in the graph, the deadlock involves the resources and *at least* all the tasks that belong to the cycles. More specifically, the deadlocked tasks are the ones in the cycles, plus all the other tasks that are blocked on some resources held by one of the tasks in the cycles. In the example of Figure 8.4, resources ρ_3, ρ_4, and ρ_5, as well as tasks τ_2, τ_3, and τ_4 are directly involved in the deadlock because they belong to the cycle. Moreover, also τ_5 is deadlocked because (due to arc C) it will wait indefinitely for ρ_5, which is held by τ_2.

Dealing with deadlock becomes more complex if, instead of considering individual resources as done so far, we divide resources into classes. In this case, we allow multiple resources to be in each class and resource requests must be directed to a class rather than to a specific resource within it. All resources in a class are considered to be identical and any of them can be used interchangeably to satisfy a resource request.

This is a scenario of practical interest because, for instance, in a paged memory system, when any task dynamically requests a page to store some data, any free page will do. Other examples of resources that can be assigned and used interchangeably include disk blocks (if we neglect access time optimization) and entries in most memory-resident operating system tables.

The definition of resource allocation graph can be extended to handle resource classes, giving rise to a general resource allocation graph [61]. However, as also proved in Reference [61], the theorem that relates cycles and deadlocks becomes

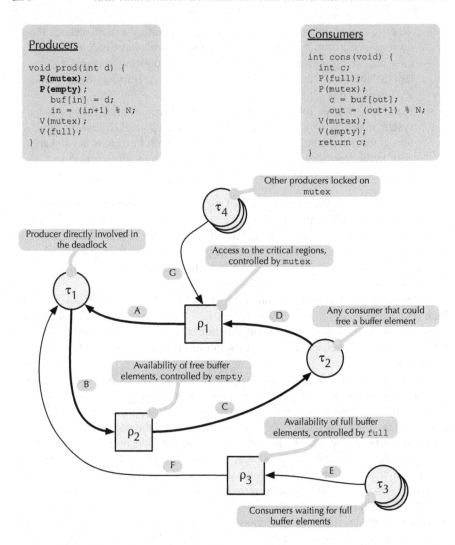

Producers

```
void prod(int d) {
  P(mutex);
  P(empty);
    buf[in] = d;
    in = (in+1) % N;
  V(mutex);
  V(full);
}
```

Consumers

```
int cons(void) {
  int c;
  P(full);
  P(mutex);
    c = buf[out];
    out = (out+1) % N;
  V(mutex);
  V(empty);
  return c;
}
```

Other producers locked on mutex

Producer directly involved in the deadlock

Access to the critical regions, controlled by mutex

Any consumer that could free a buffer element

Availability of free buffer elements, controlled by empty

Availability of full buffer elements, controlled by full

Consumers waiting for full buffer elements

FIGURE 8.5   A simple deadlock condition.

weaker. Namely, the presence of a cycle is still a necessary condition for a deadlock to take place, but it is no longer sufficient. The theorem is still useful in practice, though, because it can still be used to *deny* the presence of a deadlock in the system if there are no cycles in the general resource allocation graph.

8.2.2   DEADLOCK IN THE PRODUCERS–CONSUMERS PROBLEMS

We are now in the position to analyze the deadlock condition we found in the incorrect solution to the producers–consumers problem we have discussed at the beginning of Section 7.3. The corresponding code is shown at the top of Figure 8.5 and the questionable statements in the producers' code are highlighted in bold. To analyze the system with respect to the Havender–Coffman condition stated previously, we shall first of all identify the following three resources in the system:

- The right to access the critical regions that manipulate the circular buffer. Access is controlled by the `mutex` semaphore.
- The availability of at least one empty buffer element, determined by the `empty` semaphore.
- Symmetrically, the availability of at least one full buffer element, determined by the `full` semaphore.

All resources can be used by at most one task at a time. For the critical regions, this constraint is enforced by the corresponding mutual exclusion semaphore. Similarly, empty and full buffer elements are acquired for exclusive use by producers and consumers, respectively. They can be reused over time, but not used concurrently by multiple tasks. As a consequence, the system satisfies the *mutual exclusion* condition.

Depending on the circumstances, producers and consumers can also satisfy the *hold and wait* condition. This fact can be confirmed by looking at their code and observing that both contain multiple, potentially blocking `P()` primitives in sequence without any intervening `V()`. When a certain task blocks on the second `P()`, it clearly holds a resource (because it went past the first `P()`) and waits for another (because it is blocked on the second).

Due to the way their code has been designed, producers and consumers use resources in a non-preemptive way. For instance, the producer cannot be forced to release the mutual exclusion semaphore before it gets some empty buffer space. Similarly, the consumer cannot be forced to release some empty buffer space without first passing through its critical region, controlled by the mutual exclusion semaphore. Therefore, the system also satisfies the *non-preemption* condition.

Last, but most importantly, as soon as a producer blocks on `P(empty)` and a consumer is blocked on `P(mutex)`, there is also a *circular wait* in the system. In other words, the producer waits for the consumer to release the resource "availability of at least one empty buffer element," an event represented by `V(empty)`, and the consumer waits for the resource "critical region access," whose release is marked by a `V(mutex)`. The fact that all four conditions are satisfied leads us to formally conclude there is deadlock in the system.

We can draw the same conclusion by examining the resource allocation graph at the time of the deadlock, shown in the lower part of Figure 8.5. The graph has been constructed according to the following reasoning:

- The producer τ_1 that tries to store a data item in a buffer that is completely full acquires the mutual exclusion semaphore with `P(mutex)` to get access to its critical region (arc A in the figure), and blocks on `P(empty)` waiting for a buffer element to become free (arc B). These two resources are denoted as ρ_1 and ρ_2.

- Any consumer like τ_2 that does not block on P (full) could free a buffer element (arc C) and signal this event by means of V(empty) but, in order to do so, it must wait for the right of entering its critical region (arc D) by performing a P (mutex).
- As consumers continue to arrive, full.v continues to decrease. When it reaches zero, consumers start blocking on P (full). As τ_3 does in the figure, they start waiting for the resource "availability of full buffer elements," represented by ρ_3 (arc E). This resource is owned by τ_1 (arc F) because it is the only task that could possibly store a new data item in the buffer.
- All the other producers wait on ρ_1, that is, for the right of entering their critical region (arc G). As said previously, resource ρ_1 is owned by τ_1 (arc A) because τ_1 is the task that is in its critical region.

The presence of the cycle $\tau_1 \xrightarrow{B} \rho_2 \xrightarrow{C} \tau_2 \xrightarrow{D} \rho_1 \xrightarrow{A} \tau_1$ indicates there is a deadlock. The deadlock involves tasks τ_1 and τ_2 because they are part of the cycle. Moreover, it also involves all tasks like τ_3, due to $\tau_3 \xrightarrow{E} \rho_3 \xrightarrow{F} \tau_1$, and all tasks like τ_4, due to $\tau_4 \xrightarrow{G} \rho_1 \xrightarrow{A} \tau_1$. Both paths denote that these tasks are waiting for resources owned by τ_1, which is part of the deadlock cycle.

On the contrary, it is also easy to note that, if we rearrange P (mutex) and P (empty) in the producers' code and put them in the right order (P (empty) first and P (mutex) next) the cycle cannot exist anymore. This is because the presence of arc A prevents the existence of arc B (because a producer that has acquired mutex cannot wait on empty afterwards) and vice versa (because a producer waiting on empty cannot own mutex).

8.2.3 DEADLOCK PREVENTION

Generally speaking, according to an idea originally proposed by Havender [55], it is possible to prevent deadlocks by making sure that at least one of the necessary conditions identified in Section 8.2.1 can never be satisfied in the system. This constraint is enforced by establishing appropriate software design and implementation rules.

1. Generally, the *mutual exclusion* condition cannot be removed by working directly on the resources involved because it often depends on some hardware characteristics of the resources themselves. For instance, a printer cannot easily be modified to allow multiple tasks to print concurrently and, even if this were possible, the output would be illegible. However, it is often possible to achieve the same result by means of an additional software layer, often called *spooler*, positioned between the tasks and the device.

   In its simplest form, a spooler is a task that, on one side, has permanent, exclusive ownership of a device, like the printer in our example. It is easy to prove that, if a resource is assigned permanently to a task, a circular wait that involves the resource may not occur, because other tasks may not wait for it. Therefore, a deadlock may not occur on the device side of the spooler.

On the other side, the spooler collects print requests from other tasks in the system, and carries them out one at a time. Even if the spooler still sends documents to the printer one at a time, in order to satisfy the printer's mutual exclusion requirement, it is also able to accept multiple, concurrent print requests because it makes use of other resources, like memory buffers or mass storage, to collect and temporarily store the documents to be printed. The overall result is that, from the point of view of the requesting tasks, the mutual exclusion constraint has been lifted and deadlock may not occur on their side.

Deadlocks can still occur within the spooler, though. This may happen, for instance, if the spooler has to compete with other tasks and possibly wait to acquire the memory or disk buffers it needs for temporary storage. Unless the spooler is designed properly, there is therefore the possibility of merely moving the risk of deadlock from one part of the system to another. Another, even more important issue with spooling techniques is the limited range of devices they can be applied to. For example, it is likely impossible to use a spooler on a graphics card.

2. We can falsify the *hold and wait* condition by working on either of its two parts. For instance, we can prevent tasks from *waiting* for resources by assigning them all the resources they may possibly need during their execution right from the beginning, when tasks are created. In a general-purpose system, this goal may be difficult to achieve because it is hard to know in advance what resources a task will need during its execution. For example, the amount of memory needed by a word processor is highly dependent on which document the user is working on at the moment, and this is hard to predict in advance.

Even when it is possible to identify the set of resources to be assigned at task creation, the efficiency of resource utilization will usually be low with this method. This is because it forces system designers to request resources well in advance, on the basis of *predicted* or *potential* needs rather than facts. For instance, when following this approach, the word processor would immediately request an amount of memory large enough for the biggest document it can handle, although in reality the user is only going to write a one-page letter. As a result, a potentially large quantity of memory will stay unused, but still unavailable to other tasks, until the word processor terminates.

In relatively simple real-time systems, however, these disadvantages may no longer be a limiting factor because many resources, like hardware devices, are naturally tied to certain tasks. For example, allocating an analog-to-digital converter (ADC) to the task that acquires, validates, and filters the data it produces when the task is created is likely to be perfectly acceptable. Early allocation and low utilization are not an issue in this case, because no other tasks in the system would be capable of using the ADC anyway, even if it were available. Moreover, accepting that tasks may *wait* for resources in a hard real-time system implies being able to calculate an upper bound on the waiting time, a non-trivial task in many cases.

Alternatively, tasks could be constrained to release all the resources they own before requesting new ones, thus invalidating the *hold* part of the condition. The new set of resources being requested can include, of course, some of the old ones if they are still needed, but the task must accept a temporary loss of resource ownership anyway. Although it is correct in theory, this approach is rarely used in practice because many resources are stateful. If they are lost and then reacquired, the resource state is lost, too.

3. In order to falsify the *non-preemption* condition for a resource we must introduce a resource preemption mechanism, that is, a way of forcibly take away a resource from a task. Like for the mutual exclusion condition analyzed previously, the possibility of doing this heavily depends on the resource and is often impractical. For instance, in Chapter 3 we saw that preempting a processor is relatively straightforward and is done quite efficiently by many operating systems.

   Instead, going back to the printer we considered in previous examples, preempting a print operation sounds awkward at best, even if the operation is later resumed from the right place. Nevertheless, this technique was actually used in the past by the THE operating system [45, 43]. It might preempt its printers on a page-by-page basis and a patient operator was supposed to put the pieces together before handing users their printouts.

4. The most popular way of attacking the *circular wait* condition consists of defining a total order relation on resource classes and demanding that tasks follow the order when allocating resources. Namely, if there are N resource classes in the system and we uniquely label each resource class as $\rho_i, 1 \leq i \leq N$, a suitable total order relation is the one that orders resources by increasing index: $\rho_1, \ldots, \rho_N$.

   We can then express the rule on resource allocation order by stating that, if a task already owns a resource in class ρ_i, it can request another resource in ρ_j if and only if $i < j$. It can easily be proven by contradiction that a circular wait may not occur in the system if all tasks obey this rule [112].

   Let us assume that, although all tasks followed the rule just stated, there is a circular wait in the system. Without loss of generality, let us also assume that the circular wait involves M tasks, $\tau_1, \ldots, \tau_M$, and M resource classes $\rho_{q_1}, \ldots, \rho_{q_M}$, with $1 < M \leq N$, so that:

   τ_1 owns a resource in ρ_{q_1} and waits for a resource in ρ_{q_2},

   $$\cdots$$

   τ_k owns a resource in ρ_{q_k} and waits for a resource in $\rho_{q_{k+1}}, \quad 1 < k < M \quad (8.1)$

   $$\cdots$$

   τ_M owns a resource in ρ_{q_M} and waits for a resource in ρ_{q_1}.

   If tasks $\tau_1, \ldots, \tau_{M-1}$ followed the rule, it must be $q_k < q_{k+1}, 1 \leq K < M$ because each of these tasks owns a resource in class ρ_{q_k} and is waiting for a resource in class $\rho_{q_{k+1}}$. If τ_M also followed the rule, it must be $q_M < q_1$. We can therefore write the following chain of inequalities:

   $$q_1 < \ldots < q_M < q_1 \quad (8.2)$$

   Due to the transitive property of inequalities, we come to the contradiction $q_1 < q_1$, thus disproving the presence of a circular wait.

In a complex system, the main shortcoming of this method is the difficulty of enforcing the resource allocation rules, and then verifying whether they have been followed or not. For instance, the FreeBSD operating system kernel uses this approach but, even after many years of handmade improvements, a relatively big number of "lock order reversals" (that is, situations in which locks are actually requested in the wrong order) were still present in the kernel code. To address the issue, a special tool was specifically designed to help programmers detect them [19].

8.2.4 DEADLOCK AVOIDANCE

In the previous section we saw that the most important feature of deadlock prevention algorithms, that is, the fact they operate at system design time, is both an advantage and a disadvantage. On the positive side, deadlock prevention methods do not cause any direct run-time overhead. On the other hand, they impose some design rules that may be difficult or inconvenient to follow.

Deadlock *prevention* algorithms take a different approach because they entirely work at run time, rather than design time. Their underlying idea is to check resource allocation requests one by one, as tasks submit them, and determine whether or not they are safe for what concerns deadlock. In order to do this, a deadlock prevention algorithm has to store and maintain some data structures that represent the current resource allocation state of the system and provide a way to predict its future evolution, as resource allocation and releases continue to arrive. Moreover, most algorithms also need a certain amount of advance information about possible tasks behavior to work properly.

Unsafe requests, that is, requests that could bring the system into a deadlock, are postponed even though the resources being requested are in fact free, and the requesting task is blocked. Postponed requests are reconsidered at a later time, and eventually granted when the deadlock prevention algorithm can prove their safety. By intuition, this usually happens when other tasks release system resources, thus moving the system into a more favorable resource allocation state.

Among all the deadlock avoidance algorithms, we will describe in detail the *banker's algorithm*, originally proposed by Dijkstra for a single resource class [42] and later extended by Habermann to multiple resource classes [54].

In the following we will use capital letters to denote matrices and boldface to denote vectors. Moreover, we will sometimes treat the j-th column of a matrix M as a column vector and write it as $\mathbf{m_j}$. To simplify the notation, we also introduce a weak ordering relation between vectors. In particular we state that, given two vectors of the same length, $\mathbf{v}$ and $\mathbf{w}$:

$$\mathbf{v} \leq \mathbf{w} \iff \forall i \ v_i \leq w_i. \tag{8.3}$$

Informally speaking, a vector $\mathbf{v}$ is less than or equal to another vector $\mathbf{w}$ of the same length if and only if all its elements are less than or equal to the corresponding elements of the other one. Analogously, the strict inequality is defined as

$$\mathbf{v} < \mathbf{w} \iff \mathbf{v} \le \mathbf{w} \wedge \mathbf{v} \ne \mathbf{w}. \tag{8.4}$$

If there are n tasks and m resource classes in the system, the banker's algorithm must maintain the following data structures:

- A column vector $\mathbf{t}$ of length m, which represents the total number of resources of each class initially available in the system:

$$\mathbf{t} = \begin{pmatrix} t_1 \\ \vdots \\ t_m \end{pmatrix} \tag{8.5}$$

The i-th element of $\mathbf{t}$, denoted by t_i, indicates the number of resources of the i-th class initially available in the system. The vector $\mathbf{t}$ is assumed to be constant, thus implying (somewhat unrealistically) that resources never break up or become unavailable for use for any other reason.
- A matrix C, with m rows and n columns, that is, a column for each task and a row for each resource class:

$$C = \begin{pmatrix} c_{11} & \cdots & c_{1n} \\ \cdots\cdots\cdots \\ c_{m1} & \cdots & c_{mn} \end{pmatrix} \tag{8.6}$$

The elements of C represent the current resource allocation state. More specifically, the element c_{ij} indicates how many resources of class i are allocated to the j-th task at the moment. As a consequence, the column vector $\mathbf{c_j}$ summarizes how many resources of each class are currently allocated to the j-th task. Initially, $\forall i, j \; c_{ij} = 0$, because it is assumed that no resources are allocated when the system starts up, and then the contents of C change as the system evolves.
- A matrix X, also with m rows and n columns:

$$X = \begin{pmatrix} x_{11} & \cdots & x_{1n} \\ \cdots\cdots\cdots \\ x_{m1} & \cdots & x_{mn} \end{pmatrix} \tag{8.7}$$

This matrix represents an example of the auxiliary information about tasks behavior needed by this kind of algorithms, because is specifies the maximum number of resources that each task may possibly require, for each resource class, during its whole lifetime. In other words, it is assumed that each task τ_j will declare in advance its worst-case resource needs by means of a column vector $\mathbf{x_j}$:

$$\mathbf{x_j} = \begin{pmatrix} x_{1j} \\ \vdots \\ x_{mj} \end{pmatrix} \tag{8.8}$$

Overall, matrix X can be built by placing all the vectors $\mathbf{x_j}$ provided by tasks $\tau_j, j = 1, \ldots, n$ side by side. Moreover, tasks are not allowed to change their mind about their needs unless they have no resources currently allocated to them, that is, τ_j cannot ask the banker to change its $\mathbf{x_j}$ unless its $\mathbf{c_j}$ is zero. A further constraint on the value of each $\mathbf{x_j}$ is that it must be $\forall j \ \mathbf{x_j} \le \mathbf{t}$, otherwise τ_j could never get all the resources it needs and complete its job, even if it were executed alone in the system.

- A matrix N, representing the worst-case future resource needs of the tasks. It has the same size as C and X, and can readily be calculated as:

$$N = X - C \qquad (8.9)$$

Since C changes over time, N also does. In addition, since C, X, and N are clearly dependent on each other, real-world implementations of the banker's algorithm may store only two of them explicitly if they prefer storage efficiency to execution efficiency.

- A column vector $\mathbf{r}$, which represents the resources remaining in the system at any given time:

$$\mathbf{r} = \begin{pmatrix} r_1 \\ \vdots \\ r_m \end{pmatrix}. \qquad (8.10)$$

The elements of $\mathbf{r}$ can be calculated from other data structures as follows:

$$r_i = t_i - \sum_{j=1}^{n} C_{ij} \ \forall i = 1, \ldots, n. \qquad (8.11)$$

As it happens for N, this gives implementations the choice of not storing $\mathbf{r}$ explicitly, at the expense of execution efficiency.

Informally speaking, equation (8.11) simply means that r_i (the number of remaining resources in class i) is given by t_i (the total number of resources in class i) minus the resources in that class currently allocated to any task. This value can be calculated by summing up the i-th row of C, whose elements C_{ij} indicate how many resources in class i have been allocated to task τ_j.

For uniformity, resource requests issued by tasks are also represented by vectors. Namely, a request coming from task τ_j is represented as a column vector $\mathbf{q_j}$ defined as:

$$\mathbf{q_j} = \begin{pmatrix} q_{1j} \\ \vdots \\ q_{mj} \end{pmatrix}. \qquad (8.12)$$

Within $\mathbf{q_j}$, the element q_{ij} indicates how many resources in class i task τ_j is requesting. A value of zero is allowed and means that τ_j is not requesting any resource of class i.

Whenever it receives a new request q_j from τ_j, the banker executes the following algorithm to determine whether or not the request shall be granted immediately, or the requesting task must wait.

1. It verifies that the request is *legitimate*, that is, τ_j is not trying to exceed x_j, which contains the maximum number of resources τ_j itself declared it needs. Since the j-th column of N, denoted by n_j represents the worst-case future resource needs of τ_j, this test can be written as:

$$q_j \leq n_j \qquad (8.13)$$

If the test is satisfied, the banker proceeds with the next step of the algorithm. Otherwise, it returns an error indication to the calling task. This indication is not deadlock-related, but it has to do with an erroneous behavior of the task.

2. It compares the request with resource *availability*, to check if there are enough free resources in each class to satisfy the request immediately, without letting τ_j wait. Given that r represents the resources that are currently free in the system, this test can be written as:

$$q_j \leq r \qquad (8.14)$$

If the test is not satisfied the request cannot be granted immediately, not for deadlock-related reasons but because, quite simply, there are not enough resources available at the moment. In this case, the banker blocks the requesting task until the system evolves more favorably. Otherwise, the banker continues with the next step of the algorithm, in which it will check whether or not granting the request would put the system at risk of deadlock.

3. The general idea of this and the next steps is that the banker *simulates* the effects of the request at hand on the resource allocation state, and then checks whether or not the simulated state is *safe* for what concerns deadlock.

   To carry out the simulation, the banker calculates a new state and, in particular, new task-specific resource allocation information c_j' and future needs n_j', as well as a new resource availability vector r':

$$
\begin{aligned}
c_j' &:= c_j + q_j \\
n_j' &:= n_j - q_j \\
r' &:= r - q_j
\end{aligned}
\qquad (8.15)
$$

4. To assess the safety of the simulated state computed in the previous step, the banker tries to build a *safe* sequence of tasks. To be safe, a sequence must include all the n tasks in the system and allow each task, in turn, to reach its worst-case resource requirement, and hence, successfully conclude its work. To build the safe sequence, the banker makes temporary use of two additional data structures:

   - A column vector w of m elements, one for each resource class. It is initially set to the currently available resources (that is, $w = r'$) and tracks the evolution of the available resources as the safe sequence is being constructed.

- A row vector $\mathbf{f}$, of n Boolean elements. The j-th element of the vector, f_j, corresponds to task τ_j. Element f_j is false if and only if task τ_j has not yet been inserted into the safe sequence. At the beginning, all elements of $\mathbf{f}$ are false, because the safe sequence is still empty.

The construction of the safe sequence proceeds according to the following steps:

a. Try to find a task suitable for being appended to the safe sequence being constructed. In order to be a fitting candidate, task τ_j must not be part of the safe sequence already, and must be able to reach its worst-case resource need given the currently available resources. That is, it must be:

$$f_j = false \qquad (\tau_j \text{ is not in the safe sequence yet})$$
$$\wedge \qquad\qquad\qquad\qquad\qquad\qquad\qquad (8.16)$$
$$\mathbf{n_j}' \leq \mathbf{w} \qquad (\text{there are enough resources to satisfy } \mathbf{n_j}')$$

If no suitable candidates can be found, the algorithm proceeds to step 4c.

b. Append the candidate task τ_j to the safe sequence. At this point, the correct termination of τ_j is guaranteed by definition (because it will be able to get all the resources it needs to conclude its work). Upon termination, τ_j will release all the resources it holds and the banker can update its notion of available resources as it extends the safe sequence:

$$f_j := true \qquad (\tau_j \text{ belongs to the safe sequence now})$$
$$\mathbf{w} := \mathbf{w} + \mathbf{c_j}' \quad (\tau_j \text{ releases its resources upon termination}) \qquad (8.17)$$

Then, the algorithm goes back to step 4a, to further extend the sequence with additional tasks.

c. At this stage, the safe sequence has been extended as much as possible. If $\forall j \; f_j = true$, then all tasks belong to the safe sequence and safety verification succeeds, otherwise it fails.

Failing the safety verification does not necessarily imply that a deadlock is going to occur. This is because the algorithm considers the worst-case resource requirements stated by each task, and it is therefore being conservative. Even if a system state is unsafe according to the banker's definition, all tasks could still be able to conclude their work without deadlock if, for example, they never actually request the maximum number of resources they declared.

It should also be remarked that the preceding algorithm can arbitrarily choose any suitable task in step 4a and does not need to backtrack when there were multiple candidates at some steps, but the safe sequence it has constructed does not comprise all tasks. A theorem proved in Reference [54] guarantees that, in this case, no safe sequence exists. The most important consequence of this property is that is considerably reduces the computational complexity of the algorithm.

5. If the simulated state has been confirmed to be safe, then the request is granted and the simulated state becomes the new, actual state of the system:

$$\begin{aligned} \mathbf{c_j} &:= \mathbf{c_j}' \\ \mathbf{n_j} &:= \mathbf{n_j}' \\ \mathbf{r} &:= \mathbf{r}' \end{aligned} \qquad (8.18)$$

Otherwise, the simulated state is discarded and the requesting task has to wait, even though the resources it requested are in fact available.

The banker may force tasks to wait as they ask for additional resources, at steps 2 and 5 of the algorithm just presented. This may happen because either there are not enough free resources to satisfy a request, or granting the request would bring the system into an unsafe state.

We should therefore explain when and how the banker evaluates their requests again, and possibly grant some of them. By intuition, it is pointless to do so as the banker grants further resource allocation requests made by other tasks because, in this case, the state of the system becomes even worse from the perspective of the waiting tasks. It can actually be proven that, if the banker postponed a certain re- source allocation request, a re-evaluation of the same request with even less free resources in the system would necessarily lead to the same result.

On the contrary, the banker goes back to examine the waiting tasks' situation when a task releases some resources. When a task τ_j releases some of the resources it owns, it presents to the banker a release vector, $\mathbf{l_j}$. This column vector has one element for each resource class and its i-th element l_{ij} indicates how many resources of the i-th class the j-th task wants to release. As for resource requests, if a task does not want to release any resource of a given class, it leaves the corresponding element of $\mathbf{l_j}$ at zero. Upon receiving a resource release request from τ_j, the banker executes the following algorithm:

1. It checks if the request is legitimate, that is, the task is trying to release only resources it legally acquired in the past. More specifically, it must be $\mathbf{l_j} \leq \mathbf{c_j}$, otherwise the banker gives an error indication to the calling task.
2. It updates its state variables to reflect that the resources indicated in $\mathbf{l_j}$ have been freed, as follows:

$$\begin{aligned} \mathbf{c_j} &:= \mathbf{c_j} - \mathbf{l_j} \\ \mathbf{n_j} &:= \mathbf{n_j} + \mathbf{l_j} \\ \mathbf{r} &:= \mathbf{r} + \mathbf{l_j} \end{aligned} \qquad (8.19)$$

Unsurprisingly, this update is symmetric with respect to the one performed in (8.15) to simulate resource allocation. The only difference is that, in this case, the update is performed directly and immediately on the state variables because no safety checks are necessary.
3. In this step, the banker reconsiders the pending requests from tasks that were blocked in step 2 (due to insufficient resources) or step 5 (deadlock-related safety considerations) of the resource allocation algorithm. In order to do this, the banker follows the same algorithm as for newly arrived requests, with the exception of step 1 because pending requests have already been proven to be legitimate.

The order in which pending requests are considered in step 3 of the resource release algorithm just described affects the time-related properties of the tasks in- volved, a first-in, first-out (FIFO) policy being a common choice because it guaran- tees that no tasks will ever be forced to wait for an unbounded amount of time.

The complexity of the banker's algorithm is $\mathcal{O}(mn^2)$, where m is the number of resource classes, and n is the number of tasks in the system. Its complexity is dominated by the safety assessment (step 4 of the resource request algorithm) because it requires up to n iterations, each composed of a number of vector operations, in contrast with all the other steps that are made up of a fixed sequence of vector operations on vectors of length m, each having a complexity of $\mathcal{O}(m)$. This overhead is incurred on every resource allocation and release because, in the latter case, the banker has to reconsider any waiting requests and this is done by performing again most of the resource request algorithm.

In the safety assessment algorithm, the banker builds the safe sequence one task at a time, without backtracking. In order to do this, it must check at most n candidate tasks at the first step, then up to $n-1$ at the second step, and so on. The maximum length of the sequence is n, and hence, the worst-case number of checks is:

$$n+(n-1)+\ldots+1 = \frac{n(n+1)}{2} \tag{8.20}$$

Each individual check (8.16) is made of a comparison between two scalars, as well as a comparison between two vectors of length m, leading to a complexity of $\mathcal{O}(m)$. Combining this result with the worst-case number of checks given by (8.20) leads to a total complexity of $\mathcal{O}(mn^2)$ for the whole inspection process.

Appending a task to the safe sequence (8.17), an operation performed at most n times, does not worsen the complexity because the complexity of one insertion is $\mathcal{O}(m)$, thus giving a complexity of $\mathcal{O}(mn)$ for n insertions.

So far, the banker's algorithm has been discussed assuming that the number of tasks n is constant. However, as seen in Chapter 5, most operating system support the dynamic creation and termination of tasks. Luckily, the banker's algorithm can be easily extended to deal with this scenario. In particular:

- The creation of a new task τ_{n+1} requires the extension of matrices C, X, and N with an $(n+1)$-th column. The additional column of C must be initialized to zero because a newly created task does not own any resources. The additional column of X must contain the maximum number of resources of each class the new task will possibly need during its lifetime. Finally, the initial value of the additional column of N must be $\mathbf{x_{j+1}}$, according to the definition of N given in (8.9).
- Similarly, when task τ_j terminates the corresponding j-th column of matrices C, X, and N must be suppressed—or, alternatively, it can be reused for a new task at a later time—after confirming that the current value of $\mathbf{c_j}$ is zero. If this is not the case, it means τ_j has concluded its execution without releasing all the resources that have been allocated to it. Therefore, the banker must release those resources forcibly, to enable other tasks to use them again in the future.

8.2.5 DEADLOCK DETECTION AND RECOVERY

Although in the previous sections we discussed only a couple of specific deadlock prevention and one deadlock avoidance algorithms, they all share the same general characteristics. Namely, deadlock prevention algorithms often impose significant restrictions on system designers. Avoidance algorithms require information about tasks behavior that may or may not be readily available and have a significant run-time overhead.

In general-purpose operating systems, a third approach is possible, which acts even later than deadlock avoidance algorithms. With this approach the system may enter a deadlock condition, but a deadlock *detection* algorithm is able to recognize it. The system then reacts with a deadlock *recovery* action. Therefore, these algorithms are collectively known as deadlock detection and recovery algorithms.

If there is only one resource in each resource class, a straightforward way to detect a deadlock condition is to maintain a resource allocation graph as described in Section 8.2.1 and update it whenever a resource is requested, allocated, and released. With the help of an efficient underlying data structure, all these updates are not computationally expensive and can be performed in constant time, because they only involve adding and removing arcs from the graph.

Then, the resource allocation graph is examined at regular intervals, looking for cycles. Due to the theorem discussed in Section 8.2.1, the presence of a cycle is a necessary and sufficient indication that there is an ongoing deadlock in the system. Moreover, it provides information about the tasks and resources involved in the deadlock, which proves valuable in the subsequent deadlock recovery action.

If there are multiple resources belonging to the same resource class, this method cannot be applied because the presence of a cycle in the resource allocation graph is still necessary, but no longer sufficient, to identify a deadlock. However, other algorithms serve the same purpose. In the following, we will describe an algorithm similar to the banker's algorithm and due to Shoshani and Coffman [111, 37]. The algorithm makes use of the following data structures:

- A matrix C that represents the current resource allocation state. As for the banker's algorithm, the j-th column of C, denoted by c_j, indicates how many resources of each class task τ_j currently owns.
- A column vector r, indicating how many resources are currently available in the system.
- For each task τ_j in the system, the column vector:

$$s_j = \begin{pmatrix} s_{1j} \\ \vdots \\ s_{mj} \end{pmatrix}$$

indicates for how many resources of each class τ_j is currently waiting, if any. A zero in element s_{ij} means that τ_j is not waiting for any resource of the i-th class. When τ_j is not waiting for resources at all, all the elements of its s_j are zero.

All these data structures must be updated whenever a task requests, receives, and releases resources because they must give a faithful representation of the resource wait and allocation state at any time. However, all of them can be maintained in constant time. The deadlock detection algorithm is then based on the following steps:

1. Define an auxiliary column vector $\mathbf{w}$, with an element for each resource class, and initialize it with the vector that represents the currently available resources, that is, $\mathbf{w} := \mathbf{r}$.
2. Define an auxiliary row vector of Booleans $\mathbf{f}$, with an element for each task, and initialize its elements to false. This vector plays the same role as the vector with the same name in the banker's algorithm, that is, it records which tasks have successfully concluded as the algorithm progresses.
3. Find a task τ_j that has not been marked in $\mathbf{f}$ yet and whose pending resource request can be satisfied, that is:

$$
\begin{aligned}
f_j &= false && (\tau_j \text{ has not been marked yet}) \\
&\wedge \\
\mathbf{s_j} &\leq \mathbf{w} && (\text{there are enough resources to satisfy its request})
\end{aligned}
\tag{8.21}
$$

Tasks that are not currently waiting for resources have their $\mathbf{s_j}$ at zero. For them, the condition $\mathbf{s_j} \leq \mathbf{w}$ is always satisfied regardless of $\mathbf{w}$ and they can always be picked.
4. Mark the task τ_j picked in step 3 and simulate its successful termination by returning the resources it holds to the pool of available resources:

$$
\begin{aligned}
f_j &:= true && (\text{mark } \tau_j) \\
\mathbf{w} &:= \mathbf{w} + \mathbf{c_j} && (\text{release its resources})
\end{aligned}
\tag{8.22}
$$

Then, go back to step 3 until no more tasks can be picked.

It can be proved that a deadlock exists if, and only if, there are unmarked tasks— in other words, at least one element of $\mathbf{f}$ is still false—at the end of the algorithm. This algorithm bears a strong resemblance to the state safety assessment part of the banker's algorithm presented in Section 8.2.4 and has the same computational complexity. There are also two important differences, though:

- The deadlock detection algorithm works on *actual* resource requests that tasks perform as they proceed with their execution, represented by the $\mathbf{s_j}$ vectors. Instead, the banker's algorithm relies on a priori statements (or forecasts) that tasks make about their *worst-case* resource needs, when they communicate their $\mathbf{x_j}$ vectors.
  As a consequence, the banker's algorithm results are conservative and, also depending on how accurate the $\mathbf{x_j}$ vectors are, may pessimistically mark a state as unsafe even though a deadlock will not necessarily ensue. On the contrary, the deadlock detection algorithm just presented provides exact on-the-spot indications.

- Although the complexity of *one execution* of the banker's algorithm and of the deadlock detection algorithm is the same, there is a crucial difference, very important from the practical standpoint. The banker's algorithm must necessarily be invoked whenever tasks request and release any resource, and hence, how often the banker's algorithm is executed only depends on tasks behavior. On the contrary, the execution frequency of execution of the deadlock detection algorithm can be chosen at will.

  Therefore, it can be adjusted to obtain the best trade-off between conflicting system properties, such as the maximum deadlock detection overhead that may be imposed on the system and the reactivity to deadlocks of the system itself, that is, the maximum time that may elapse between the onset of a deadlock and its detection.

The major problem after detecting a deadlock is to decide how to recover from it. A very simple recovery principle, proposed in Reference [37], consists of aborting the tasks that have been caught in the deadlock. Less aggressive strategies abort tasks one at a time, until the additional resources made available in this way allow the remaining tasks to exit from the deadlock. More sophisticated algorithms, one example of which is also given in References [37, 111], forcibly remove resources from deadlocked tasks based on a cost function.

In both cases, assigning a cost to the abortion of a task or to the forced removal of a resource from a task may be a daunting proposition because it depends on several factors, like the role and importance of a task in the system, its relationship with the others, and its ability to recover from resource preemption.

Other recovery techniques act on resource *requests*. In order to recover from a deadlock, they deny one or more pending resource requests and give an error indication to the corresponding tasks. In this way, they force some of the s_j vectors to become zero and bring the system in a more favorable state with respect to deadlock. The choice of the most appropriate requests to deny is still subject to cost considerations similar to those already discussed.

The same technique can also be used with deadlock avoidance algorithms. In this case, potential deadlocks are still detected on a request-by-request basis, by means of a deadlock safety assessment algorithm like the one presented in Section 8.2.4. However, instead of forcing tasks to wait if granting their request would bring the system into an unsafe state, they immediately receive an error indication, thus shifting the burden of reacting to a potential deadlock from a general algorithm that is part of the operating system into a specifically designed part of the application logic. This is the approach taken by the POSIX standard [68], in which some primitives may optionally fail and return an `EDEADLK` error indication to the caller when they detect that executing the primitive would cause a deadlock.

Another important aspect of deadlock avoidance based on passive wait and automatic deadlock recovery is that these methods have hard to predict, adverse effects on task timeliness and may make them violate their deadlines. Moreover, some deadlock recovery strategies—like the possibility of aborting a task chosen automatically at an arbitrary point if its execution—may be plainly unacceptable in a real-time system. Deadlock avoidance based on error reporting, like in POSIX, is more favorable from this point of view because each task has an explicit, well-defined deadlock recovery strategy in place, and its effects on task timings can be accurately predicted and measured.

8.3 SUMMARY

In this chapter, we described two important issues that may arise when lock-based task interactions are designed or implemented incorrectly, namely, unbounded priority inversion and deadlock. One of these issues, unbounded priority inversion, is especially relevant in real-time systems because it may affect task timings in a subtle and time-dependent way and should be solved by introducing ad-hoc methods, known as protocols, to properly manage the priority of tasks engaged in a critical region.

Deadlocks can also be prevented, avoided, or at least detected by means of a variety of methods, which offer different trade-offs between runtime overhead and added design complexity or programmer's discipline. Operating systems themselves may offer, at least for some kinds of synchronization device, facilities to detect and report incumbent deadlocks.

9 IPC Based on Message Passing

This chapter complements the previous one by presenting *message passing*, an IPC mechanism that does not implicitly assume the availability of shared memory for data transfer. Message passing primitives do provide inter-task data exchange besides synchronization, but the details of how it takes place are left to the underlying implementation and are transparent to their users.

In principle, this approach paves the way to a unified IPC technique that is applicable to a whole range of system architectures, from relatively simple concentrated systems with a single processor, in which the availability of shared memory can be taken for granted, to large distributed systems whose nodes are connected by a real-time network and shared memory is not generally available.

As is done in the rest of the book, after presenting the mechanisms from the theoretical point of view, we describe how programmers can use them in RTEMS. Although this chapter focuses on message passing as an IPC mechanism among tasks residing on a tightly coupled set of cores or processors, the theoretical concepts presented here also apply and serve as a general introduction to Chapters 10 and 11, which discuss network-based communication in detail.

9.1 UNIFIED SYNCHRONIZATION AND DATA EXCHANGE

The inter-task synchronization and communication methods introduced in Chapter 7 are able to convey *synchronization* signals from one task to another by themselves, but must rely on shared memory for *data exchange*. Like the producers and consumers depicted in Figure 7.6, tasks read and write information from and to a shared memory buffer. The data transfer is meaningful because a set of semaphores ensures that tasks perform read and write operations at the appropriate time, but those semaphores play no other role in the data transfer. Even the value of a semaphore, which could be seen as a small item of information shared among tasks, is often not directly accessible. Even when it is, like in POSIX semaphores, tasks can only read the value but cannot change it at will. In other words, these methods depend on *two* distinct mechanisms:

- *Semaphores,* to pass synchronization signals from one task to another.
- *Shared objects,* to support data transfer.

Message passing takes a radically different approach to inter-task synchronization and communication by providing *one* single mechanism that implements both synchronization and data transfer at the same time and with the same set of primitives. In this way, the mechanism works at a higher level of abstraction and becomes easier to use. Even more importantly, it can be adopted with minimal updates on systems where shared memory is not available, for instance, in distributed systems in which communicating tasks may be executed by distinct computers linked by a communication network. As will be discussed in Chapters 10 and 11, virtually all network protocol stacks export their services through a message passing interface.

In its simplest and most abstract form, a message passing mechanism is based on two basic primitives, defined as follows:

- A *send* primitive, which transfers a *message* from one task to another. Besides control information, the message encloses a certain amount of data provided by the sender. In addition to data transfer, a send operation may imply a synchronization action that blocks the sender until the data transfer can take place.
- A *receive* primitive, which allows the caller to retrieve the contents of a message sent to it by another task. Also in this case, the primitive may block the calling task if the message it is looking for is not immediately available, thus synchronizing the receiver with the sender.

Although this definition still lacks many important lower-level details that will be discussed later in this chapter, it is already clear that the most apparent result of message passing primitives is to transfer a certain amount of information from the sending task to the receiving one. At the same time, message passing primitives also take care of synchronization because they may block the caller when needed.

Moreover, mutual exclusion is not a concern with message passing because messages are never shared among tasks. On the contrary, the message passing mechanism works "as if" the message were atomically transferred from the sender to the receiver—even though real implementations do their best to avoid actually copying messages to improve performance. Along with message contents, the logical ownership of a message is also passed from the sender to the receiver when message passing takes place. In this way, even if the sender modifies its local copy of a message after sending it, this does not affect the message it sent. Symmetrically, the receiver may modify a message it received without affecting the sender's local copy at all.

Real-world message passing schemes comprise a number of variations around the basic theme outlined so far. The main design choices left open by our summary description and to be discussed in the next sections are:

1. The *synchronization model*, that is, under which circumstances communicating tasks shall be blocked, and for how long, when they are engaged in message passing.
2. How many *message buffers,* to hold messages already sent but not yet received, the operating system shall provide, if any.
3. How to identify the intended recipient of a message when sending and, symmetrically, how to specify intended senders when receiving a message. In other words, a suitable *naming scheme* for message passing must be defined.
4. Whether the message passing system should be aware of and preserve *message boundaries* during the transfer, or treat messages as mere sequences of bytes to be sent one after another.

9.2 MESSAGE PASSING SYNCHRONIZATION MODELS

Message passing incorporates both data transfer and synchronization within the same communication primitives. The data transfer mechanism is straightforward and is invariably accomplished by moving a message from the source to the destination task. However, synchronization aspects are more complex and subject to variations, giving rise to several different *synchronization models*.

The most basic synchronization constraint, which is always supported, stipulates that the `receive` primitive must be able to block the caller and wait for a message to arrive, if it is not already available. In the *asynchronous* model, this is the only synchronization constraint in effect. This gives rise to two possible scenarios, both depicted in Figure 9.1:

- If the receiving task τ_2 executes `receive` before the sending task τ_1 has sent the message, it blocks and waits for the message to arrive. The message transfer will take place when τ_1 eventually sends the message and τ_2 will also continue at that time.
- If the sending task τ_1 sends the message before the receiving task τ_2 has performed a matching `receive`, the system buffers the message—that is, stores it somewhere temporarily—and lets τ_1 continue immediately. The message becomes available for reception so that a `receive` later performed by τ_2 will be immediately satisfied.

In this figure and in the other figures of this chapter, time flows vertically from top to bottom and thick vertical lines depict tasks evolution, namely, a solid line indicates that a task is ready or running, while a dashed line denotes that a task is blocked. Rectangles represent message passing primitives and thin arrows portray

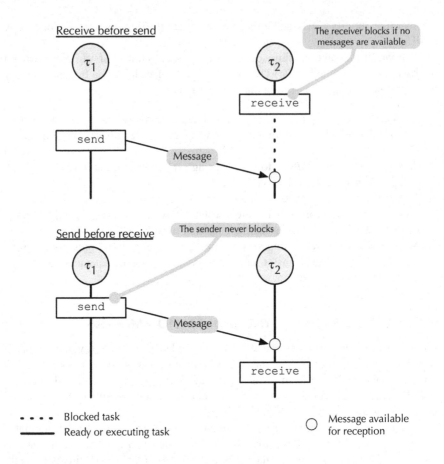

FIGURE 9.1   Asynchronous communication model with unlimited buffer.

message flow. Small circles mark the instant in time at which a message becomes available for reception.

One important aspect to keep in mind about the asynchronous model is that when τ_2 eventually receives a message from τ_1 it does not also acquire any significant timing information about τ_1. This is because, due to the asynchronous nature of the send operation, the sender might already be executing well beyond it and, in principle, could even have sent more than one message that were all duly buffered by the system. In other words, an asynchronous message transfer may convey *outdated* information to the receiver. Symmetrically, τ_1 does not acquire any information about the timings of τ_2 either, for the same reasons.

Although the asynchronous model is very useful from the theoretical point of view, also because it is used as the building block of more complex communication models, it has one important shortcoming that hinders its direct practical applicability. Namely, its implementation may require a large, and potentially *unbounded* number of buffers, which is hardly feasible.

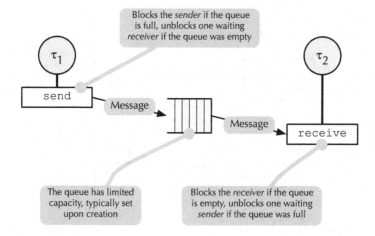

FIGURE 9.2 Asynchronous communication model with limited buffer.

To visualize this scenario it is sufficient to go back to Figure 9.1 and imagine a sender τ_1 that periodically performs a send and a receiver τ_2 that does not perform any receive at all or, more subtly, systematically receives and consumes message at a lower rate. In order to fulfill the asynchronous communication model requirements, the system would need a number of buffers that grows indefinitely with time. As a side effect, the messages received by τ_2 would become more and more outdated, and probably less useful, as time goes by.

For this reason, the variant of the asynchronous model with *limited* buffer illustrated in Figure 9.2 is the one provided by RTEMS and specified by the POSIX standard. The most important differences with respect to the original model are:

- A *queue* is interposed between the sender and the receiver. The system uses the queue to buffer messages that have been sent but not yet received. Its capacity is fixed and is defined when the queue itself is created.
- As in the original model, the receiving task (τ_2 in the figure) still blocks in its receive operation when no messages are available, that is, the queue is empty.
- Moreover, the sending task (τ_1 in the figure) also blocks in the send operation if the queue is completely full. The system will unblock the task when some space in the queue becomes available, so that the message being sent can be enqueued successfully.

Despite the additional synchronization constraint just introduced, if we consider the case in which the queue is neither empty nor full, messages still flow asynchronously from the sender to the receiver. No synchronization actually occurs because neither of them is blocked by the message passing primitives it invokes.

In most real-world implementations—including the one provided by RTEMS and the one specified by the POSIX standard—there are also non-blocking variants of receive and send. A non-blocking receive simply checks whether a message is

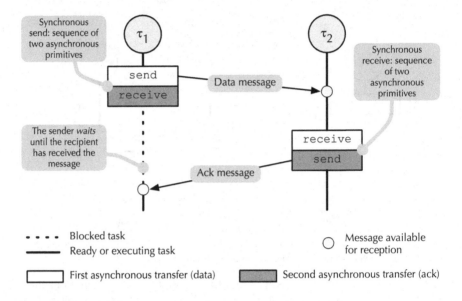

FIGURE 9.3 Synchronous communication model.

available and, in that case, retrieves it, but fails without ever waiting if it is not. Similarly, a non-blocking `send` fails immediately when asked to enqueue a message in a queue that is completely full. Even though these variants may sometimes be useful from the software development point of view, they will not be further discussed here because they simply remove a synchronization constraint.

Another widespread variant, especially in real-time systems, is a *timed* version of `send` and `receive`, in which it is possible to specify the maximum amount of time these primitives are allowed to block the caller before failing. If the operation cannot be completed within the allotted time, they unblock the caller anyway and return an error indication.

When necessary, more elaborate forms of synchronization can be implemented using the asynchronous model as a building block. As shown in Figure 9.3, a sequence of two asynchronous transfers going in opposite directions can be used to implement a *synchronous* communication model, also called *rendezvous*. More specifically:

- In the first part of the synchronous transfer the sender τ_1 sends its data message to the recipient τ_2 by means of an asynchronous transfer. As described previously, τ_1 is never blocked as a result of this transfer, whereas τ_2 may or may not be blocked depending on whether its `receive` preceded the matching `send` or not. The figure illustrates the second case.
- In the second part of the synchronous transfer τ_1 waits for an *acknowledgment* (ACK) message from τ_2. As shown in the figure, τ_2 sends this message immediately after receiving the data message.

As a result, in a synchronous transfer the task τ_1 that sends the data message is blocked until the message recipient τ_2 has received the message. Moreover, as in an asynchronous transfer, the recipient is blocked until a message becomes available.

A peculiarity of the second part of the synchronous transfer, corresponding to the dark gray primitives in Figure 9.3, is that it does not actually move data between tasks—the transferred message may, in fact, be empty—but it is important only for its synchronization semantics. Its purpose is to block τ_1 until τ_2 has successfully received the data message and has sent the acknowledgment.

The differences between the asynchronous and the synchronous models have an important side effect for what concerns message buffering and performance, too. Since in a rendezvous the sender is forced to wait until the receiver has received the message, the system must not necessarily provide any form of temporary buffer to handle the transfer. The message can simply be kept by the sender until the receiver is ready and then copied directly from the sender to the receiver address space, thus saving one memory-to-memory copy operation. For this reason in some programming languages, for instance, Promela [62], the synchronous communication model is expressed as an asynchronous communication model with a zero-size buffer.

A *remote invocation* message transfer, also known as *extended rendezvous*, enforces even stricter synchronization constraints between communicating tasks. More specifically, when task τ_1 sends a remote invocation message to task τ_2, it is blocked until a reply message is sent back from τ_2 to τ_1. Symmetrically—and here is the difference with respect to a synchronous transfer—τ_2 is blocked until that reply has been successfully received by τ_1.

As the name suggests, this synchronization model imitates a function call, or method invocation, using message passing. As in a regular function call, the requesting task τ_1 prepares the arguments of the function it wants task τ_2 to execute. Then, it encapsulates them into a request message and sends the message to τ_2, which will be responsible to execute the requested function. Then, τ_1 performs two `receive` operations in sequence. The first one waits for an ACK message from τ_2 and ensures that τ_2 has successfully received the request, while the second one blocks τ_1 until function results become available.

Meanwhile, τ_2 has received the request and, through a local computation, executes the requested function and eventually generates some results. When results are ready, τ_2 encapsulates them in a reply message, sends it to τ_1, and unblocks it. Afterwards, τ_2 blocks until the reply has been received by τ_1, as confirmed by a second ACK message that τ_1 sends back to τ_2.

Although, as has been shown previously, asynchronous message transfers can be used as a "building block" for constructing the most sophisticated ones—and are therefore very flexible and popular in real-world operating systems—they have some drawbacks as well. As has been remarked, for instance, in Reference [29] the most important concern is probably that asynchronous message transfers give "too much freedom" to the programmer, somewhat like the "goto" statement of unstructured sequential programming.

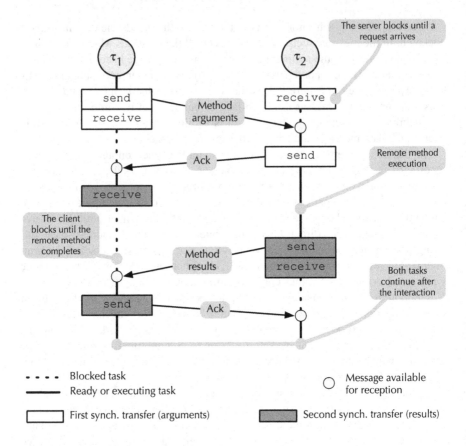

FIGURE 9.4 Remote method invocation.

The resulting programs are therefore more complex to understand and check for correctness, also due to the proliferation of explicit message passing primitives in the code. This observation leads us to highlight the importance of good programming practice to avoid using those message passing primitives directly if at all possible, but encapsulate them within higher-level communication functions, which implement the stricter, but more structured, semantics presented in this section. This trend is further encouraged by the availability of portable, open-source libraries, like Open MPI [48], which go in the same direction.

9.3 DIRECT AND INDIRECT NAMING

The message transfer diagrams drawn in Figures 9.1–9.4 leave two important and related points open, that is, how message senders and recipients identify each other, and whether the `send` and `receive` primitives are symmetric or not. This is the role played by the *naming scheme* in message passing.

Direct naming scheme

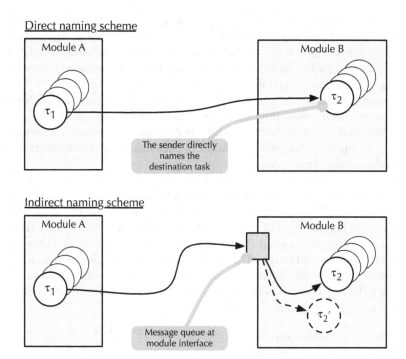

Indirect naming scheme

FIGURE 9.5   Direct versus indirect naming in message passing.

Concerning the first point, the most straightforward and intuitive approach is for the sending task to name the receiver *directly*, for instance, by passing its task identifier to send as an argument. For instance, as shown in the upper part of Figure 9.5, if τ_1 would like to send a message to τ_2, it directly names τ_2 when it invokes the send primitive.

More complex naming schemes are also possible and useful, though. In particular, when the software gets more complex, it may be more convenient to adopt an *indirect* naming scheme. With this naming scheme, the send and receive primitives are not associated based on task identities, but because they both name the same intermediate entity, as shown in the bottom part of Figure 9.5.

In the figure, this intermediate entity is represented by a gray rectangle and called *message queue*, because this is the name used by both the POSIX specification and RTEMS. Other operating systems and programming languages sometimes use other terms as well, for instance, *port* (in the Mach microkernel) or *channel* (in the Promela programming language).

The figure highlights that an indirect naming scheme is advantageous to software modularity and integration. If, for example, a software module *A* wants to send a message to another module *B* using a direct naming scheme, the task τ_1 of module *A* responsible for the communication must know the identity of the intended recipient task τ_2 within module *B*.

In other words, module A becomes dependent not only on the *interface* of module B, which would be perfectly acceptable, but also on its internal design and implementation. For instance, if the internal architecture of B is later changed, so that the intended recipient becomes τ_2' instead of τ_2, module A must be updated accordingly, or communication would no longer be possible.

Moreover, sometimes the unique task identifier used to name a task becomes known only after the task itself has been created, and there is no way to guarantee that the identifier of a certain task will stay the same, for instance, across reboots. In this case, module B becomes responsible of publishing in some way the task identifiers other modules may need.

On the contrary, when using an indirect naming scheme, module A and its task τ_1 must only know the name of the message queue that module B is using for incoming messages. The message queue can then be viewed as located at module B's boundary and becomes part of the external interface of the module itself. Hence, it will likely stay the same even if B's implementation and internal design change with time, unless the external interface of the module is radically redesigned, too.

Another consequence of indirect naming is that the relationship among communicating tasks may become more complex. For both direct and indirect naming, two kinds of relationship are already possible:

1. A *one-to-one* relationship, like the one depicted in Figures 9.1–9.4, in which one task sends messages to another.
2. A *many-to-one* relationship, in which many tasks send messages to a single recipient.

With indirect naming, since multiple tasks can receive messages from the same message queue, a *one-to-many* relationship becomes possible, too. In this case, a task sends a message to a group of many tasks, all receiving from the same message queue, without caring about or specifying which of them will actually get the message. A further generalization, encompassing multiple potential senders, leads to a *many-to-many* relationship.

This is often useful to conveniently handle concurrent processing on the server side in a client–server communication paradigm. For instance, the server may be implemented as a set of *worker* tasks, all able to process one request at a time. This keeps the internal structure of workers simple, because they do not have to take concurrency into account.

All workers wait for requests by performing a `receive` from the same message queue. When a request arrives, one of the workers will get it, process it, and provide an appropriate reply to the client. Meanwhile, the other workers will still be waiting for additional requests and may start working on them concurrently. As long as requests are independent from each other, no worker-to-worker communication is required, thus avoiding the introduction of undue timing relationships among requests.

A further, important aspect of naming schemes is their *symmetry* or *asymmetry*:

- In a symmetric scheme the sender task names either the recipient task or the destination message queue, depending on whether naming is direct or

TABLE 9.1
RTEMS Message Queue Primitives, Classic API

Function	Purpose
rtems_message_queue_create	Create a message queue
rtems_message_queue_ident	Find the identifier of a message queue given its name
rtems_message_queue_delete	Delete a message queue
rtems_message_queue_send	Send a message, enqueuing at the rear
rtems_message_queue_urgent	As above, but enqueuing at the front
rtems_message_queue_broadcast	Send a message to all tasks blocked on a message queue
rtems_message_queue_receive	Receive a message, possibly blocking the caller
rtems_message_queue_flush	Remove all messages from a message queue
rtems_message_queue _get_number_pending	Retrieve the number of messages in a message queue

indirect. Symmetrically, the receiver names either the sending task or the source message queue.

- In an asymmetric scheme the sender still names either the recipient task or the destination message queue. However, the receiver does not name the source of the message in any way. Instead, it accepts messages from any source, and it is often informed by the system of where the received message comes from.

The asymmetric scheme fits the client–server paradigm better because, in this case, the server is usually willing to accept requests from any of its clients and may not even know their identity in advance.

9.4   RTEMS API FOR MESSAGE PASSING

9.4.1   CLASSIC API

Message queue creation and attributes

The RTEMS component in charge of implementing message queues is the *Message Manager*. Table 9.1 summarizes the primitives it provides. The function rtems_message_queue_create creates a new message queue:

```
rtems_status_code rtems_message_queue_create(
    rtems_name name,
    uint32_t count,
    size_t max_message_size,
    rtems_attribute attribute_set,
    rtems_id *id);
```

TABLE 9.2

RTEMS Message Queue Attributes, Classic API

Attribute	Meaning
RTEMS_LOCAL	Local message queue
RTEMS_GLOBAL	Globally accessible message queue
RTEMS_FIFO	Use a first-in, first-out (FIFO) queuing policy when receiving
RTEMS_PRIORITY	Use a priority-based queuing policy when receiving

This function creates a message queue with the given name and, if successful, stores its unique identifier into the location pointed by id. This identifier must be used to refer to the message queue in all ensuing operations on it, except rtems_message_queue_ident. The message queue can store up to count messages, each consisting of up to max_message_size bytes. The argument attribute_set allows the caller to select some aspects of the message queue to be created and consists of the bitwise OR of the constants listed in Table 9.2.

The first two attributes, RTEMS_LOCAL and RTEMS_GLOBAL, are mutually exclusive and determine the visibility and accessibility of the message queue. They have the same meaning as the semaphore attributes with the same name described in Section 7.4.

The attributes RTEMS_FIFO and RTEMS_PRIORITY are also mutually exclusive. They establish the queuing policy of tasks that are waiting to receive a message from an empty queue, and hence, the order in which they will obtain a message and be unblocked when messages arrive.

More specifically, RTEMS_FIFO (the default) indicates that task are enqueued in the same order as they perform the blocking receive operation on the message queue and will be unblocked in first-in, first-out (FIFO) order. Instead, RTEMS_PRIORITY specifies that tasks must be enqueued according to their priority. When there are multiple tasks blocked on the same, empty message queue, a message sent to that message queue will be received by the highest-priority blocked task.

The function rtems_message_queue_create returns the status code RTEMS_SUCCESSFUL if it was able to create the message queue correctly. Otherwise, it returns one of the following status codes to indicate failure:

RTEMS_INVALID_NAME The name of the message queue given in the name argument is invalid.

RTEMS_INVALID_ADDRESS The id argument is a NULL pointer, and hence, it does not point to a valid location where the new message queue identifier could be stored.

RTEMS_INVALID_NUMBER The number of messages specified in the count argument is invalid.

RTEMS_INVALID_SIZE The maximum message size specified in the `max_message_size` argument is invalid.

RTEMS_TOO_MANY The new message queue could not be created because the maximum number of local message queues or global objects in the system (depending on the kind of message queue being created) has been reached. These limits are set by means of the RTEMS configuration system, as described in Section 2.4. For instance, the value of `RTEMS_CONFIGURE_MAXIMUM_MESSAGE_QUEUES` determines the maximum number of local message queues.

RTEMS_UNSATISFIED The message queue could not be created because the system could not allocate the number of message buffers that have been requested.

RTEMS_MP_NOT_CONFIGURED The attribute set given as argument asked for a global messages queue, that is, `RTEMS_GLOBAL` was set, but multi-node support has not been enabled in the system configuration.

Although the message queue creation function just described stores the message queue identifier into the location pointed by the `id` argument, the identifier can also be retrieved at a later time, starting from the message queue name, by means of the function:

```
rtems_status_code rtems_message_queue_ident(
    rtems_name name,
    uint32_t node,
    rtems_id *id);
```

This function looks for a message queues whose name is given by the `name` argument on node `node` and stores its identifier into the location pointed by `id`. If `node` is the special value `RTEMS_SEARCH_ALL_NODES`, the function searches all nodes in the system, starting from the local node, that is, the node where the calling thread resides. If multiple message queues have the same name, the function is guaranteed to return the identifier of a message queue with a matching name, but exactly which one is left unspecified.

Upon failure, `rtems_message_queue_ident` returns one of the following status codes:

RTEMS_INVALID_NAME The function did not find any message queue with the requested `name` within the scope of its search.

RTEMS_INVALID_NODE The `node` identifier is invalid.

RTEMS_INVALID_ADDRESS The `id` argument is a `NULL` pointer, and hence, it does not point to a valid location where the message queue identifier could be stored.

The function:

```
rtems_status_code rtems_message_queue_delete(
    rtems_id id);
```

deletes the message queue whose unique identifier is `id`. The message queue may have been created by another task, but must reside on the same node as the calling thread. The deletion of a remote message queue, that is, a queue that resides on a remote node, is not supported. After a successful call to this function, the identifier is no longer valid and shall no longer be used.

When a message queue is deleted, any tasks that are blocked on it because of a pending receive operation are unblocked and receive the RTEMS_OBJECT_WAS_DELETED status code. If the queue is not empty, all the messages it contains are deleted. In both cases, the system reclaims all the memory previously allocated to the message queue and its message buffers.

Besides RTEMS_SUCCESSFUL, which indicates that the message queue has been successfully deleted, rtems_message_queue_delete may return the following status codes:

RTEMS_INVALID_ID The message queue identifier is invalid.

RTEMS_ILLEGAL_ON_REMOTE_OBJECT The message queue could not be deleted because it does not reside on the same node as the calling task.

Message queue operations

The concrete counterparts of the abstract send and receive operations described in Section 9.2 are rtems_message_queue_send and _receive, plus a couple of useful RTEMS-specific variants also listed in Table 9.1. The function:

```
rtems_status_code rtems_message_queue_send(
    rtems_id id,
    const void *buffer,
    size_t size);
```

sends the message referenced by buffer and consisting of size bytes, to the message queue identified by id.

If the queue is currently empty and there is at least one task blocked on a receive operation this function picks one of them, according to the enqueuing policy established when the message queue was created, copies the message into the blocked task's buffer, and unblocks it.

Otherwise, it copies the message into a message buffer and places it at the rear of the message queue. If the message queue is full, that is, it already contains the maximum number of messages declared upon creation, rtems_message_queue_send fails and immediately returns to the caller an appropriate status code. The function may return RTEMS_SUCCESSFUL or one of the following status codes:

RTEMS_INVALID_ID The message queue identifier is invalid.

RTEMS_INVALID_ADDRESS The buffer argument is NULL, and hence, it cannot refer to a valid message in memory.

RTEMS_INVALID_SIZE The size of the message is invalid, for instance, because it exceeds the maximum message size declared upon message queue creation.

RTEMS_TOO_MANY The message queue is full, that is, it already contains the maximum number of messages specified upon creation.

RTEMS_UNSATISFIED The system ran out of message buffers.

The function:

```
rtems_status_code rtems_message_queue_urgent(
    rtems_id id,
    const void *buffer,
    size_t size);
```

has the same signature as rtems_message_queue_send, the same possible status codes, and similar semantics, the only difference being that it places the message at the *front* of the message queue, rather than the *rear*.

Although the RTEMS Classic API—unlike the POSIX API to be discussed next—does not provide the ability to assign a priority to individual messages, a mindful use of rtems_message_queue_urgent (to send high-priority messages) and rtems_message_queue_send (to send low-priority messages) on the same message queue is a straightforward alternative way of implementing a two-priority message passing scheme.

Another variant of the send operation broadcasts copies of the same message to *all tasks* waiting on a message queue:

```
rtems_status_code rtems_message_queue_broadcast(
    rtems_id id,
    const void *buffer,
    size_t size,
    uint32_t *count);
```

As for the other functions described so far, id identifies the target message queue, while buffer and size locate the buffer that contains the message to be sent and its length in bytes. When successful, the function returns RTEMS_SUCCESSFUL and stores into the variable pointed by count the number of tasks it unblocked.

If no tasks are blocked on the message queue at the moment, the function stores zero in the location pointed by count and does *not* store the message into the message queue for later use. Upon failure, rtems_message_queue_broadcast returns one of the following status codes:

RTEMS_INVALID_ID The message queue identifier is invalid.
RTEMS_INVALID_ADDRESS Either buffer or count (or both) are NULL pointers, and hence, cannot refer to a valid memory location.
RTEMS_INVALID_SIZE The size of the message is invalid, for instance, because it exceeds the maximum message size declared upon message queue creation.

It is worth noting that rtems_message_queue_broadcast never returns RTEMS_TOO_MANY or RTEMS_UNSATISFIED because it never stores the message into the message queue.

The function:

```
rtems_status_code rtems_message_queue_receive(
    rtems_id id,
```

TABLE 9.3

RTEMS Message Queue Receive Blocking Rules, Classic API

Argument		Meaning
option_set	timeout	
RTEMS_NO_WAIT	—	Never block the caller
RTEMS_WAIT	RTEMS_NO_TIMEOUT	No upper limit on blocking time
RTEMS_WAIT	k	Block the caller for up to k ticks

```
void *buffer,
size_t *size,
rtems_option option_set,
rtems_interval timeout);
```

receives a message from the message queue specified by id. Upon successful completion, it stores the message into the buffer pointed by buffer and writes its size, in bytes, into the location pointed by size. The buffer must be big enough to contain a maximum-length message, that is, the length specified in the max_message_size argument when the message queue was created.

The behavior of the function when the message queue is empty is controlled by the arguments option_set and timeout, as summarized in Table 9.3. In more details:

- If option_set is set to RTEMS_NO_WAIT the function fails immediately, without blocking the caller, and returns the status code RTEMS_UNSATISFIED.
- If option_set is set to RTEMS_WAIT and timeout is set to the special value RTEMS_NO_TIMEOUT the function blocks the caller potentially forever, until it either successfully receives a message or an error occurs.
- If option_set is set to RTEMS_WAIT and timeout is set to a timeout k expressed in ticks, the function blocks the caller only for up to k ticks. If no message has been successfully received before the timeout expired, the function returns the status code RTEMS_TIMEOUT.

When the calling task blocks on an empty message queue, the enqueuing order is determined by the message queue attributes specified upon creation, namely, RTEMS_FIFO (first-in, first-out order) or RTEMS_PRIORITY (priority order).

Besides RTEMS_SUCCESSFUL, which indicates successful completion, the function rtems_message_queue_receive may return one of the following status codes:

RTEMS_INVALID_ID The message queue identifier is invalid.

RTEMS_INVALID_ADDRESS Either buffer or size (or both) are NULL pointers, and hence, cannot refer to a valid memory location.

RTEMS_UNSATISFIED The message queue was empty and option_set specifies RTEMS_NO_WAIT.

RTEMS_TIMEOUT The option_set specifies RTEMS_WAIT and no messages could be received before the timeout specified in timeout expired.

The following functions removes (flushes) all messages from the message queue identified by id:

```
rtems_status_code rtems_message_queue_flush(
    rtems_id id,
    uint32_t *count);
```

Upon successful completion, the function stores into the location pointed by count the number of messages it removed and returns RTEMS_SUCCESS. Flushing an empty queue is not an error and makes the function store zero into the location pointed by count. Upon failure, rtems_message_queue_flush returns one of the following status codes:

RTEMS_INVALID_ID The message queue identifier is invalid.

RTEMS_INVALID_ADDRESS The argument count is a NULL pointer.

The last function in this group queries a message queue and returns the number of *pending* messages, that is, the number of messages currently waiting to be received. In particular:

```
rtems_status_code rtems_message_queue_get_number_pending(
    rtems_id id,
    uint32_t *count);
```

stores into the location pointed by count the number of pending messages in the message queue identified by id. If the message queue is empty, it stores zero into the location pointed by count. Besides RTEMS_SUCCESSFUL, which denotes a successful completion, the function may return one of the following status codes to indicate failure:

RTEMS_INVALID_ID The message queue identifier id is invalid.

RTEMS_INVALID_ADDRESS The argument count is a NULL pointers.

9.4.2 POSIX API

Unlike for semaphores, the POSIX standard specifies only one kind of message queue, which works according to the asynchronous communication scheme with limited buffer discussed in Section 9.2. Message queues are *named* global objects,

TABLE 9.4

RTEMS Message Queue Primitives, POSIX API

Function	Purpose
mq_open	Create/open a message queue
mq_close	Close a message queue
mq_unlink	Mark a message queue for destruction
mq_getattr	Retrieve message queue attributes
mq_setattr	Modify message queue attributes
mq_send	Send a message
mq_timedsend	Timed variant of mq_send
mq_receive	Receive a message
mq_timedreceive	Timed variant of mq_receive
mq_notify	Request an asynchronous notification of message arrival

are represented by the opaque data type mqd_t, and are handled like the named semaphores presented in Section 7.4.2. However, there is an important practical difference with respect to named semaphores:

- A named semaphore is represented by the data type sem_t. The function sem_open returns a pointer to a semaphore object, that is, a sem_t *, while the function sem_init initializes an object of type sem_t provided by the user. After initialization, a *pointer* to the semaphore object must be used to refer to the semaphore.
- A (named) message queue is represented by the data type mqd_t. The function mqd_open returns an object of that type and the object itself, *not a pointer to it*, must be used to refer to the message queue in all the other message queue-related functions.

Message queue creation and destruction

The POSIX message functions available in RTEMS are summarized in Table 9.4. The function:

mqd_t mq_open(**const char** * name, **int** flags, ...);

can be used to get access to an existing message queue or to create a new one. In both cases, the message queue is identified by a unique character string referenced by the name argument and, upon successful completion, the function returns a value of type mqd_t that represents the message queue. The flags argument is the bitwise OR of the flags listed in Table 9.5. They affect some aspects of the function's behavior and of the subsequent send and receive operations on the message queue.

TABLE 9.5
RTEMS Message Queue Flags, POSIX API

Flag	Meaning
O_CREAT	Create the message queue if it does not exist
O_EXCL	Fail is the message queue already exists
O_RDONLY	Open the queue only for receiving messages
O_WRONLY	Open the queue only for sending messages
O_RDWR	Open the queue both for receiving and sending
O_NONBLOCK	Make send and receive operations non-blocking

The first two flags, O_CREAT and O_EXCL, determine whether mq_open should create the message queue if it does not exist already, and if it should fail when attempting to create an already existing message queue:

O_CREAT This flag, when set, allows mq_open to create the message queue if it does not exist already. New message queues are initially empty, that is, they contain no messages. When O_CREAT is set, mq_open takes two additional arguments in place of the ellipses . . . shown in the prototype:
- mode_t mode determines the access permissions of the newly created message queue, and
- struct mq_attr *attr points to a data structure that contains the message queue *attributes*, to be discussed later.

O_EXCL This flag shall only be set together with O_CREAT. When set, it causes mq_open to fail when trying to create a message queue that already exists. The existence check and the creation are carried out in an atomic step with respect to other tasks doing the same.

The second set of flags, comprising O_RDONLY, O_WRONLY, and O_RDWR, determines which kind of operations the calling thread would like to perform on the message queue. More specifically:

O_RDONLY Open the message queue only for receiving messages (that is, "reading" from the message queue).

O_WRONLY Open the message queue only for sending messages (that is, "writing" into the message queue).

O_RDWR Open the message queue for both receiving and sending messages.

The last flag, O_NONBLOCK, controls whether a receive operation from an empty queue, and a send operation to a full queue, should block the caller or fail immediately. When set, it makes these operations non-blocking.

A struct mq_attr contains the fields listed in Table 9.6. The fields used by mq_open while creating a message queue are:

TABLE 9.6

RTEMS Message Queue Attributes, POSIX API

Field	Meaning
long mq_flags	Message queue flags (the O_NONBLOCK flag is read-write)
long mq_maxmsg	Maximum number of messages the queue can hold (read-only after creation)
long mq_msgsize	Maximum size of individual messages (read-only after creation)
long mq_curmsgs	Number of messages in the queue (read-only)

mq_maxmsg The maximum number of messages the message queue can store.
mq_msgsize The maximum size of each message, in bytes.

A call to mq_open may fail for a variety of reasons. Upon failure, the function returns the special value (mqd_t) -1 instead of a valid message queue descriptor and sets errno to a status code that provides more information about the error it encountered. The most common status codes are:

ENOENT The message queue does not exist, but flags did not include O_CREAT.
EEXIST The message queue already exists, but both O_CREAT and O_EXCL were set in flags.
EINVAL The given name is not a valid name for a message queue, or the flags include O_CREAT and the value of mq_maxmsg or mq_msgsize specified in the structure pointed by attr was zero or negative.
EACCES The message queue could not be accessed or created due to permission issues.

Other status codes, like EMFILE, ENFILE, ENOMEM, and ENOSPC indicate that the system lacks various kinds of resources it needs to create the message queue.

Tasks should close a message queue when they no longer intend to use it, by means of the function:

int mq_close(mqd_t mqd);

The only parameter of this function is mqd, the descriptor of the message queue to be closed. No further operations on mqd shall be attempted after a successful mq_close. The function returns zero when it succeeds and -1 when it fails. In this case, it also sets errno to a status code. The only possible status code is:

EBADF The message queue descriptor mqd is invalid.

By itself, mq_close only breaks the association between the message queue and its descriptor mqd, without destroying the message queue itself, unless it has been previously marked for destruction by means of the function:

int mq_unlink(**const char** *name);

This function takes the *name* of a message queue rather than a descriptor, and hence, it can be used even if the caller does not hold any valid message queue descriptor. For instance, a task can mark a message queue for destruction without opening it first. The execution of mq_unlink has two possible outcomes, depending on how many active references to the message queue exist. More specifically:

- If no tasks have the message queue open at the moment, mq_unlink destroys the message queue immediately.
- Otherwise, the message queue is marked for destruction. Further attempts to open the message queue fail. Any attempt to create a message queue with the same name succeeds, but creates a new message queue distinct from the previous one. The message queue will be destroyed as soon as the number of active references to it drops to zero.

The function mq_unlink returns zero if it succeeds. Otherwise, it returns -1 and sets errno to a status code. Possible reasons for failure include:

ENOENT No message queues named name exist in the system.
EACCES The caller lacks the permission to destroy the message queue.

Message queue attributes

As reported in the description of mq_open, some of the message queue attributes listed in Table 9.2 are set when the message queue is created. They can all be retrieved at a later time by means of the function:

int mq_getattr(mqd_t mqd, **struct** mq_attr *attr);

The function mq_getattr, given a message queue descriptor mqd, retrieves the message queue attributes and stores them into the data structure pointed by attr. It returns zero upon successful completion. Upon failure, the function returns -1 and sets errno to the status code:

EBADF The message queue descriptor mqd is invalid.

Besides mq_maxmsg and mq_msgsize, which have already been discussed previously and cannot be changed after the message queue has been created, the struct mq_attr also contains the following fields:

mq_curmsgs is the number of message currently in the message queue. Unlike mq_maxmsg and mq_msgsize, this attribute changes as messages are sent and received through the queue.
mq_flags is the set of flags given to mq_open when the message queue descriptor was created, and possibly modified afterwards by means of the mq_setattr function.

Unlike all the other attributes, mq_flags is associated with the message queue *descriptor* rather than the message queue itself, and affects only the functions that operate on the message queue through that specific descriptor.

The function mq_setattr retrieves the current attributes of the message queue identified by the descriptor mqd and stores them into the data structure pointed by old_attr (like mq_getattr does). In addition, it modifies the attributes according to what is specified in the data structure pointed by new_attr:

```
int mq_setattr(mqd_t mqd,
    const struct mq_attr *restrict new_attr,
    struct mq_attr *restrict old_attr);
```

In the current edition of the standard, only the O_NONBLOCK flag in mq_flags, plus other implementation-defined flags, can be modified, whereas the other fields of *new_attr are ignored. Being mq_flags an attribute associated with the message queue *descriptor*, it takes effect on subsequent message queue operations invoked on the descriptor mqd, without affecting any other descriptors associated with the same underlying message queue.

Like most other functions in this group, mq_setattr returns zero to indicate successful completion. When it fails, it leaves the message queue descriptor flags unchanged, returns -1, and sets errno to the following status code:

EBADF The message queue descriptor mqd is invalid.

Send and receive operations

The function:

```
int mq_send(mqd_t mqd, const char *buffer, size_t size,
    unsigned priority);
```

sends a message to the message queue represented by the message queue descriptor mqd. The message to be sent is copied from a user-provided buffer consisting of size bytes starting at memory address buffer. Unlike the RTEMS classic API, which lets callers decide whether to enqueue the message at the rear or the front of the message queue, the POSIX API lets callers assign a priority to each message they send.

The priority is given by the unsigned integer argument priority, whose value must be greater than or equal to zero and less than MQ_PRIO_MAX. Higher values correspond to higher enqueuing priorities, that is, higher-priority messages are enqueued in front of lower-priority messages. The minimum required value of MQ_PRIO_MAX is 32, which is also the minimum required number of execution scheduling priorities.

The function may exhibit three different behaviors depending on the current state of the queue:

1. If the message queue is empty and there is at least one thread waiting to receive a message, the message is transferred directly to one of the waiting threads and

the queue remains empty. Provided the *Priority Scheduling* POSIX option is supported (as is the case in virtually all POSIX-based real-time operating systems), the receiving thread is selected based on its scheduling priority. More specifically, the highest-priority waiting thread is selected and, in case of a tie, the thread that has been waiting the longest. The `priority` argument does not play any role in this scenario.

2. If there are no threads waiting on the message queue and it is not full, the message is enqueued based on its `priority` and `mq_send` returns without blocking the caller, thus realizing a completely asynchronous message passing operation.

3. If the message queue is full, that is, it already contains a number of messages equal to the value of `mq_maxmsg` specified upon message queue creation, the function's behavior depends on the setting of the `O_NONBLOCK` flag of the message queue descriptor `mqd`. As discussed previously, this flag is set when the message queue is opened and can later be modified by means of `mq_setattr`.

    If `O_NONBLOCK` is not set (the default), `mq_send` blocks the caller until it becomes possible to enqueue the message. Namely, when space becomes available in the message queue, one of the threads waiting to send a message is selected, according to its scheduling priority. This thread is allowed to enqueue its message in the message queue and then continue past `mq_send`.

    Instead, if `O_NONBLOCK` is set, `mq_send` fails and immediately returns to the caller without sending the message, after setting `errno` to `EAGAIN`.

Upon successful completion, `mq_send` returns zero. Otherwise, it returns -1 and sets `errno` to one of the following status codes:

EBADF The message queue descriptor `mqd` is invalid, or send operations are not allowed on it because neither `O_WRONLY` nor `O_RDWR` was specified when the message queue was opened.

EINVAL The `priority` argument is outside the valid range. It must be greater than or equal to zero and less than `MQ_PRIO_MAX`.

EMSGSIZE The message is too long to be sent, that is, the `size` argument exceeds the `mq_msgsize` attribute specified upon message queue creation.

EAGAIN The message queue was full and the `O_NONBLOCK` flag is set in the message queue descriptor, calling for a non-blocking send.

A timed variant of the send operation also exists:

```
int mq_timedsend(mqd_t mqd,
    const char *buffer, size_t size, unsigned priority,
    const struct timespec *abstime);
```

Compared to `mq_send`, the timed variant has one additional argument, `abstime`. It indicates the absolute instant in time when any wait initiated by the function when the target message queue is full must end and the function must return to the caller, even though the message could not yet be sent. As for semaphores (see Section 7.4.2), the reference clock of `abstime` is `CLOCK_REALTIME`.

In all other cases, mq_timedsend behaves identically to mq_send. The two functions are equivalent and the abstime argument is ignored when:

- The message queue is empty and there are threads waiting to receive a message from it.
- There are no threads waiting on the message queue and it is not full.
- The O_NONBLOCK flag is set in the message queue descriptor.

The function mq_timedsend shall fail and set errno to the following status codes, besides the ones already described for mq_send:

EINVAL The function had to block the calling thread, but the abstime argument was invalid.

ETIMEDOUT The timeout specified by abstime expired before the send operation could be concluded.

A thread can receive a message from a message queue and possibly block until one becomes available, by calling the function:

```
ssize_t mq_receive(mqd_t mqd,
    char *buffer, size_t size, unsigned *priority);
```

The function operates on the message queue referenced by mqd. Upon successful completion it stores the received message in the user-provided buffer pointed by buffer and its priority in the unsigned variable pointed by priority. Moreover, it returns the length of the received message in bytes that, by definition, is a non-negative integer.

Upon failure, it returns -1 and sets errno to one of the status codes to be discussed later. The argument size is the actual length of the user-provided message buffer and is used to avoid overflowing the buffer when storing the received message into it.

As for mq_send, the behavior of mq_receive depends on the state of the underlying message queue. Two scenarios are possible:

1. If the message queue is not empty, mq_receive extracts the highest-priority message from it and immediately returns to the caller, without blocking. If more than one message with the highest priority is available, the function extracts the oldest one, that is, the one that has been in the queue for longest. The extraction of a message from a full message queue may also unblock one of the threads waiting to send a message to it, if any, as mentioned in the description of mq_send.

2. If the message queue is empty, mq_receive may or may not block the caller depending on the setting of the O_NONBLOCK flag associated with the message queue descriptor mqd. If O_NONBLOCK is not set (the default), mq_receive blocks the caller until a message becomes available for reception.

   Whenever a message is sent to the message queue, one of the blocked threads is selected to receive the message and continue, in scheduling priority order. In case of a tie, the system selects the highest-priority thread that has been blocked for the longest time. Instead, if O_NONBLOCK is set, mq_receive fails without receiving any message and returns to the caller after setting errno to EAGAIN.

The function mq_receive shall fail for the following reasons:

EBADF The message queue descriptor mqd is invalid, or receive operations are not allowed on it because neither O_RDONLY nor O_RDWR was specified when the message queue was opened.

EMSGSIZE The user-provided buffer is not big enough to hold a maximum-size message coming from the message queue, that is, the size argument is less than the mq_msgsize attribute specified upon message queue creation.

EAGAIN The message queue was empty and the O_NONBLOCK flag is set in the message queue descriptor, calling for a non-blocking receive.

In addition, some implementations may perform additional consistency checks on the received message and also report:

EBADMSG The message was corrupted.

Threads willing to block while waiting for a message to become available, but not indefinitely, may make use of the timed variant of mq_receive:

```
ssize_t mq_timedreceive(mqd_t mqd,
    char * restrict buffer, size_t size,
    unsigned * restrict priority,
    const struct timespec * restrict abstime);
```

The function mq_timedreceive behaves identically to mq_receive, except when the underlying message queue is empty and it must block the caller. In this case, the additional argument abstime comes into play.

As for mq_timedsend, this argument indicates the absolute instant in time when any wait initiated by mq_timedreceive because the message queue was empty must end and the function must return to the caller even though no message could be received. Also in this case, the reference clock of abstime is CLOCK_REALTIME. Due to this additional semantics, mq_timedreceive shall also fail and set errno to the following status codes, besides the ones already described for mq_receive:

EINVAL The function had to block the calling thread, but the abstime argument was invalid.

ETIMEDOUT The timeout specified by abstime expired before the receive operation could be concluded.

Asynchronous notifications

Message queues support asynchronous notifications, configured by means of a struct sigevent data structure as described in Section 6.5. Asynchronous notifications are managed with the function:

```
int mq_notify(mqd_t mqd,
    const struct sigevent *notification);
```

Depending on the value of the notification pointer, this function activates or deactivates asynchronous notifications for message queue mqd. More specifically:

- If the notification argument is NULL, the function deactivates asynchronous notifications for the given message queue.
- If the notification argument is not NULL, the function activates asynchronous notifications for the given message queue, according to the method specified in the struct sigevent structure it points to.

When asynchronous notifications are active, a notification is sent only when the queue transitions from the empty to the non-empty state, that is, when the queue changes state. In particular, no notifications are sent if the message queue is already non-empty when mq_notify is called to activate them. Moreover, if a message is sent to an empty queue while asynchronous notifications are active, but there is also a pending receive operation on the queue, the receive operation takes precedence. That is, the message satisfies the receive operation and no notifications are sent. Another aspect worth noting is that, in any case, asynchronous notifications are automatically deactivated when a notification is sent, and must be explicitly activated again as needed.

The function mq_notify returns zero upon successful completion. Otherwise, it returns -1 and sets errno to one of the following error codes:

EBADF The message queue descriptor mqd is invalid.

EBUSY The notification argument is not NULL, but notifications are already active for the message queue.

Moreover, the function may optionally perform additional checks on the notification status and also return the following status code when appropriate:

EINVAL The notification argument is NULL, but notifications are not active for the message queue.

9.5   SUMMARY

This chapter described message passing, an inter-task communication mechanism that is equally suited to systems with and without shared memory because its primitives encompass both task synchronization and data transfer.

The first part of the chapter, that is, Sections 9.1–9.3 introduced key concepts like the most common synchronization models for message passing primitives and the problem of identifying message passing endpoints by means of a suitable naming scheme.

Instead, the second part of the chapter, Section 9.4 was devoted to more practical considerations and provided a thorough description of the RTEMS Classic and POSIX message passing APIs, highlighting their analogies and differences.

Part IV

Network Communication

10 Network Communication in RTEMS

CONTENTS

In this chapter, we outline how RTEMS implements its TCP/IP network communication facilities, by means of suitable protocol stacks. More specifically, we will describe the internal structure of the RTEMS networking code and highlight the most important aspects of operating system–protocol stack integration, such as synchronization and the device driver interface. A detailed discussion of the network communication application programming interface (API) will be the topic of the next chapter.

10.1 INTERNAL STRUCTURE OF THE RTEMS NETWORKING CODE

At the time of this writing, RTEMS support two distinct TCP/IP protocol stacks:

- The standard protocol stack is part of the core RTEMS distribution and derives from a snapshot of the FreeBSD protocol stack forked in 1998. The FreeBSD code has been frozen and cannot be updated easily from upstream.
- A port of a newer version of the FreeBSD, which replaces the standard one, is available in the *RTEMS LibBSD* project []. Unlike the previous one, this new port is meant to be updated by pulling from the FreeBSD kernel sources directly.

Both ports are compatible with the POSIX sockets API, and hence, they can be used interchangeably at the application level, provided the application restricts itself to use that interface. However, they differ significantly for what concerns features.

In this chapter, we will recall the internal organization of the standard RTEMS protocol stack, mainly focusing on its layered structure and the synchronization mechanisms used among layers. Both derive directly from the original TCP/IP protocol stack of the "Berkeley UNIX" operating system [86]. This has several purposes:

1. Describe one of the very first implementations of the socket concept, which was used as a starting point by many other protocol stacks nowadays found in popular open source real-time operating systems [91, 103].
2. Discuss how a complex piece of software, most importantly a real-world protocol stack in widespread use, has been designed and implemented.
3. Highlight how the original, historical inter-level synchronization mechanisms have been adapted to a modern, real-time operating system with minimal changes.

10.2 PROTOCOL STACK ORGANIZATION

Figure 10.1 summarizes the internal structure of the protocol stack. There are four main software layers, to be discussed in top-down order:

1. At the top, a set of small functions sometimes called *stubs* enables user-level code to invoke the system calls corresponding to the socket functions. These functions are part of the C runtime library and their execution implies a trap into the operating system kernel.

   Besides causing the processor to transition from unprivileged to privileged mode, the execution of the system call trap also puts in effect a peculiar form of mutual exclusion typical of traditional single-processor BSD kernels. In these kernels, as soon as a process enters the kernel by means of a system call, it cannot be preempted by other processes until it exits from the kernel or blocks within it. Similarly, the operating system scheduler also enforces mutual exclusion when a process resumes execution in the kernel after blocking.

   For example, the system call recv is implemented by means of a user-level library function with the same name, which adapts the arguments to the kernel calling conventions if needed, and executes a trap into the operating system. The system call trap handler then dispatches the processor to the in-kernel implementation of the system call, embodied by the soreceive protocol stack function in this case.
2. The second layer from the top is executed with interrupts fully enabled for the most part. It is responsible for the implementation of the socket-level portion of the system calls. For instance, the soreceive function is responsible for moving data from protocol stack buffers into user buffers.

   To enforce mutual exclusion and ensure data consistency, the functions in this layer temporarily raise the interrupt priority level (IPL) of the processor to mask off some interrupts when they need to access data shared with lower layers.

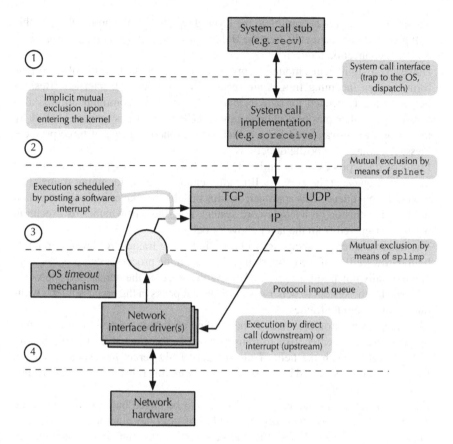

FIGURE 10.1 Simplified structure of the BSD 4.4 TCP/IP protocol stack.

More specifically, when they need to execute in mutual exclusion with respect to the functions in the layer immediately below, they raise the IPL to `splnet`. When an even wider-range mutual exclusion is needed, for example, including also the bottom software level, the IPL will be raised to an even more restrictive value, `splimp`. The traditional names of these IPLs derive from the names of the functions used in the kernel to actually change the IPL.

3. The third layer is further subdivided into multiple sub-layers. The sub-layers follow the typical layered structure of network protocols and each protocol module is generally responsible of implementing a specific protocol. As an example, Figure 10.1 shows the most well-known protocol modules typically implemented by a TCP/IP protocol stack: TCP, UDP, and IP.

   Functions in this layer can be executed, as described previously, as a consequence of a direct call from the upper layer. For instance, a connection request made by an application that calls the `connect` function on a TCP socket will ultimately result in a call to the function within the TCP layer that is responsible of initiating a connection. In turn, according to the layered structure of the protocol stack, this will give rise to related calls into the IP layer, to transmit the messages needed to initiate a connection request.

However, a significant amount of protocol stack processing must also be performed when incoming frames are received from a network interface. Besides being obviously needed for data exchange, bidirectional communication is also required for other purposes. For instance, TCP connections are established by means of a three-way handshake, which implies the exchange of three protocol messages going in opposite directions.

Protocol stack processing is initiated by the bottom layer by posting a software interrupt request, at the `splnet` IPL, after enqueueing the incoming frames in a protocol input queue. This approach ensures that the third layer is always executed in mutual exclusion on a single-core system, regardless of whether its execution has been triggered from the layer above or below it.

The use of a queue is necessary to buffer incoming frames, retrieved at relative high hardware IPL of the network interface, until the processor can process them at the usually much lower `splnet` IPL. With respect to the unbuffered approach, this enables the system to better sustain transient peaks in the processor load without losing incoming frames.

Moreover, many protocols need to be notified of elapsing time to perform time-based internal activities, like re-transmissions in the case of TCP. This feature is implemented with the help of the standard BSD kernel *timeout* mechanism. Elapsing time is measured by means of a hardware timer that generates interrupts, or *ticks*, periodically.

Kernel functions can arrange for a function to be called back by the operating system timer handler after a certain number of ticks has elapsed. The call takes place at the `splsoftclock` IPL, and hence, the called function must explicitly change its IPL to `splnet` before proceeding to execute any other functions in the third layer.

In order to reduce the number of callbacks submitted to the kernel timeout mechanism and, above all, keep their number constant regardless of the number of active protocol modules, the protocol stack registers by itself two callbacks, called *fast* and *slow* timeout, which are invoked every 200 ms and 500 ms, respectively.

Within the protocol stack, each protocol module can define its own slow and fast timeout functions. Upon each activation, the main callbacks will invoke these functions for all active protocol modules. The two periods have been chosen to suit the needs of most network protocols. Although nothing prevents a protocol module from having its own additional callbacks, none of the RTEMS protocol modules makes use of this feature.

4. The bottom layer contains the network interface drivers. Their execution is triggered either by a synchronous, direct call from the layer above, or by a hardware interrupt. As described previously, these interrupts all have an IPL lower than `splimp`, so the upper layers can ensure mutual exclusion with respect to all network interface drivers by temporarily raising the IPL to this value. The `splimp`

IPL is guaranteed by design to be higher (more restrictive) than `splnet`.
Figure 10.1 also includes the operating system timeout mechanism in this layer, although its callbacks are initially executed at a different IPL (`splsoftclock` instead of `splimp`).

An important aspect of kernel mutual exclusion has to do with blocking and synchronization of different parts of the protocol stack. Even though the mutual exclusion enforced upon kernel entry, plus the appropriate manipulation of the processor IPL, are adequate for mutual exclusion as long as the processes involved are continuously ready for execution, a different mechanism is needed when a process needs to *block* within the kernel.

This is very common occurrence in the protocol stack, since most socket primitives may wait until an event of interest takes place. For instance, the application-level function `connect`, which initiates a connection request, may block the caller and wait until connection has been established. Similarly, the functions that receive data from a socket may wait until some data arrive.

In this case, the synchronization mechanism should allow other processes to execute and, possibly, let one of them enter the kernel as well. When the blocked process is eventually unblocked, the same mechanism shall also guarantee that it is not executed until no other process is also executing within the kernel.

To implement this aspect of synchronization, the BSD protocol stack makes widespread use of *wait channels*, the traditional synchronization mechanism of the BSD operating system kernel. Two abstract primitives are defined on a wait channel:

- The `sleep` primitive blocks the caller in a passive wait on the channel and schedules another process for execution. In doing this, it also implicitly releases the kernel mutual exclusion constraint, so that another process is also allowed to enter the kernel if it executes a system call trap.
- The `wakeup` primitive awakens all processes blocked on the channel. Unlike semaphores, wait channels have *no memory*, hence a `wakeup` performed before a matching `sleep` does not prevent `sleep` from blocking, thus leading to the loss of the event signified by the execution of `wakeup`.

Since this mechanism has no equivalent counterpart in other operating systems, its emulation constitutes a significant part of the effort needed to port the BSD protocol stack, as discussed in Section 10.4.

10.3 MAIN DATA STRUCTURES

The Berkeley Sockets implementation is based on several key data structures, with most of them summarized in Figure 10.2.

- The *domain* data structure `struct domain` (not shown in the figure) holds all information about a communication domain. For instance, it contains the symbolic address family identifier assigned to the communication domain (like `AF_INET`, which identifies the Internet communication

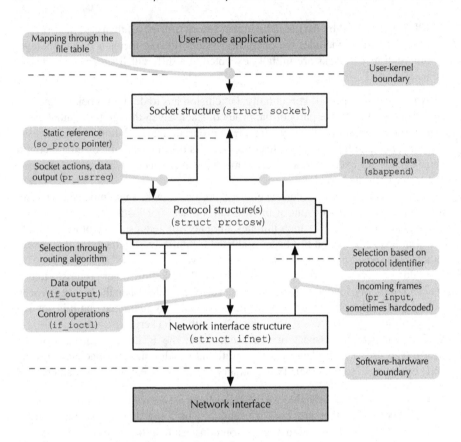

FIGURE 10.2   Main BSD 4.4 protocol stack data structures.

domain), its human-readable name. Most importantly, it also contains a pointer to an array of protocol switch structures, one for each protocol supported by the communication domain, to be described next.

In addition, it contains a set of pointers to domain-specific routines for the management and transfer of access rights to a socket, and for routing initialization. The socket implementation maintains a globally-accessible table of domain structures, one for each communication domain known to the system and configured for use. In RTEMS, the table contains two entries, one for the Internet communication domain, the other for the abstract *route* domain, used to control and configure the Internet routing function of the protocol stack.

- The *protocol switch* data structures `struct protosw`, shown in the middle of Figure 10.2, describe protocol modules. Among other information, a protocol switch specifies the type of sockets it supports, the communication domain it belongs to, and a unique protocol number used throughout the protocol stack to refer to the protocol itself. This information is used to choose the right protocol when creating a new socket.

Moreover, it holds a set of pointers that indirectly reference the externally accessible functions, or entry points, of the protocol module. The set of pointers is the same for all protocol modules, although some of them are optional and may be left unimplemented. Hence, they can be used as a uniform interface to access any protocol module. This gives to the protocol modules an object-oriented design, in which these entry points can be seen as abstract methods although they are implemented in C, a language that intrinsically does not support objects.

Instead, the main interface between the topmost protocol module and the socket level, is the entry point referenced by the `pr_usrreq` pointer in the protocol switch. This entry point is invoked by the topmost level of the protocol stack, with an appropriate request code and additional arguments, whenever the protocol module must perform an action on their behalf. For example, the `pr_usrreq` entry point is invoked with the `PRU_SEND` request code when a `send` operation is invoked at the socket level. In turn, this asks the protocol module to send some data and may result in further requests directed to other protocol modules below it.

This is because protocol modules can be, and usually are, stacked on top of each other to build a whole protocol stack. For instance, the TCP protocol module is invariably stacked on top of the IP protocol module. Since in this case the association between protocols is fixed, the interface between modules takes place with direct calls instead of passing through function pointers, to improve efficiency. For instance, the TCP protocol module directly calls the IP output function (`ip_output`) by name whenever it needs to transmit an IP datagram.

An alternate, more efficient interface towards a protocol module is also possible. It makes use of a structure `struct pr_usrreqs` linked to the protocol switch. The structure contains a set of entry points, one for each request code, thus eliminating the need for explicitly passing the request code as argument and checking its value within the `pr_usrreq` entry point.

A different entry point, referenced by the `pr_input` pointer in the protocol switch, is used to push incoming data through the protocol stack. These data originate at the lowest level of the stack and are forwarded to upper layers, possibly after consulting and removing some lower-level protocol headers. In some cases, when the name of the input function to be called is fixed and known in advance, the function is called directly instead of indirectly. This is the case, for instance, of the `ip_input` function, which is called directly by the `splnet` software interrupt handler for incoming IP traffic.

- The socket data structure `struct socket` represents a communication endpoint. It contains information about the type of socket it supports and its state. In addition, it provides buffer space for data coming from, and directed to, the process that owns the socket and may hold a pointer to a chain of protocol state information.

Upon creation of a new socket, the table of domain structures and the table of protocol switch structures associated with each domain are scanned, looking for a protocol switch entry that matches the requirements set forth by the arguments passed to the socket creation function. That entry is then linked through the so_proto pointer of the socket data structure, and is used as the only interface point between the top-level socket layer and the communication protocol.

Within the socket data structure there are two data queues, one for transmission and the other for reception. These queues are manipulated through a uniform set of utility functions. For example, the sbappend function appends a chunk of data to a queue and is therefore invoked whenever a new data message is received from the lower levels.

At the user application level, the association between a socket descriptor and the corresponding socket structure is carried out by means of a table that, in general-purpose operating systems, often coincides with the in-kernel file table. On simpler systems, it can simply be a per-process array of pointers to socket structures, in which the socket descriptor (a small integer) is used as an index.

- The network interface data structure, struct ifnet, represents a network interface module, with which a hardware device is usually associated, and to its device driver. It provides a uniform interface to all network devices that may be present on a host, and insulates the upper layers of software from the implementation details of each device and its corresponding device driver code. Uniform interfacing is realized in the same way as for the protocol switch, that is, by means of a set of function pointers to the externally visible entry points of the network interface device driver.

The main purpose of a network interface module is to interact with the corresponding hardware device, in order to send and receive data-link level packets. Within a struct ifnet, the main entry point pointers are: if_output, which is responsible for data output through the interface, and if_ioctl, which performs all control operations on the interface. On networks that provide for network-layer routing, for example the Internet, the selection of the correct output interface for a certain IP frame is usually carried out by the local routing algorithm. Therefore, there is no static link between a struct protosw and a struct ifnet, contrary to what happens between the higher layers. For incoming data, the selection of the correct pr_input entry point to be called is based on the incoming protocol identifier, usually contained in the data-link level header. A list of struct ifaddr data structures, each representing an interface address in possibly different communication domains, is linked to the main struct ifnet structure.

Another data structure not shown in Figure 10.2, the struct mbuf, is used whenever dynamic memory allocation is needed. Its implementation makes it particularly suitable to prepend and append further data to an existing buffer, an operation frequently used in communication protocol for encapsulation and de-encapsulation.

10.4   RTEMS PORT AND ADAPTATION LAYER

With respect to the description of the general structure of the BSD protocol stack given in Section 10.2, most of the porting effort is concentrated in a few key areas:

- The implicit *mutual exclusion* upon entering a single-processor BSD kernel through a system call has no counterpart in most real-time operating systems, including RTEMS, and must be implemented explicitly by means of a mutual exclusion device.
- The *sleep/wakeup* mechanism on a wait channel does not exist in RTEMS. It must therefore be emulated by means of one of the synchronization devices RTEMS provides, the closest one being the *events* discussed in Section 7.6.
- A significant amount of processing, within layer 2 of the protocol stack depicted in Figure 10.1, takes place in *software interrupt handlers*, whose execution is triggered by the lower layer upon incoming traffic.
- Similarly, a non-negligible amount of processing concerning incoming and outgoing frames through network interfaces is performed within *hardware interrupt handlers*. Since hardware and software interrupt handlers escape the normal operating system scheduling policy, their use is discouraged in a real-time execution environment.
- The protocol stack relies on the BSD kernel *timeout* mechanism to schedule internal, periodic activities for execution. This mechanism, also based on software interrupts, has no direct counterpart in RTEMS.

10.4.1   MUTUAL EXCLUSION AND SLEEP/WAKEUP

The specific synchronization device chosen to emulate kernel mutual exclusion is the RTEMS *recursive mutex* outlined in Section 7.4.1. As shown in the top-right part of Figure 10.3, the uppermost layer of the protocol stack has been modified to call the RTEMS adaptation layer functions `rtems_bsdnet_semaphore_obtain` and `rtems_bsdnet_semaphore_release` where the system call boundaries to enter and exit from the kernel were originally placed. These functions are simple wrappers around the RTEMS recursive lock acquisition and release functions `rtems_recursive_mutex_lock` and `rtems_recursive_mutex_unlock` described in Section 7.4.1.

In order to work correctly, the mutual exclusion mechanism must be properly integrated with the sleep/wakeup mechanism, so that mutual exclusion is released when a task sleeps, and then acquired again before the task continues after having been woken up.

Per se, the sleep/wakeup mechanism relies on RTEMS *system events*. To integrate it with mutual exclusion, the blocking call to the system event receive function has been surrounded by calls to `rtems_bsdnet_semaphore_release` and `rtems_bsdnet_semaphore_obtain`. Besides this, two additional expedients are however required to emulate the original mechanism properly:

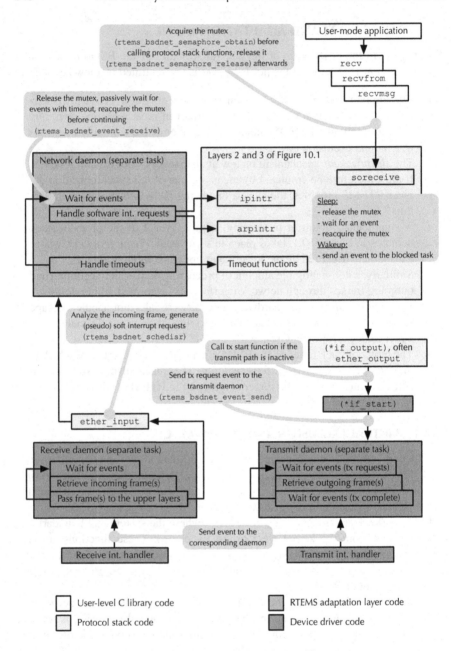

FIGURE 10.3   RTEMS network adaptation layer and device driver.

- RTEMS events have memory, that is, if a task looks for an event that has already been generated in the past, it receives the event and then continues without blocking. On the contrary, a wakeup call that precedes a matching sleep call shall have no effect and the subsequent sleep shall still block the caller.

  Therefore, the implementation of the sleep function must flush all pending events before releasing the mutual exclusion semaphore and calling the system event receive function `rtems_event_system_receive`. With this approach, the boundary between past and future events with respect to the sleep call is set to the instant in which the mutual exclusion semaphore is released: All events generated before that instant are discarded, whereas any event generated afterwards will wake up the calling task.

- Since the call to `rtems_event_system_receive` and then to `rtems_bsdnet_semaphore_obtain` are not performed as an indivisible unit, there is the possibility that further events are generated after `rtems_event_system_receive` returned, but before the task could obtain the mutual exclusion semaphore again and continue. Conceptually, these events have been generated before the sleep operation was completed, and hence, should be taken into account.

  To this purpose, the sleep function performs a second, non-blocking system event receive operation after re-acquiring the mutual exclusion semaphore and combines the events received in this way with the ones received with the first call. The fact that other tasks are allowed to generate wakeup events only if they hold the mutual exclusion semaphore themselves prevents further events from being generated after the re-acquisition of the mutual exclusion semaphore.

Strictly speaking, a minor difference between the original sleep/wakeup mechanism and the one provided by RTEMS remains. Namely, a wakeup operation should wake up *all* tasks sleeping on the matching wait channel. Instead, in the RTEMS emulation, only *one* task is woken up. This is a direct consequence of the fact that RTEMS events must be directed to one specific task.

However, this difference has no practical consequence since in the BSD protocol stack this "broadcast" feature of the wakeup function is not used, as long as at most one task is allowed to sleep on a certain socket at any given time.

10.4.2 SOFTWARE INTERRUPTS AND NETWORK DAEMON

In order to eliminate software interrupts, the RTEMS adaptation layer implements a *network daemon* as a separate task. This task provides a proper execution context for the protocol stack code formerly executed within software interrupt handlers. To support the orderly coexistence of the network daemon with other real-time task, its scheduling priority and affinity mask can be configured at will by the user.

Software interrupt requests are emulated by means of the same event mechanism already used to emulate the sleep and wakeup operations. Namely, the network daemon consists of an infinite loop in which it waits for an event to arrive. An event towards the network daemon is generated by the `rtems_bsdnet_schedisr` when a software interrupt request would have been posted in the original code.

No changes to the protocol stack are needed in order to do this because the original protocol stack consistently invokes the function `schedisr` to post a software interrupt, and this function is directly mapped onto `rtems_bsdnet_schedisr` by the RTEMS adaptation layer.

After being woken up by an event, the network daemon analyzes the event to determine which software interrupt has been requested and call the appropriate protocol stack function. As shown in Figure 10.3, the most important software interrupt entry points are `ipintr` (for incoming IP traffic) and `arpintr` (for incoming ARP traffic).

In order to ensure mutual exclusion, the function used by the network daemon (and by all the other tasks playing a similar role to be discussed in the following) to wait for an event, `rtems_bsdnet_event_receive`, releases the mutual exclusion semaphore that protects the protocol stack before waiting, and re-acquires it afterwards.

Moreover, the function `rtems_bsdnet_newproc` used to create the network daemon ensures that the network daemon acquires the mutual exclusion semaphore before it starts executing its main entry point, by means of a suitable trampoline.

This approach realizes a different, and somewhat stricter, form of mutual exclusion than in the original protocol stack, in which the software interrupt handlers and other parts of the protocol stack could execute concurrently unless these other parts raised the IPL to at least `splnet`.

This is safe because the scope of the mutual exclusion enforced by the RTEMS adaptation layer has indeed a wider scope than the original, and also enables the removal of all IPL changes within the protocol stack. An additional benefit is that the mutual exclusion mechanism also works correctly on multi-core systems.

10.4.3 TIMEOUT EMULATION

The network daemon is also responsible of emulating the *timeout* mechanism outlined in Section 10.2, used by the protocol stack to schedule the future execution of its internal activities. The implementation leverages the timeout feature of the RTEMS system event wait function. More specifically:

- When it is about to block for events, the network daemon calculates the timeout to be passed to the system event wait function by consulting the list of pending timeout requests. The timeout is calculated so that, even in absence of events, the system event wait function returns to the network daemon when the first pending timeout is due to expire. If there are no pending timeouts, an infinite timeout is used.
- After calling the appropriate software interrupt entry points triggered by the incoming events, if any, the network daemon checks the elapsed time

before blocking again and, if any pending timeouts expired, calls the corresponding timeout handling function within the protocol stack.

Also in this case, mutual exclusion is guaranteed without additional provisions, because the timeout handling functions are always called while the network daemon holds the protocol stack mutual exclusion semaphore.

10.4.4 DEVICE DRIVER ORGANIZATION

The idea of replacing execution in interrupt handlers (and the shortcomings associated with it) with execution in a task context, which the RTEMS adaptation layer adopted for software interrupts, has been extended to hardware interrupts as well.

As shown at the bottom of Figure 10.3, the activities performed within the interrupt handlers of a network device shall be reduced to a minimum by design. In other words, interrupt handlers must perform only time-critical device operations that cannot be postponed and delegate everything else to two tasks, the *receive* and *transmit daemons*, by sending appropriate events to them.

These two tasks are responsible for retrieving incoming frames from the network interface and push them up through the protocol stack, and to initiate the transmission of outgoing frames coming down from the protocol stack, respectively.

The structure of the receive daemon, at the bottom left of the figure, is similar to the one of the network daemon and makes use of the same mutual exclusion and synchronization mechanisms:

- The receive daemon consists of an infinite loop in which it waits for events and processes them. The wait is carried out by calling `rtems_bsdnet_event_receive`, the same function used by the network daemon, to ensure proper mutual exclusion with other parts of the protocol stack. This approach also enables the removal of all IPL changes to the `splimp` level in the protocol stack code.
- Events are generated by the hardware interrupt handler upon arrival of incoming frames. Unless the interface itself is capable of direct memory access (DMA), the receive daemon must retrieve each frame from the network interface, encapsulate it into an appropriate chain of `struct mbuf`, and enqueue it in the appropriate protocol input queue depending on the network-level protocol while also removing the data link-level header. After that, the receive daemon must trigger further protocol stack activities by posting the appropriate software interrupt request, which also depends on the network-level protocol.

  For Ethernet devices, most of these activities except frame retrieval, which depends on the characteristics of the underlying hardware device, are performed by the `ether_input` function of the protocol stack.

Referring back to Figure 10.3, the `struct ifnet` that describes a network interface has two entries related to frame transmission:

- The function referenced by the if_output function pointer is called by the protocol stack when a frame shall be enqueued for transmission. It receives as argument a chain of struct mbuf that contains the IP datagram to be transmitted.

  This function is responsible for adding an appropriate data link-layer header to the datagram and enqueueing the resulting frame in the interface transmission queue, which is also linked to the struct ifnet. If the transmit path of the interface was idle, this function must also call the function pointed by if_start to start it.

  For Ethernet interfaces, all these functions are performed by the function ether_output, provided by the protocol stack itself. For a device driver to use it, it just needs to link it to the if_output field of the struct ifnet.

- Instead, the function referenced by the if_start fuction pointer is typically implemented on a case-by-case basis by the device driver. Its main purpose is to send an appropriate event to the transmit deamon, so that it can pull the outgoing frames from the interface transmission queue and transfer them to the hardware. The details of this operation depend on the underlying hardware and its level of sophistication.

The transmit deamon itself consists of an infinite loop in which it waits for events that either request it to enqueue some outgoing frames for transmission or signal that the transmission of some frames has been completed. In the first case, the events come from the if_start function, while in the second case they are generated by the transmit interrupt handler. Mutual exclusion with respect to other parts of the protocol stack are handled as for the network and receive daemons.

10.5 SUMMARY

In this chapter we outlined how network communication is implemented in RTEMS. The discussion began with Section 10.1, in which we briefly discussed the two TCP/IP protocol stacks that RTEMS supports. Then, in Sections 10.2 and 10.3, we described the internal organization of one of those protocol stacks and its main data structures.

In the next section, Section 10.4, we focused instead on the adaptation layer between the protocol stack and the RTEMS operating system. There, we provided more details on the most important aspects of this layer, namely, the mutual exclusion and passive wait mechanisms, the interface toward network device drivers, and the emulation of some synchronization mechanisms needed by the protocol stack but not directly available in RTEMS.

11 POSIX Sockets API

CONTENTS

This chapter is entirely devoted to the application programming interface for network communication specified by the POSIX standard. By means of this interface, widely available on a variety of operating systems, applications gain access to network communication in a system- and protocol-independent way, to the benefit of portability.

11.1 MAIN FEATURES

The *sockets* API, nowadays backed by the POSIX international standard [68], was first introduced in the "Berkeley Unix" operating system many years ago [86]. It is now available on virtually all general-purpose operating systems and most real-time operating systems that support network communication. Generally speaking, the main advantage of the *sockets* API with respect to a custom interface, tailored to a specific network or protocol, is that it supports in a *uniform* way *any kind* of communication network, protocol, naming conventions, and hardware.

In this sense sockets represent abstract communication endpoints that are largely agnostic with respect to the underlying networks and protocols. Semantics of communication and naming are captured by communication *domains* and socket *types*, both specified upon socket creation. Namely, communication domains are used to distinguish between IP-based network environments and other kinds of network, while the socket type determines whether communication will be stream-based or datagram-based and also implicitly selects which network protocol a socket will use.

Additional socket characteristics can be set up after creation through abstract *socket options*. For example, a socket option provides a uniform, implementation-independent way to set the amount of receive buffer space associated with a socket, without requiring any prior knowledge about how buffers are managed by the underlying implementation of the communication layers. As often happens, the price to be paid for these advantages is mainly related to *efficiency* in terms of execution time and memory footprint, as well as the expressive power of the API, which are usually lower for POSIX *sockets* with respect to other, less general approaches.

For instance, the OSEK VDX operating system specification [73, 93], focused on automotive applications, specifies a communication environment (OSEK/VDX COM) less general than *sockets* and oriented to real-time message-passing networks, such as the Controller Area Network (CAN) [72]. The API that this environment provides is more flexible and efficient because it allows applications to easily set message filters and perform out-of-order receives, thus enhancing their timing behavior. Neither of these functions is straightforward to implement with *sockets*, because they do not fit well within the more general and abstract socket paradigm, although it is definitely possible [75, 114].

Table 11.1 summarizes the main functions made available by the POSIX *sockets* API, divided into functional groups that will be described in more detail in the following.

11.2   COMMUNICATION ENDPOINT MANAGEMENT

The first thing to do in order to use the *sockets* API for network communication, is to create at least one communication endpoint, known as *socket*. This is accomplished by means of the `socket` function:

`int socket(int domain, int type, int protocol);`

Its three arguments indicate:

1. An *address family* identifier, which uniquely specifies the network communication domain the socket belongs to and operates within.
2. A *socket type* identifier, which specifies the communication model that the socket will use and, as a consequence, determines which communication properties will be available.
3. A *protocol identifier*, to select which specific protocol stack, among those suitable for the given protocol family and socket type, the socket will use—if more than one protocol is available.

A communication domain groups together sockets with common communication properties, for example, their endpoint addressing scheme or the underlying communication protocols they use. It also implicitly determines a communication boundary because data exchange can take place only among sockets belonging to the same domain. At the time of this writing, the default protocol stack of RTEMS supports only the `AF_INET` address family, which identifies the Internet communication domain.

The communication domain and the socket type are first used together to determine a set of communication protocols that belong to the domain and obey the

TABLE 11.1
Main Primitives of the POSIX *Sockets* API

Function	Purpose
Communication endpoint management	
socket	Create a communication endpoint
shutdown	Shutdown a connection, in part or completely
close	Close a communication endpoint
getsockopt	Retrieve a socket option
setsockopt	Set a socket option
fcntl	Make socket operations blocking or non-blocking
ioctl	Perform assorted control operations on a socket
Local socket address	
bind	Assign a well-known local address to a socket
getsockname	Retrieve the local address of a socket
Connection establishment	
connect	Initiate a connection
listen	Prepare a socket to listen for incoming connections
accept	Accept an incoming connection from a listening socket
getpeername	Retrieve the address of a connected peer
Data transfer	
send	Send a message from a connected socket
sendto	Send a message to a given destination
sendmsg	Gather a message from memory and send it to a given destination
recv	Receive a message from a connected socket
recvfrom	Receive a message and its source address information
recvmsg	Receive a message and scatter it into memory
write	Send data from a connected socket
read	Receive data from a connected socket
Synchronous I/O multiplexing	
select	Simple interface for synchronous I/O multiplexing
pselect	Improved select
poll	General interface for synchronous I/O multiplexing
FD_ZERO	Initialize an empty file descriptor set
FD_SET	Add a file descriptor to a set
FD_CLR	Remove a file descriptor from a set
FD_ISSET	Check whether or not a file descriptor belongs to a set

communication model the socket type indicates. Then, the protocol identifier is used to narrow the choice down to a specific protocol within this set.

The special protocol identifier 0 (zero) specifies that a suitable default protocol, selected by the underlying socket implementation, shall be used. In most cases, this is not a source of ambiguity because most protocol families support at most one protocol for each socket type.

When it completes successfully, socket returns to the caller a non-negative integer, known as *socket descriptor*, which shall be passed to all other socket-related functions, in order to refer to the socket itself. Instead, the negative value −1 indicates that the function failed and no socket has been created. In this case, like for most other POSIX functions, the errno variable conveys to the caller additional information about the reason for the failure:

EAFNOSUPPORT The system does not support the address family given in the domain argument.

EPROTOTYPE The system does not support the socket type given in the type argument.

EPROTONOSUPPORT The system does not support the communication protocol specified in the protocol argument, or the protocol is unsuitable for the given combination of domain and type.

EMFILE The limit on the number of open file descriptors that the calling process is allowed to have has been reached.

ENFILE The system-wide limit on the number of open file descriptors has been reached.

Optionally, the socket function may also check the following additional error conditions and fail if it encounters them:

EACCESS The calling process does not have sufficient privileges to create the socket.

ENOBUFS The system does not have sufficient resources to create the socket.

ENOMEM The system does not have enough memory to create the socket.

The default RTEMS protocol stack currently supports three different socket types:

1. The SOCK_STREAM socket type provides a connection-oriented, bidirectional, sequenced, reliable transfer of a byte stream. The underlying protocol does *not* necessarily keep track of message boundaries, and hence, a message sent as a single unit may be received as two or more separate pieces. The opposite is also possible, that is, multiple messages may be grouped together at the receiving side and received as a single unit.

2. The SOCK_DGRAM socket type supports a bidirectional data flow, but does not provide any guarantee of sequenced or reliable delivery. In other words, messages sent through a datagram socket may be duplicated, discarded, or received in an order different from the transmission order, with no indication about these facts being conveyed to the user. Message boundaries are preserved though.

TABLE 11.2

Supported Raw Protocols in the AF_INET Family

Name	Acronym	RFC	Protocol ident.
Internet Control Message Protocol	ICMP	RFC 792 [98]	IPPROTO_ICMP
Internet Group Management Protocol	IGMP	RFC 3376 [32]	IPPROTO_IGMP
Resource Reservation Protocol	RSVP	RFC 2205 [22]	IPPROTO_RSVP
IP Encapsulation within IP	IPIP	RFC 2003 [95]	IPPROTO_IPIP
Raw access to all IP traffic	—	—	IPPROTO_RAW

3. The SOCK_RAW socket type is similar to SOCK_DGRAM, but gives applications direct access to lower-level protocols, further specified by the protocol identifier argument of socket. The format of datagrams sent and received through this type of socket depends on the underlying protocol and on the protocol stack implementation.

The POSIX standard also specifies a fourth socket type, SOCK_SEQPACKET. It is similar to SOCK_STREAM, but it also preserves message boundaries. On the transmitting side, a single message can be built incrementally with a sequence of send operations, marking the last one in a special way to indicate the end of a message.

Similarly, a single message may be returned in one or more chunks at the receiving side, but the end of the message is still marked, thus giving the receiver the ability to recognize message boundaries. Although the default RTEMS protocol stack supports this socket type in principle, it does not implement any underlying protocols able to provide its features unless suitably extended.

For what concerns SOCK_STREAM sockets, the only protocol supported in the AF_INET address family is TCP [100], and hence, the protocol identifier argument to socket is ignored in this case. Similarly, the only protocol supported for SOCK_DGRAM sockets is UDP [97], still in the AF_INET address family. As a consequence, their protocol identifiers, IPPROTO_TCP and IPPROTO_UDP respectively, are seldom used.

Instead, the protocol identifier becomes important for SOCK_RAW sockets, which provide access to several other Internet protocols, listed in Table 11.2 along with the Request For Comments (RFC) documents that define them and their protocol identifiers within the AF_INET family. Moreover, using the pseudo-protocol identifier IPPROTO_RAW grants unfiltered access to all the underlying IP traffic.

The default RTEMS protocol stack also supports the PF_ROUTE domain, which will not be further discussed here. Opening a socket in this domain and sending/receiving messages through it allows a thread to manipulate the protocol stack routing tables.

The function close closes and destroys a socket. Its only argument is descriptor of the socket to be closed, formerly returned by socket:

```
int close(int fildes);
```

This function is the same function also used to close regular files and must be used to reclaim system resources—for instance, memory buffers—assigned to a socket when it is no longer in use. It is important to highlight that—unless the default behavior is changed by setting the SO_LINGER socket option, to be discussed in Section 11.7—the close function is designed to return to the caller as soon as possible.

More specifically, the function may return to the caller before all the data sent through a socket have actually reached their destination. In this case, there is no guarantee that data transmission will eventually be successful, even though the underlying socket type (like SOCK_STREAM) promises reliable data delivery.

In addition, when using the TCP protocol, the close function may abort the connection instead of closing it gracefully (by means of a TCP RST instead of the usual FIN) if there are unread data in the socket receive buffers. This is done to make the peer at the other side of the connection aware that some of the data it sent has not been received and processed. Interested readers should refer to [104] for more information about these important TCP-related topics.

The close function returns zero when successful, otherwise it returns −1 and sets errno to an error code. The function shall fail if:

EBADF The given file descriptor fildes is invalid.
EINTR The function was interrupted by a signal.

The error code EINTR can also be returned by many other socket-related functions, for the same reason. In the descriptions that follow, it will be omitted for conciseness, unless it is particularly important for the function at hand.

The standard also specifies a way to shutdown a socket only *partially*, by disabling further send and/or receive operations, but without destroying the socket itself. This is done by means of the shutdown function:

```
int shutdown(int socket, int how);
```

The how argument specifies which direction of the data transfer shall be shutdown:

SHUT_RD Further receive operations will no longer be allowed.
SHUT_WR Further send operations will no longer be allowed. In the case of TCP, this request enqueues the transmission of a FIN.
SHUT_RDWR Combines SHUT_RD and SHUT_WR, that is, further send and receive operations will no longer be allowed.

The shutdown function returns zero when successful, otherwise it returns −1 and sets errno to an error code. The function shall fail if:

EBADF The given file descriptor socket is invalid.
ENOTSOCK The given file descriptor is valid, but does not refer to a socket.
EINVAL The how argument does not have one of the values previously mentioned.

ENOTCONN The socket is not connected.

In addition, shutdown may also fail if:

ENOBUFS The system lacks the resources needed to carry out the operation.

11.3   LOCAL SOCKET ADDRESS

When it is initially created by means of the socket function, a socket has no local network address associated with it. However, a socket must have a unique local address to be actively engaged in data transfer. This is because the local address is used as the source address for outgoing data frames, and incoming data are conveyed to sockets by matching their destination address with local socket addresses.

The exact address format and its interpretation may vary depending on the communication domain. Within the Internet communication domain, addresses consist of a 4-byte IP address and a 16-bit port number, assuming that IPv4 [99] is in use.

For connection-oriented sockets used to initiate a connection, the explicit assignment of a local address is unnecessary since the system implicitly binds the socket to an appropriate, unique local address when given an unbound socket. The address consists of the IP address of the outgoing network interface and a unique *ephemeral* port, usually taken from a system-dependent range of port numbers located above port 1024. In this case, the application is not concerned at all with local address assignment but it also has no control on it.

On the contrary, explicit address assignment is mandatory in order to use a connection-oriented socket as a listening point for incoming connection requests, because the peer must obviously know the target address to initiate a connection. For the same reason, connectionless sockets are also explicitly given a local address in virtually all cases.

The bind function explicitly assigns a local address to a socket:

```
int bind(int socket, const struct sockaddr *address,
    socklen_t address_len);
```

As usual, the socket is identified by means of its socket descriptor socket. The data structure that contains the address is referenced by the pointer address and the argument address_len indicates the size of the structure, in bytes.

An IPv4 address is stored in a struct sockaddr_in data structure. According to the general address format described previously, it contains the following fields:

- The field sin_family, of type sa_family_t, denotes the address family to which the address belongs and must be set to AF_INET.
- The IP address is held in the field sin_addr, of type struct in_addr. In turn, this structure contains a field called s_addr, of type in_addr_t, which is equivalent to a 32-bit integer and holds the IP address.
- The field sin_port, whose type in_port_t is equivalent to a 16-bit integer, holds the port number.

To make them compatible with all possible communication domains and their address formats, `bind` and all other functions that take a network address as argument use a pointer to a generic `struct sockaddr` to refer to it.

When dealing with an address belonging to a specific communication domain, for instance, an IPv4 address, programmers must explicitly cast its `struct sockaddr_in` pointer into a `struct sockaddr` pointer when passing it as argument, to avoid compiler warnings. Even more importantly, it is the programmer's responsibility to ensure that the address given to `bind` or any other function belongs to the same address family as the socket they operate upon.

Another common source of mistakes comes from the fact that both the IP address and the port number—stored in the fields `sin_addr.s_addr` and `sin_port`, respectively—must be written in *network* byte order. For historical reasons, this is *big-endian* for IPv4, and hence, does not coincide with the native byte order of many computer architectures in common use nowadays.

A portable conversion from host to network byte order and vice versa can be performed by means of the following functions:

- The functions `htons` and `htonl` take a 16-bit (also called *short*) or a 32-bit (also called *long*) integer as argument, respectively, convert it from host to network byte order, and return the result.
- The functions `ntohs` and `ntohl` do the opposite, that is, they convert a 16-bit or a 32-bit integer from network to host byte order and return the result.

Th `bind` function may fail for a variety of reasons. In this case, like most other *sockets* functions, it returns -1 to the caller instead of zero, and sets `errno` to an error code that better explains the error:

EBADF The given file descriptor `socket` is invalid.

ENOTSOCK The given file descriptor is valid, but does not refer to a socket.

EOPNOTSUPP The socket does not support the `bind` operation.

EAFNOSUPPORT The address referenced by `address` is not valid for the address family of the given `socket`.

EADDRNOTAVAIL The address is not available on the local machine. This may happen, for instance, if the IP address given in the network address does not belong to any of the network interfaces configured on the machine.

EADDRINUSE The address is already in use. No two sockets can be bound to the same local address at the same time.

EALREADY Another address assignment request is already in progress for the same socket. This is possible because sockets are shared among all threads belonging to the same process and they can operate on them concurrently.

EINVAL The socket has previously been bound to an address and the underlying protocols do not support rebinding it to a new address, or the socket has been shutdown.

ENOBUFS The system lacks the resources needed to carry out the operation.

Regardless of the way the local socket address has been assigned, it can be retrieved by means of the `getsockname` function:

```
int getsockname(int socket,
    struct sockaddr *restrict address,
    socklen_t *restrict address_len);
```

Upon successful completion, the function stores the local address of the given `socket` into the data structure pointed by `address`. When invoked on a socket that has not been bound to any local address, the value stored into the data structure is unspecified.

Addresses belonging to different communication domains may have different lengths and it may be difficult to precisely determine in advance what the address family of the address to be returned should be. To avoid buffer overflows, the variable of type `socklen_t` referenced by `address_len` has a twofold meaning:

- The caller must initialize it to the size in bytes of the buffer referenced by `address` before calling `getsockname`.
- Then, `getsockname` updates the value to reflect the actual size of the address it stored in the buffer. If the buffer size is insufficient to store the whole address, the function truncates it.

Upon failure, `getsockname` returns -1 to the caller instead of zero, and sets `errno` to an error code that better explains the error:

EBADF The given file descriptor `socket` is invalid.
ENOTSOCK The given file descriptor is valid, but does not refer to a socket.
EOPNOTSUPP The socket does not support the `getsockname` operation.

Moreover, `getsockname` may also fail if:

EINVAL The socket has been shutdown and no longer has a local address.
ENOBUFS The system lacks the resources needed to carry out the operation.

11.4 CONNECTION ESTABLISHMENT

A connection-oriented socket of type `SOCK_STREAM` implements a bidirectional, point-to-point communication between two peers. As its name itself says, it must therefore be connected to another socket before data transfer can take place between them. This is done by means of the `connect` function:

```
int connect(int socket, const struct sockaddr *address,
    socklen_t address_len);
```

The first argument, `socket`, is a socket descriptor that identifies the local communication endpoint. The other two arguments, `address` and `address_len` provide

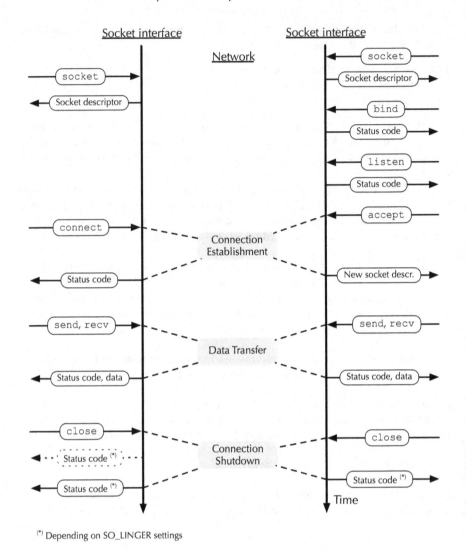

(*) Depending on SO_LINGER settings

FIGURE 11.1 Lifetime of a SOCK_STREAM connection.

the address of the target communication endpoint, specified in the same way as for the bind function.

As shown in the top left part of Figure 11.1, when invoked on a connection-oriented socket, as those using the TCP protocol are, it sends out a connection request directed toward the target address specified in the second and third argument. If the local socket is currently unbound, the system also selects and binds an appropriate local address to it beforehand, as described in Section 11.3.

If the function succeeds, it associates the local and the target sockets, and data transfer can begin. Otherwise, it returns to the caller an error indication. In order to be a valid target for a `connect`, a socket must satisfy two conditions:

1. It must have a well-known address assigned to it—because the communication endpoint willing to connect has to specify this address in the second and third argument of the `connect` function. This is usually obtained by means of a `bind` operation.
2. It must have been marked as willing to accept connection requests, by means of the `listen` function to be discussed next. Newly created sockets do not accept incoming connection requests by default.

Upon failure, `connect` returns −1 to the caller instead of zero, and sets `errno` to an error code that better explains the error:

EBADF The given file descriptor `socket` is invalid.

ENOTSOCK The given file descriptor is valid, but does not refer to a socket.

EALREADY Another connection establishment attempt is already in progress for the same socket.

EADDRNOTAVAIL The connection attempt failed because the socket has been bound to a local address that is invalid, or the system ran out of ephemeral local ports.

EAFNOSUPPORT The target address referenced by `address` is not valid for the address family of the local `socket`.

EPROTOTYPE The address referenced by `address` corresponds to a socket of a different type than the local `socket`.

EISCONN The local `socket` is already connected.

ECONNREFUSED The target address refused the connection, for instance, because it was not listening for incoming connections.

ENETUNREACH The target network is unreachable.

ETIMEDOUT The connection establishment process timed out before the connection could be established.

Moreover, `connect` may also fail if:

EINVAL The target address family or the address length are invalid.

EADDRINUSE The connection attempt failed because it tried to use address/port combinations the were already in use.

ENETDOWN The local network interface that should have been used to reach the target is down.

EHOSTUNREACH The target host is unreachable.

ECONNRESET The remote host reset the connection

EOPNOTSUPP The socket is listening for incoming connection requests, and hence, it cannot be used to send outgoing connection requests.

ENOBUFS The system lacks the resources needed to carry out the operation.

The `listen` function marks a socket as willing to accept incoming connection requests:

```
int listen(int socket, int backlog);
```

The first argument of this function is, as usual, the socket descriptor to be acted upon. Informally speaking, the second argument is an integer that specifies the maximum number of outstanding connection requests that can be waiting acceptance on the given socket, known as *backlog*. The special value zero, or a negative value, leaves the underlying implementation free to choose a suitable backlog size. It should also be noted that, even if the user specifies a non-zero value, it is treated as a hint by the socket implementation, which is free to reduce it if necessary.

If a new connection request is received while the queue of outstanding requests is full, the connection can either be refused immediately or, if the underlying protocol implementation supports this feature, the request can be retried at a later time. Reference [104] contains more in-depth details about how the backlog mechanism works.

When successful, the listen function returns zero to the caller. Otherwise, it returns −1 and sets errno to an error code. The function shall fail for the following reasons:

EBADF   The given file descriptor socket is invalid.

ENOTSOCK   The given file descriptor is valid, but does not refer to a socket.

EOPNOTSUPP   The underlying protocol does not support the listen operation. This may happen, for instance, when calling listen on a connectionless socket.

EINVAL   The socket is already connected, and hence, it cannot be used to listen for incoming connection requests.

EDESTADDRREQ   The socket is not bound to a local address and the underlying protocol, like TCP, does not support listening for incoming connection requests on an unbound socket.

Moreover, listen may also fail if:

EINVAL   The socket has been shutdown and can no longer be used for listening.

ENOBUFS   The system lacks the resources needed to carry out the operation.

A successful execution of bind and listen is necessary, but not yet sufficient, to listen for incoming connection requests and establish new connections. The accept function must also be used, after listen, to wait for the arrival of a connection request on a given socket. The whole process is summarized in the top right part of Figure 11.1. The accept function has the following prototype:

```
int accept(int socket, struct sockaddr *restrict address,
    socklen_t *restrict address_len);
```

The function blocks the caller until a connection request arrives, then accepts it, creates a *new* socket and returns its descriptor to the caller. It returns −1 upon failure, but this is not a source of ambiguity because socket descriptors are non-negative integers by definition. The new socket is connected to the socket that originated the connection request, sometimes called the *peer* socket, while the original socket passed to `accept` as argument is still available to wait for and accept further connection requests, if used with `accept` again.

In addition, if the `address` argument is not a `NULL` pointer, `accept` stores the address of the socket that originated the connection request into the location pointed by it. The `address_len` argument is handled in the same way as the `getsockname` function does. Namely, the location it points to is used both to indicate the size of the buffer pointed by `address` and the actual length of the address that `accept` stored into it.

The `accept` function shall fail for the following reasons:

EBADF The given file descriptor `socket` is invalid.

ENOTSOCK The given file descriptor is valid, but does not refer to a socket.

EINVAL The given socket is not accepting connections.

EOPNOTSUPP The underlying protocol does not support the `accept` operation.

ECONNABORTED The connection has been aborted.

EMFILE The limit on the number of open file descriptors that the calling process is allowed to have has been reached.

ENFILE The system-wide limit on the number of open file descriptors has been reached.

ENOBUFS The system lacks the buffer space needed to carry out the operation.

ENOMEM The system does not have enough memory to carry out the operation.

The address of the peer socket can also be retrieved at a later time by means of the `getpeername` function, which should be invoked on a connected, connection-oriented socket:

```
int getpeername(int socket,
    struct sockaddr *restrict address,
    socklen_t *restrict address_len);
```

Also in this case, the location referenced by `address_len` has a twofold meaning, as explained previously. The `getpeername` function returns zero upon successful completion, or −1 upon failure. It shall fail for the following reasons:

EBADF The given file descriptor `socket` is invalid.

ENOTSOCK The given file descriptor is valid, but does not refer to a socket.

ENOTCONN The given socket is not connected, and hence, it does not currently have a peer.

EOPNOTSUPP The underlying protocol does not support the `getpeername` operation or the concept of peer.

In addition, the function may fail if:

ENOBUFS The system lacks the buffer space needed to carry out the operation.

Figure 11.1 on page 362 summarizes the steps that the two peers must take in order to establish a connection between them. More specifically:

1. Both peers must create a new connection-oriented socket by calling `socket`. The two sockets must necessarily belong to the same communication domain and be of the same type.
2. The peer that will passively wait for incoming connection, shown on the right of the figure, must call `bind` to give its socket an address that is well-known to the other peer.
3. Then, it must mark the socket as willing to accept connections by invoking `listen` on it.
4. Finally, it must call `accept` to block and wait for incoming connection requests.
5. The peer that will actively initiate the connection request, shown on the left of the figure, does not need to call `bind` because the system will automatically pick a local address for its socket as needed.
6. Instead, it can proceed directly to call `connect`, which will block the caller until the connection is established.
7. After a successful connection establishment, both peers have a socket available for data transfer: The peer that initiated the connection will use the socket returned by its call to `socket`, and the peer that waited for an incoming connection request will use the socket returned by `accept`.
8. The socket marked with `listen` is still available to `accept` further incoming connection requests if so desired.

11.5 CONNECTIONLESS SOCKETS

Once connected to a peer, a connection-oriented socket implements a point-to-point communication channel between two peers. The establishment of a connection is a prerequisite for data transfer. Instead, *connectionless* communication does not require any form of connection negotiation or establishment before data transfer can take place. It is typical of datagram sockets, such as the ones using the UDP protocol.

Since in datagram communication each message is routed independently from any other, connectionless sockets readily support multi-point communication in which a single socket is used to send messages to multiple targets. Symmetrically, one single socket can also receive messages from multiple sources.

Socket creation proceeds in the same way as it does for connection-oriented sockets, and `bind` can be used to assign a specific, well-known local address to a socket. Moreover, if a data transmission operation is invoked on an unbound socket, the socket is implicitly bound to an available local address before transmission takes place. Instead, due to the lack of need for connection establishment, `listen` and `accept` may not be used on a connectionless socket.

The `connect` function can still be used, albeit with different semantics. When invoked on a connectionless socket, `connect` does not send any connection request—it would not make sense—but simply associates the given destination address with the socket. In this way, it becomes possible to use the socket with data transmission functions that do not explicitly indicate the destination address, for instance, `send`.

Moreover, after a successful `connect`, only data received from that remote address will be delivered to the user.

The `connect` function can be used multiple times on the same connectionless socket, but only the last address specified remains in effect. Unlike for connection-oriented sockets, in which `connect` implies a certain amount of network activity, connect requests on connectionless sockets return almost immediately to the caller, because they simply perform a local operation. The only way to send data through a connectionless socket without using `connect` is by means of a function that allows the caller to specify the destination address on a message-by-message basis, such as `sendto`.

11.6   DATA TRANSFER

The functions `send`, `sendto`, and `sendmsg` send data through a socket, with different trade-offs between expressive power and interface complexity. The `send` function is the simplest one. The most important underlying assumption of its interface is that the destination address is already known to the system. This is true in two cases:

- If the function is invoked on a connection-oriented socket that has been successfully connected to a remote peer in the past and has not been disconnected yet.
- On a connectionless socket, the use of `send` is allowed if a remote address has been associated with the socket by means of a `connect` call.

On the contrary, `send` cannot be used, for instance, on a connectionless sockets on which no `connect` has ever been performed. Its four arguments specify the socket to be used, the position and size of a memory buffer containing the data to be sent, and a set of *flags* that may alter the semantics of the function:

```
ssize_t send(int socket, const void *buffer,
    size_t length, int flags);
```

More specifically:

- The `socket` argument is a socket descriptor
- The `buffer` argument points to the in-memory data buffer and `length` gives its length, in bytes.
- The `flags` argument contains the bitwise OR of the flags.

Of all the flags available to the caller, here we will only discuss `MSG_OOB` and `MSG_NOSIGNAL`, which are the most relevant to real-time systems:

MSG_OOB is used on sockets that support out-of-band communication to send a (usually limited) quantity of out-of-band data. These data are sent and received independently of normal data and may have a higher delivery priority. The TCP protocol supports one single outstanding byte of out-of-band data at any given time. The presence of out-of-band data waiting to be delivered is signaled to the remote peer, but they are nevertheless sent in sequence with respect to normal data.

MSG_NOSIGNAL is used to suppress the `SIGPIPE` signal that would normally be generated if `send` is called on a stream-oriented socket that is no longer connected to its remote peer. When this flag is set, `send` fails and returns the `EPIPE` error code instead.

The `send` function returns the non-negative number of bytes successfully transferred from the user buffer given as argument into network buffers to be sent. If there is no buffer space available, `send` blocks the caller until some space becomes available. The function may also succeed partially, that is, it may return a number of bytes lower than `length`.

It is important to remark that a success indication from `send` (or from any other data transmission primitives that will be discussed in the following) does *not* guarantee in any way that any data has actually been delivered to the remote peer. This is true even though the underlying network protocol, like TCP, promises *reliable delivery*. In fact, a successful `send` does not even imply that the system has started sending those data, because there may still be older data in the queues of the network transmission data path.

The only true meaning of a successful `send` is that no local errors that would prevent the transmission have been detected, the data to be transmitted have been successfully moved into suitable network buffers, and the system will do its "best effort" to deliver them in due course. If a timely delivery confirmation is needed, it must be implemented by means of a higher-level protocol above the socket layer. For instance, a very simple approach could be to have the remote peer send back an acknowledgment message whenever it successfully received a valid data message.

Upon failure, `send` does not send any data, returns −1 to the caller, and sets `errno` to an error code. The `send` function shall fail when:

EBADF The given file descriptor `socket` is invalid.

ENOTSOCK The given file descriptor is valid but does not refer to a socket.

ENOTCONN The given socket is connection oriented but it is not currently connected to a peer.

EDESTADDRREQ The given socket is connectionless and no predefined destination address has been set with `connect`.

EOPNOTSUPP The underlying protocol does not support some of the `flags`.

EMSGSIZE The `length` of the message is too large.

ECONNRESET The remote peer closed the connection.

EPIPE This error code is returned when the conditions to raise a `SIGPIPE` signal are satisfied, but signal generation is disabled by the `MSG_NOSIGNAL` flag.

In addition, `send` may also detect and report the following additional error conditions:

EACCESS The calling process does not have sufficient privileges to send data.

ENETDOWN The local network interface that should have been used to reach the target is down.

ENETUNREACH The target network is unreachable.

ENOBUFS The system lacks the buffer space needed to carry out the operation.

With respect to `send`, the `sendto` function is more flexible because it enables the caller to explicitly specify a destination address by means of the `dest_addr` and `dest_len` arguments, making it suitable for connectionless sockets on which `connect` has not been used:

```
ssize_t sendto(int socket, const void *message,
    size_t length, int flags,
    const struct sockaddr *dest_addr, socklen_t dest_len);
```

When invoked on a connection-oriented socket, `sendto` silently ignores the destination address given as argument. When used on a connectionless socket on which `connect` has never been used, `sendto` simply sends the message to the address given as argument. Instead, when `connect` has previously been used on the socket, the implementation has two choices:

- Send the message to the address given as argument anyway, thus overriding the one set by `connect`.
- Fail and report the `EISCONN` error code.

As usual in the POSIX *sockets* interface, the destination address is specified by means of two additional arguments:

- A pointer `dest_addr` to the data structure that holds the address.
- The size of the data structure, `dest_len`, in bytes.

The other arguments have the same meaning as for the `send` function. As `send` does, `sendto` also returns the number of bytes of data that have successfully been moved into network buffers for transmission, or −1 to indicate an error. The `sendto` function shall detect and report the same error conditions as `send`, plus:

EAFNOSUPPORT The target address referenced by `dest_addr` is not valid for the address family of the local `socket`.

EISCONN The `sendto` function has been invoked on a connectionless socket on which `connect` has been used, and the implementation opted to fail in this scenario.

The `sendto` function may also detect and report the same additional error conditions as `send`, plus:

ENOMEM The system does not have enough memory to carry out the operation.

The `sendmsg` function is the most general of the group. The most important difference with respect to the previous ones is that it can gather the data to be sent from a set of memory buffers of various sizes, instead of accepting as argument a single, monolithic buffer. This makes application-level memory management more

TABLE 11.3

Fields of a `struct msghdr`, POSIX API

Field	Purpose
msg_name	Pointer to a communication endpoint address
msg_namelen	Size of the address pointed by msg_name
msg_iov	Pointer to an array of struct iovec, used to gather/scatter the message to be sent/received
msg_iovlen	Number of elements in the array pointed by msg_iov
msg_control	Pointer to ancillary data, beyond the scope of this book
msg_controllen	Length of the ancillary data referenced by msg_control
msg_flags	Flags on received messaged, unused when sending

efficient because, for instance, the application can leverage the advantages of fixed-size rather than variable-size dynamic memory allocation, and at the same time it also avoids extra memory-to-memory copy operations, which are often expensive, especially on low-end microcontrollers.

To avoid having too many arguments, only the `flags` are passed as an argument on their own, while most of the information `sendmsg` needs must be stored in a structure of type `struct msghdr`, to be allocated by the caller. Then, the address of this structure shall be passed to `sendmsg` as a single argument, `message`:

```
ssize_t sendmsg(int socket,
    const struct msghdr *message, int flags);
```

The `struct msghdr` includes the fields listed in Table 11.3. Besides the fields `msg_name` and `msg_namelen`, which hold a pointer to the destination address and its size, respectively, the data structure also contains the two fields `msg_iov` and `msg_iovlen`. As illustrated in Figure 11.2, these two fields give to `sendmsg` the information it needs to gather the data to be sent from a number of non-contiguous buffers in memory.

In particular, the `msg_iov` field of the `struct msghdr` points to an array of `struct iovec`. The `msg_iovlen` field contains the number of *elements* of the array, not to be confused with its size in bytes. Each element of the array corresponds to, and describes, a contiguous memory buffer that holds part of the data to be sent. As shown in the figure, each `struct iovec` contains two fields:

- The `iov_base` field points to the *base* (the lowest memory address) of the buffer.
- The `iov_len` indicates its *length*, or size, in bytes.

In order to gather the data to be sent, the `sendmsg` function scans the array referenced by `msg_iov` starting from the first element and logically concatenates the data it finds in the memory buffers (depicted as dark gray rectangles in Figure 11.2).

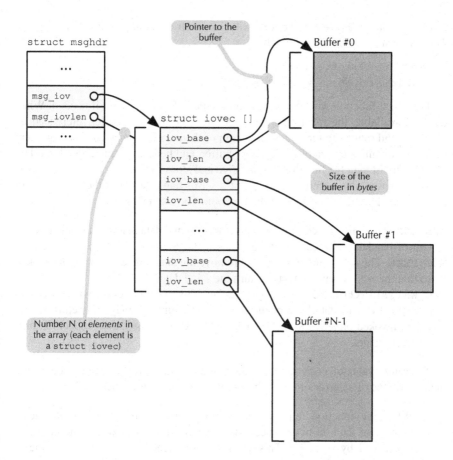

FIGURE 11.2 Data gathering and scattering through a `struct iovec`.

It is important to remark that this is a *logical* operation, which may or may not correspond to a physical copy of the data into a single, contiguous memory buffer before transmission. As recalled previously, this is a computationally expensive operation that protocol stack implementations try to avoid if possible.

The other fields of the `struct msghdr` are related to the transmission and reception of ancillary data, whose meaning is highly dependent on the underlying communication protocols and are therefore beyond the scope of this book. The `sendmsg` function returns zero when successful. Otherwise, it returns −1 after setting `errno` to an error code. The error codes are the same as for `sendto`. In addition, `sendmsg` shall fail and report the `EMSGSIZE` error code if the number of elements of the array pointed by `msg_iov` exceeds `IOV_MAX`, whose value is guaranteed to be at last 16 across POSIX-compliant implementations.

Symmetrically, the `recv`, `recvfrom`, and `recvmsg` functions allow a process to wait for and receive incoming data from a socket. These functions have different levels of expressive power and complexity, like their data transmission counterparts. The simplest one is the `recv` function:

```
ssize_t recv(int socket,
    void *buffer, size_t length, int flags);
```

The `recv` function waits for data to be available from `socket`, if needed. When data are available, the function stores them into the memory buffer referenced by `buffer`, and returns to the caller the length of the data just received. The argument `length` contains the size of the memory buffer, and hence, sets the upper bound on the quantity of data the function may receive and store into it.

The function also accepts as argument a set of flags, stored in the `flags` argument, which alter the semantics of the function. The main flags are:

MSG_OOB The `recv` function will receive out-of-band data instead of ordinary data. Out-of-band data are sent by giving the same flag to `send`, `sendto`, or `sendmsg`.

MSG_PEEK The `recv` function will not remove the data it received from the socket, so that the next receive will still return the same data

MSG_WAITALL On a `SOCK_STREAM` socket, the `recv` function will wait until `length` bytes of data have been received. The function may still return earlier if it is invoked on a `SOCK_DGRAM` socket, if the peer closes the connection, or if an error occurs.

The return value of `recv` indicates the amount of data that has been received and the occurrence of an error or other abnormal conditions. More specifically:

- When the function returns a non-zero positive value, it represents the amount of data, in bytes, successfully received and stored into the buffer referenced by `buffer`. This value cannot exceed `length`, that is, the length of the buffer itself.
- A return value of zero indicates that the peer shutdown the connection in a normal way and no more data were available to be received. This corresponds to the "end of file" indication for regular files.
- The value −1 is an error indication. In this case, the function also sets `errno` as described below to provide more information about the error.

For a `SOCK_STREAM` socket, even if no errors occur, the `recv` function may return after receiving more than zero, but less than `length` bytes of data, unless the `MSG_WAITALL` flags has been set in the `flags`.

It is important that the application is prepared to deal with this scenario correctly because the exact amount of data actually received by `recv` cannot be predicted. In the case of TCP it depends, among other things, on the inner workings of the segmentation mechanism of the underlying transport protocol. On the same lines, it is useful to remark once more that `SOCK_STREAM` sockets treat data as an unstructured stream of bytes that flows from the source to the destination socket and do not preserve message boundaries in any way.

In any case, although the `recv` function may arbitrarily split incoming data into pieces and return prematurely, it will never discard data. On the other hand, `SOCK_DGRAM` sockets do preserve message boundaries, but this may lead to other unexpected buffer-related issues. In particular, the `recv` function always considers the message as an indivisible unit for this kind of socket. If the message length exceeds the length of the buffer provided by the user, the extra bytes are silently discarded.

The `recv` function shall fail for the following reasons and set `errno` to the corresponding error code:

EBADF The given file descriptor `socket` is invalid.

ENOTSOCK The given file descriptor is valid but does not refer to a socket.

EOPNOTSUPP The underlying protocol does not support some of the `flags`.

ENOTCONN The given socket is connection oriented but it is not currently connected to a peer, and hence, no data can possibly be received from it.

ECONNRESET The remote peer forcibly closed the connection. In case of orderly connection shutdown, the `recv` does not report an error, but returns zero instead.

EINVAL The caller requested out-of-band data by setting the `MSG_OOB` flag, but the socket does not support out-of-band data.

ETIMEDOUT The connection establishment process timed out before the connection could be established or a transmission timeout occurred afterwards.

In addition, `recv` may also fail if:

ENOBUFS The system lacks the buffer space needed to carry out the operation.
ENOMEM The system does not have enough memory to carry out the operation.

With respect to `recv`, the `recvfrom` function has two extra arguments that enable the caller to retrieve the address of the sending socket along with data:

```
ssize_t recvfrom(int socket,
    void *restrict buffer, size_t length, int flags,
    struct sockaddr *restrict address,
    socklen_t *restrict address_len);
```

In particular, if the `address` argument is not `NULL`, `recvfrom` stores into the data structure pointed by `address` the address of the sending socket. Also in this case, the location referenced by `address_len` has a twofold meaning: It is used to specify both the size of the data structure referenced by `address` and provided by the caller (as input to the `recvfrom` function), as well as the actual length of the address that `recvfrom` stored there (as output from the `recvfrom` function itself). The meaning of `recvfrom`'s return value and the possible values of `errno` are the same as for `recv`.

The most powerful and flexible receive function is `recvmsg`:

```
ssize_t recvmsg(int socket,
    struct msghdr *message, int flags);
```

As for `sendmsg`, most of the arguments to `recvmsg` must be stored by the caller in a data structure of type `struct msghdr`, whose address is then passed as the single argument `message`. The `sendmsg` function also uses some fields of this structure to provide additional information to the caller. Only the `flags` are passed as an argument on their own. The fields of the `struct msghdr` are the ones listed in Table 11.3. They are used in a similar was as `sendmsg` does, with the following differences:

- The fields `msg_name` and `msg_namelen` are used to return the address of the sending socket, like the `address` and `address_len` arguments of `recvfrom`.
- The information referenced by `msg_iov` is used to *scatter* the receive data into multiple receive buffers, instead of gathering.
- The field `msg_flags` provides additional information about the received data, instead of being unused.

More specifically, the `msg_flags` field may contain the bitwise or of the following flags:

MSG_OOB The received data consist of out-of-band data rather than ordinary data.

MSG_TRUNC The ordinary data received from a `SOCK_DGRAM` socket were truncated because the caller-provided buffers were not big enough to hold the whole message.

The meaning of `recvmsg`'s return value are the same as for `recv` and `recvfrom`. In addition to the possible values of `errno` used by `recv`, `recvmsg` shall also fail if:

EINVAL The sum of all the `iov_len` fields of the array elements referenced by the `msg_iov` field of the `struct msghdr` given as argument cannot be represented by an `ssize_t` data type.

EMSGSIZE The number of elements of the array pointed by the `msg_iov` field of the `struct msghdr` passed as argument—given in the `msg_iovlen` field of the same data structure—exceeds `IOV_MAX`. The value of `IOV_MAX` is guaranteed to be at last 16 across POSIX-compliant implementations.

Finally, if we look at `SOCK_STREAM` sockets, we may notice some important analogies with the sequential file access model specified in POSIX. Both provide reliable access to a sequential, byte-oriented stream of data without any notion of record or message boundaries. For this reason, POSIX specifies that the file-access primitives `write` and `read`, with their standard signature, shall also be applicable to sockets:

```
ssize_t write(int fildes, const void *buf, size_t nbyte);
ssize_t read(int fildes, void *buf, size_t nbyte);
```

The mapping between `write` and `read` onto `send` and `recv`, respectively, is extremely straightforward. Namely, the function calls:

```
write(fildes, buf, nbyte);
read(fildes, buf, nbyte);
```

are implemented as:

```
send(fildes, buf, nbyte, 0);
recv(fildes, buf, nbyte, 0);
```

by the protocol stack.

11.7 SOCKET OPTIONS

Socket options are used to customize the general behavior of a socket, as well as some aspects of the underlying communication protocols. In addition, some read-only socket options provide useful information about the state of a socket. Options also have sensible default values, which are automatically applied if they are not explicitly set by the application. Default values are adequate in the majority of applications.

Socket options are organized hierarchically, in multiple *levels*, which reflect the typical layered structure of communication protocols and software. Within a level, individual options are uniquely identified by an option *name*, a preprocessor macro that expands into an integer value.

Options have a value associated with them. Since the data type of the value may change from one option to another, values are passed by reference by means of a generic void * pointer and an ad-hoc mechanism is used to convey the intended and actual size of an option value when interacting with the protocol stack. The actual mechanism is similar to the one used to accommodate variable-length network address, described in the previous sections.

Socket options are retrieved by means of the getsockopt function:

```
int getsockopt(int socket, int level, int option_name,
    void *restrict option_value,
    socklen_t *restrict option_len);
```

where:

- The socket argument is a socket descriptor and indicates which socket the function must operate on.
- The level and option_name arguments are the level and the unique name within the level of the option to be retrieved, respectively.
- The option_value points to a caller-provided memory buffer in which the function shall store the value of the option.
- The object referenced by option_len is both an input and an output argument of the getsockopt function. As an input argument, it indicates the length, in bytes, of the buffer pointed by option_value. As an output argument, getsockopt sets it to the actual length of the option it stored in the buffer.

To avoid buffer overflows, the getsockopt function silently truncates the option value if it is too big to fit in the caller-provided buffer. Since, as a consequence, the option value would most likely be unusable, this scenario should however be considered a programming error and avoided.

The return value of getsockopt indicates whether the function succeeded or failed. Namely, a return value of zero means that the function stored the value of the option into the buffer pointed by option_value as requested by the caller. Instead a return value of −1 signifies that an error occurred. In this case, getsockopt also sets errno to an error code that provides more information about the error. The getsockopt function shall fail if:

EBADF The given file descriptor socket is invalid.
ENOTSOCK The given file descriptor is valid but does not refer to a socket.
EINVAL The option specified by option_name is not valid at the given level.
ENOPROTOOPT The option option_name is valid, but either the socket or the underlying network protocols, depending on level, do not support it.

Moreover, getsockopt may also fail if:

EACCESS The caller does not have sufficient privileges to retrieve the option.
EINVAL The option cannot be retrieved because the socket has been shutdown.
ENOBUFS The system lacks the resources needed to carry out the operation.

Symmetrically, an option can be set by invoking the setsockopt function:

```
int setsockopt(int socket, int level, int option_name,
    const void *option_value, socklen_t option_len);
```

The arguments to the function have the same meaning as for the getsockopt function, the only differences being:

- The memory buffer pointed by option_value must be preset by the caller to the desired value of the option being set. The setsockopt function does not change its contents in any way.
- The option_len argument is the actual *length* in bytes of the memory buffer pointed by option_value, rather than a *pointer* to the length, because it is an input-only argument to setsockopt.

The return value of the setsockopt function has the same meaning as the one of getsockopt. The reasons why setsockopt shall fail are also the same, plus:

EISCONN The socket is connected and the given option cannot be set while the socket is in this state.
EDOM The desired value of the option is outside the allowed domain. For instance, if the option has a timeout as a value, the timeout may be excessive.

In addition, setsockopt may also fail if:

ENOMEM The system does not have enough memory to carry out the operation.

As described previously, the way of specifying a socket option in POSIX has been modeled after the typical layered structure of the underlying communication protocols and software. In particular, each option is uniquely specified by a (level, name) pair, in which:

- The level indicates the protocol level at which the option is defined. The protocol level can be one of the protocol identifiers defined by the protocol stack headers as symbolic constants, and hence, correspond to a real communication protocol. For instance, the IPPROTO_TCP symbolic constant corresponds to the TCP protocol.
  In addition, the special level identifier SOL_SOCKET is reserved for the uppermost layer of the hierarchy, that is, the socket level itself, which does not have a direct correspondence with any communication protocols.
- The name determines the option to be set or retrieved within the level. Although names are represented as integer numbers, applications should never use those numbers directly because they are system-dependent. Instead, they should make use of the symbolic constants defined in the protocol stack headers and discussed in the following.
  The name of an option implicitly determines the data type of the object referenced by the option_value argument to getsockopt and setsockopt and, as a consequence, the meaning of its contents, which represent the option value.

Tables 11.4 and 11.5 summarize the general options defined at the SOL_SOCKET level, divided into two categories: *read-write* options, which can be set and retrieved at will by applications, and *read-only* options. By definition, any attempt to set a read-only option shall fail.

These tables do not list buffer-related options, which will be described separately. Some of the options deserve further explanations in the context of the standard TCP/IP protocol stack implemented by RTEMS. They are discussed in the same order as they appear in the tables.

- In the standard RTEMS protocol stack, only the TCP protocol module has functions that record and output debugging information. They are enabled only if the protocol stack has been compiled with the compile-time option TCPDEBUG enabled, which is not the default. Unless this option is enabled, setting the SO_DEBUG option succeeds, but has no effect.
- Although the TCP protocol does not foresee a dedicated request to probe whether a connected peer is still reachable, an equivalent effect can be obtained by periodically sending an acknowledgment message containing a zero-length data segment. This is the approach taken by the RTEMS TCP/IP protocol stack when the SO_KEEPALIVE socket option is set.
- The value associated with the SO_LINGER option is a struct linger that contains two fields. The l_onoff field is an integer treated as a

TABLE 11.4

Read-Write General Options Defined at the `SOL_SOCKET` Level, POSIX API

Name	Type	Meaning
SO_BROADCAST	int [1]	Determines whether the socket is allowed to transmit broadcast datagrams, only for sockets of type SOCK_DGRAM.
SO_DEBUG	int [1]	Enables or disables the debugging of socket activities. The actual amount of debugging information provided depends on whether and to which extent the underlying protocol modules support this option.
SO_DONTROUTE	int [1]	If set, normal network-layer routing algorithms are disabled and routing is performed only based on the destination address. This implies that the destination node must be directly reachable by one of the sender's network interfaces.
SO_KEEPALIVE	int [1]	On a connection-oriented socket, determines whether or not messages meant to check whether the connection is alive (often called *keepalive* messages) should be periodically exchanged between peers. Must be supported by the underlying communication protocols.
SO_LINGER	struct linger	Determine how queued, but still unsent, data are handled when the socket is closed. See the description in the main text for more information.
SO_OOBINLINE	int [1]	If set, out-of-band data are placed together with ordinary data in the receive queue, rather than being enqueued separately.
SO_RCVTIMEO	struct timeval [2]	Specifies the maximum amount of time a receive operation from the socket is allowed to wait for incoming data before failing and returning a timeout indication. A value of zero (the default) means that receive operations never time out.
SO_SNDTIMEO	struct timeval [2]	Specifies the maximum amount of time a send operation is allowed to block the caller, in order to wait until all data can be enqueued for transmission. When this amount of time elapses the function returns to the caller anyway, although it was able to enqueue only part of the data. A value of zero (the default) means that send operations never time out.
SO_REUSEADDR	int [1]	When set, this option relaxes the algorithm that checks whether the address given to bind can be used or not, to enable an easier/faster reuse of local addresses. See the description in the main text for an explanation of how this option affects the TCP protocol.

[1] Used as a Boolean, a non-zero value meaning *true* and zero meaning *false*.

[2] Not to be confused with `struct timespec`, see Section 11.8 for more information.

TABLE 11.5

Read-Only General Options Defined at the SOL_SOCKET Level, POSIX API

Name	Type	Meaning
SO_ACCEPTCONN	int [1]	Tells if the socket is accepting connection requests, that is, it has been marked with listen.
SO_ERROR	int	Retrieves and then clears any pending error number on the socket, returns 0 if there are no pending errors.
SO_TYPE	int	Returns the socket type (SOCK_...), useful when inheriting a socket created by others.

[1] Used as a Boolean, a non-zero value meaning *true* and zero meaning *false*.

Boolean flag. A non-zero value indicates that the option is enabled. The l_linger field is an integer, too, which contains the linger time in seconds. The contents of this field are used only if the linger option is enabled according to l_onoff.

The SO_LINGER option only applies to connection-oriented sockets and affects the behavior of a close operation performed on a socket while the socket has some queued, but still unsent, data. When the option is disabled, close shall return to the caller as quickly as possible, without waiting until those data are sent. Afterwards, the protocol stack may or may not perform further attempts to transmit them. As a consequence, there is no way for the caller to know whether any of those data will eventually reach the other peer or not.

If the SO_LINGER option is enabled with a zero l_linger time, executing close aborts a TCP connection (by sending an RST) instead of closing it gracefully. As a result, close returns immediately to the caller but any queued, unsent data are discarded. Another scenario in which some TCP implementations may do the same, that is, abort the connection instead of closing it gracefully, is when close is called on a socked that has some data in its *receive* buffers. In this case, the abort makes the peer aware that some of the data it sent (the ones in the receive buffer) are not going to be received and processed.

Instead, when the option is enabled and the l_linger time is not zero, close *blocks the caller* until either all queued data have been sent and acknowledged by the peer, or the time indicated by l_linger elapsed. However, this is still not enough to guarantee that the peer *application* has actually received and processed all data—they may have been acknowledged by the peer TCP, but still be in the peer socket receive buffers. In order to ensure end-to-end, application-level data delivery an explicit, an application-level acknowledge is generally needed.

- The SO_RCVTIMEO and SO_SNDTIMEO set the timeout of blocking receive and send functions, respectively. The timeout represents the maximum amount of time these functions are allowed to block the caller if no data are available to be received, or no data can be transferred from user to system buffers for transmission. If no data at all were actually transferred before the timeout expired, the operations report a failure by returning -1 and set `errno` to EAGAIN or EWOULDBLOCK.

The timeout is expressed by means of a `struct timeval`, which holds a relative time value written as an integral number of seconds and a fractional part in microseconds. This is also the same data structure used to represent the timeout of the `select` function, to be discussed in Section 11.8.

- When a TCP connection is closed gracefully, the socket that took the initiative of closing it (the so called *active close* of a connection) ends up in the TIME_WAIT state and remains in that state for a relatively long period of time, equal to twice the TCP Maximum Segment Lifetime (MSL). In turn, the MSL represent the maximum amount of time a TCP segment could possibly be active in the network before it is discarded. It is specified to be two minutes in [100], although some implementations may use smaller values.

  Sockets that responded to a close request without actively starting it (performing a *passive close*) do not enter the TIME_WAIT state. Also, if a connection is aborted rather than closed gracefully, neither of the peers enters the TIME_WAIT state.

  If the SO_REUSEADDR option is not set, a new socket cannot be bound to a local address and port number combination (by means of the `bind` function) for which a socket in the TIME_WAIT state currently exists. Although this behavior is generally needed to avoid interpreting residual TCP segments from an old connection as segments belonging to a new connection, in some cases it may be useful to immediately reuse a local address. This can be done by setting the SO_REUSEADDR option to a non-zero value before invoking `bind`.

Another group of options, listed in Table 11.6, is related to how transmit and receive data buffers associated with the socket are handled:

- The SO_RCVBUF and SO_SNDBUF options set the size in bytes of the receive and transmit data buffers, respectively.
- The SO_RCVLOWAT and SO_SNDLOWAT set the threshold in bytes of the receive and transmit "low-water" mechanism, respectively, as discussed in the table.

The receive low-water mechanism affects both the blocking socket operations discussed so far and the non-blocking operations to be discussed together with synchronous I/O multiplexing in Section 11.8. The send low-water mechanism is of importance only when non-blocking operations are in use.

The POSIX standard does not specify any standard options for protocols below the SOL_SOCKET level, with the exception of the IPv6 protocol that the standard

TABLE 11.6

Buffer-Related Options Defined at the `SOL_SOCKET` Level, POSIX API

Name	Type	Meaning
SO_RCVBUF	int	Sets the size, in bytes, of the receive buffers associated with the socket. The default value is protocol and implementation-dependent.
SO_SNDBUF	int	Sets the size, in bytes, of the send buffers associated with the socket. The default value is protocol and implementation-dependent.
SO_RCVLOWAT	int	The value of this option determines the minimum amount of incoming data, in bytes, which a receive operation waits to be available before returning. The default value is 1 and causes receive operations to return as soon as any data are available. Higher values may improve efficiency because they reduce the number of distinct receive operations needed to receive a given amount of data. [1]
SO_SNDLOWAT	int	The value of this option determines the minimum amount of send buffer space that must be available for the synchronous I/O multiplexing function (Section 11.8) to signal the ability to transmit through the socket. It does not affect ordinary, blocking send operations that, unless an error occurs, process all the data they have been given. The default value is implementation and protocol-dependent.

[1] Receive operations may still return less data than requested by this options if an error occurs or a signal is caught.

RTEMS protocol stack does not support. Therefore, although these options may be very convenient for some classes of application, their use makes the code potentially non-portable.

It should be noted, however, that often these portability issues are not of practical concern because many commonly used TCP/IP protocol stacks (including the ones available in RTEMS) were derived from the original BSD stack [86] and retained its TCP- and IP-level options. As an example, Table 11.7 summarizes the options available for the TCP protocol (that is, at the `IPPROTO_TCP` level) in the standard RTEMS protocol stack.

TABLE 11.7

TCP-Level Options of the Standard RTEMS Protocol Stack

Name	Type	Meaning
TCP_NODELAY	int [1]	Under some circumstances that depend on the presence of sent but not yet acknowledged data, the TCP implementation temporarily gathers data coming from multiple application-level send operations to transmit them all together at a later time. This behavior improves throughput because it avoids the transmission of multiple small segments on the network, but may also increase data transfer latency significantly. This option, when set, disables the gathering algorithm. By default this option is not set, that is, the gathering algorithm is enabled.
TCP_MAXSEG	int	The value of this option determines the maximum size of outgoing TCP segments. By default, it is equal to the maximum size negotiated between the peers when the TCP connection was established. Any value set by the application must not exceed the negotiated value, otherwise setsockopt fails with the EINVAL error number.
TCP_NOOPT	int [1]	Most TCP implementations may include a number of *options* in the TCP packet headers they send. Options are a standard mechanism conceived to support various extensions to the TCP protocol by adding optional contents to its packet headers. When this option is not set (the default) the TCP implementation is allowed to send the options corresponding to the various protocol extensions it supports. If this flag is set, options are disabled in an effort to improve compatibility with other TCP implementations, although conforming implementations should ignore any option they do not recognize anyway.
TCP_NOPUSH	int [1]	When this option is not set (the default) the TCP implementation normally starts transmitting immediately after a call to any of the send functions, if allowed to do so by the protocol state. Moreover, it marks the last segment transmitted with the *push* (PSH) bit, to recommend the peer to make the data received so far available to the application. If this option is set, the TCP implementation will instead delay data transmission until the socket is closed or its send buffers are full.

[1] Used as a Boolean, a non-zero value meaning *true* and zero meaning *false*.

11.8 NON-BLOCKING I/O AND SYNCHRONOUS I/O MULTIPLEXING

The description of the *sockets* API functions in Sections 11.2–11.7 referred to their default *blocking* mode. Informally speaking, when operating in this mode, these functions block the caller until they can proceed with at least a partial success or an error occurs. For example, the recv function, when called with no special flags set, blocks the caller until there are some data available to be received from the socket (even though the amount of data is less than what the caller requested) or an error occurs.

The same functions can also operate in *non-blocking* mode. In this case, they always return to the caller immediately, without blocking it, and signal whether or not they were able to perform all or part of the operation by setting errno appropriately. For instance, invoking the recv function in this operating mode enables the caller to *poll* the socket and check whether data are immediately available to be received from it, but without blocking until they arrive.

The default behavior in which socket functions block the caller until completion is quite useful in many cases, since it allows the software to be written in a simple and intuitive way. However, it may become a disadvantage in other, more complex situations because, especially for connection establishment and data transfer in connection-oriented communication, it may tie the timings of the two communicating peers in an excessive and undesirable way. The SO_RCVTIMEO and SO_SNDTIMEO socket options (see Section 11.7 and Table 11.4) help alleviate this issue for what concerns data transfer functions, but do not affect the behavior of other kinds of function.

There are three possible ways to switch a socket to non-blocking mode: by means of the fcntl function invoked on the socket, by means of the ioctl function also invoked on the socket, or by passing the MSG_DONTWAIT flag to any function that supports flags as an argument. The most important differences among the three methods are:

- The use of the fcntl function is the only one explicitly mentioned in the POSIX standard. The other two methods are generally available, especially on protocol stacks derived from the original BSD stack [86], but neither of them is standard.
- Both fcntl and ioctl operate at the socket level, and hence, affect the behavior of all future operations performed on that socket until the setting is changed again.
- The MSG_DONTWAIT flag is finer-grained since it works on a call-by-call basis, but it can be used only with functions that support a flags argument, either directly (send, sendto, recv, and recvfrom) or indirectly through a struct msghdr (sendmsg and recvmsg).

Moreover, selecting the non-blocking behavior at the socket level by means of fcntl or ioctl overrides the MSG_DONTWAIT flag passed upon individual function calls. In other words, if a socket has been set to be non-blocking, all functions invoked on it will be non-blocking regardless of the MSG_DONTWAIT flag.

A socket can be switched to non-blocking or blocking mode by setting or clearing, respectively, the O_NONBLOCK flag associated to its socket descriptor, by means of the fcntl function:

```
int fcntl(int fildes, int cmd, ...);
```

In this particular usage pattern, the fildes argument of fcntl represents the socket descriptor and cmd must be equal to F_SETFL, which corresponds to the "set flags" command of fcntl. The third argument, corresponding to the ellipsis ... in the signature, must be an int that holds the bitwise OR of the flags. The fcntl function returns a value other than −1 if it was able to successfully set the flags. Otherwise, it returns −1 after setting errno to an error code. Reasons for failure are:

EBADF The given file descriptor fildes is invalid.
EINVAL The command cmd is invalid. It may not happen if cmd is F_SETFL because the implementation of this command is mandatory.

Since the flags associated with a socket descriptor may include other flags besides O_NONBLOCK, and some of them can be used and manipulated by the system behind the scenes, it is advisable to never forcibly set the flags to a specific value, but always work on them in a read-modify-write fashion. Therefore, in order to set the O_NONBLOCK flag without disrupting the other flags inadvertently, the application should first of all read the current flags associated with the socket descriptor, then perform a bitwise OR of the value with O_NONBLOCK, and finally set the flags according to the result.

To this purpose, the fcntl function can also be used to retrieve the flags by means of the F_GETFL command. When fcntl is used in this way, its fildes argument still represents the socket descriptor, cmd must be F_GETFL, and the function takes no additional arguments. It returns the current value of the flags, which is guaranteed to be a non-negative value, or −1 upon error. The possible values of errno are the same as described previously.

As an example, the following fragment of code sets the O_NONBLOCK flag on socket descriptor s:

```
int fl;

if((fl = fcntl(s, F_GETFL)) == -1)
{
    /* error handling */
}
else
{
    fl |= O_NONBLOCK;    /* Use fl &= ~O_NONBLOCK to clear
    the flag */
    if(fcntl(s, F_SETFL, fl) == -1)
    {
        /* error handling */
```

```
    }
}
```

The second way to switch a socket in and out of non-blocking mode is to invoke the ioctl function on the socket descriptor:

```
int ioctl(int fildes, int request, ...);
```

Like fcntl, also ioctl is a generic function, able to carry out a variety of commands, also called *requests*, on a file descriptor. They are encoded in the request argument. Depending on the command, ioctl may require additional arguments. Here we will focus only on two commands especially relevant in this context.

1. The FIONBIO command allows the caller to set or reset the O_NONBLOCK socket flag, like fcntl does. For this command, ioctl requires a third argument of type int *. The location referenced by it must contain an integer, which the function interprets as a Boolean. If its value is not zero (Boolean true) the socket is put in non-blocking mode. Instead, if the value is zero (Boolean false) the socket reverts to the default, blocking mode.

   Unlike fcntl, ioctl operates only on the non-blocking flag of the socket descriptor. Therefore, there is no need to read the current flags, modify them, and write them back.

2. The FIONREAD command lets the caller know how many bytes of data are immediately available and waiting to be received from the socket at the moment, without actually retrieving those data or removing them from the receive buffers. Also in this case, ioctl requires a third argument of type int *, which it uses to store the value requested by the user.

As for many other functions, the return value of ioctl indicates whether the function completed successfully or not. A return value of zero means successful completion, while −1 denotes that an error occurred. In the last case, the function also sets errno to provide more information about the error itself. For the specific requests just described the ioctl function shall fail if:

EBADF The given file descriptor fildes is invalid.

The third and last method is probably the simplest because it merely implies passing an additional flag (MSG_DONTWAIT) to the socket-related functions and allows the non-blocking behavior to be selected on a call-by-call basis. However, as outlined previously, it has two main shortcomings:

- Not all socket-related functions support flags as an argument. For instance, this method cannot be used in the connection establishment phase.
- Although it is widely available this method is not standardized, and hence, it should be avoided when writing code meant to be portable.

Regardless of how it is activated, non-blocking mode has a profound effect on the behavior of many socket-related functions, also affecting the possible error codes they may generate. In particular:

- The `bind` function will return −1, a failure indication, and set `errno` to `EINPROGRESS` if it is unable to perform the requested local address assignment immediately. However, this does *not* actually indicate a failure because the system will perform and complete the address assignment asynchronously.
- If no connection requests are immediately available to be accepted, the `accept` function will fail—returning −1 and setting `errno` to either `EWOULDBLOCK` or `EAGAIN`—instead of blocking the caller until there is one.
- If the `connect` function is unable to establish a connection immediately it will return −1 and set `errno` to `EINPROGRESS` instead of blocking the caller until connection establishment is complete. The system will establish the connection asynchronously.
- The receive functions will not block the caller if no data are available to be received immediately. Instead, they will return −1 and set `errno` to either `EWOULDBLOCK` or `EAGAIN`. A return value of zero is still reserved to indicate that the connection has been shutdown normally (for connection-oriented sockets).
- Similarly, the send functions will not block the caller if no data can immediately be transferred from user to system buffers. Instead, they will return −1 and set `errno` to either `EWOULDBLOCK` or `EAGAIN`.

Somewhat contrary to the general approach just discussed, putting a socket in non-blocking mode does *not* affect the blocking that may occur in the `close` function due to the lingering interval discussed in Section 11.7.

Switching sockets to non-blocking operation mode enables a single process or thread to service multiple sockets without running the risk of ever blocking on one of them, thus neglecting the others.

However, as it generally happens with many polling-based approaches to input-output, choosing the best polling frequency for each socket may be challenging because it represents a trade-off between improving reactivity (which would push toward a higher polling frequency) and keeping overheads stemming from useless polling under control (which would of course benefit from a lower polling frequency).

For this reason, the POSIX standard also supports an event-driven, rather than time-driven, approach to socket input-output. With this method, socket operations are carried out in two stages, usually enclosed in an infinite loop:

1. The thread *waits* until some event of interest occurs on a *set of sockets*. The occurrence of an event on a socket signifies that an input–output operation became possible on the socket, or the system detected an error or an exceptional condition that affects the socket. The wait operation terminates and returns to the caller as soon as at least one event of interest occurs on *any* of the sockets in the set. Therefore, incoming events are handled in a first-in, first-out fashion regardless of

which socket they relate to. In addition, the wait function can provide information on exactly which events have occurred, to facilitate further processing.

2. Based on the information provided by the wait function invoked in the previous step, the thread performs one socket operation for each event. The operations to be performed on a socket depend on the events that occurred on the socket itself. It is unnecessary to set the socket to non-blocking mode because the occurrence of an event implies that, even though the corresponding socket operation could potentially block the caller, it will not.

For instance, one of the events of interest is "ready for reading." When this event occurs on a socket, it is guaranteed that a receive function (`recv`, `recvfrom`, or `recv`) invoked on the socket would not block the caller. However, the outcome of the function does not necessarily imply any actual data transfer. The function may indeed return some data, but it may also indicate that the remote peer has closed the connection normally or that an error occurred on the socket.

Overall, this mechanism is sometimes called *synchronous input–output multiplexing* because thread activities are still performed in a synchronous way after a passive wait. Input–output operations on different sockets are multiplexed based on the order of arrival and carried out sequentially.

The simplest and, historically, the oldest event waiting function specified by the POSIX standard is `select`:

```
int select(int nfds, fd_set *restrict readfds,
    fd_set *restrict writefds, fd_set *restrict errorfds,
    struct timeval *restrict timeout);
```

The `select` function takes as arguments three pointers to three sets of socket descriptors: `readfds`, `writefds`, and `errorfds`. Each set is represented by an object of type `fd_set` and the sets may partially or completely overlap, that is, the same socket descriptor may belong to more than one set. The `nfds` argument limits the range of socket descriptors to be checked. Namely, the `select` function will check only the socket descriptor whose numerical value is between 0 and `nfds-1` included. Numerically higher socket descriptors are ignored, even though they belong to the sets previously described.

The `select` function supports three possible events of interest for each socket: "ready for reading", "ready for writing", and "exceptional condition." The way the three sets of socket descriptors are constructed indicates whether or not the caller is interested in these events. For instance, if a socket descriptor belongs to `readfds`, the caller is interested in the "ready for reading" event occurring on that socket.

The `timeout` argument specifies the maximum amount of time the `select` function is allowed to block the caller and is a pointer to an object of type `struct timeval`. A `struct timeval` has two fields that together express a time value:

time_t tv_sec represents the number of seconds, and
time_t tv_usec represents the number of microseconds.

This data type must not be confused with the struct timespec discussed in the previous chapters, in which the fractional part of a second is expressed in *nanoseconds* rather than *microseconds*. The timeout argument can be set in three possible ways:

- A NULL pointer means *no timeout*, that is, the select function will wait indefinitely, until one of the events of interest occurs or it detects an error.
- A pointer to a struct timeval that contains a time value of zero (that is, it has zero in both the tv_sec and tv_usec fields) means that the select function must still check whether the events of interest already occurred in the past and report them back appropriately, but it must not block the caller if none of them occurred yet. This is equivalent to a *polling* operation performed on multiple sockets.
- A pointer to a struct timeval that contains a non-zero time value indicates that the select function must block for at most the specified amount of time, waiting for any of the events of interest to occur. The time value is interpreted in a relative way with respect to the time of the call to select.

The select function examines the descriptors belonging to each set in order to check whether at least one of them is ready for reading, ready for writing, or has a pending exceptional condition, respectively. If the timeout argument permits, the select function may also block the caller if none of the events of interest has already occurred. Upon successful completion, the select function also updates the three sets of socket descriptors to inform the caller about which socket descriptors became ready for the corresponding kind of operation, that is, which events actually occurred.

More specifically, a socket is considered ready for reading if:

- It has been marked with the listen function and there is at least one pending connection request on it, that is, the accept function invoked on the socket would be able to accept a connection request without blocking the caller.
- It has not been marked with listen and there are some pending data in its receive buffers, or the peer has shutdown the connection. In both cases, a receive function invoked on the socket would return without blocking the caller.
- If the socket has been configured to place out-of-band data together with ordinary data (by means of the SO_OOBINLINE socket option), the socket is considered ready for reading also if there are out-of-band data waiting to be received.

A socket is considered ready for writing if:

- A pending connection attempt, initiated by calling connect with the socket in non-blocking mode, has been completed either successfully or unsuccessfully.

- A local address assignment, initiated by calling `bind` with the socket in non-blocking mode, has been completed. In this case, the socket is also marked as ready for reading.
- A send operation of at most `SO_SNDLOWAT` bytes would be able to enqueue this amount of data immediately, without blocking the caller.

Finally, a socket has an exceptional condition pending if:

- There is a pending error condition on the socket.
- A receive operation with the `MSG_OOB` flag set would return a certain amount of out-of-band data without blocking the caller, or out-of-band data are present in the ordinary data receive queue.

Although protocol stacks are free to add additional conditions to the previous lists, they are beyond the scope of the POSIX standard and their use would make the code no longer portable on other systems.

Upon successful completion, the return value of the `select` function is the (non-negative) total number of events that have been reported to the caller, that is, the number of socket descriptors that `select` inserted in the sets of descriptors pointed by `readfds`, `writefds`, and `errorfds`. Otherwise, `select` returns −1 and sets `errno` to indicate which error occurred. The `select` function shall fail if:

EBADF At least one of the file descriptors in the sets referenced by `readfds`, `writefds`, and `errorfds` is invalid.

EINVAL The timeout interval referenced by `timeout` or the value of `nfds` are invalid.

EINTR The function was interrupted by a signal before any of the events of interest occurred.

The file descriptor sets used with the `select` function must be allocated by the caller and initialized before use by means of the `FD_ZERO` function:

```
void FD_ZERO(fd_set *fdset);
```

The function takes as argument a pointer `fdset` to a file descriptor set and "zeroes" it out, that is, initializes it to the empty set. This function, as well as the other functions that manipulate file descriptor sets, can be implemented as a macro on systems where macro expansion is more efficient than inline function expansion. It is important to remember that macros may evaluate their arguments more than once. This may lead to an unpredictable behavior if their arguments are expressions with side-effects, for instance, pre- or post-increments and decrements.

Afterwards, individual file descriptors can be added to, or removed from, a set by means of the functions:

```
void FD_SET(int fd, fd_set *fdset);
void FD_CLR(int fd, fd_set *fdset);
```

Both functions take a pointer `fdset` to a file descriptor set and a file descriptor `fd` as argument, and update the file descriptor set in-place. The numeric value of the file descriptor being added or removed must be in the range from zero to `FD_SETSIZE-1` included, but no error checks are performed. Using an out-of-range file descriptor leads to undefined behavior. Instead, adding a file descriptor to a set that already contains it (or, symmetrically, removing a file descriptor from a set that does not contain it) is allowed and produces no effects.

The last function in this group, `FD_ISSET`, queries whether a file descriptor belongs to a set or not:

```
int FD_ISSET(int fd, fd_set *fdset);
```

As before, the two arguments `fdset` and `fd` are a reference to a set and a file descriptor, respectively. The function returns an integer value to the caller, which must be interpreted as a Boolean value. More specifically, the return value is not zero (true) if file descriptor `fd` belongs to the set referenced by `fdset`, otherwise it is zero (false).

Besides `select`, the POSIX standard specifies two other functions with a similar purpose, `pselect` and `poll`. Although they are not supported by RTEMS at the time of this writing, they will be briefly discussed here anyway for the sake of completeness. The `pselect` function:

```
int pselect(int nfds, fd_set *restrict readfds,
    fd_set *restrict writefds, fd_set *restrict errorfds,
    const struct timespec *restrict timeout,
    const sigset_t *restrict sigmask);
```

is similar to `select`, although there are three important differences:

- The timeout is represented by means of a `struct timespec` instead of a `struct timeval`. This is more in line with other POSIX functions, like the ones discussed in Chapters 5, 7, and 9, which also accept a timeout as argument. As for `select`, a `NULL` pointer denotes an infinite timeout, whereas a pointer to a `struct timespec` that contains zero indicates that `pselect` must immediately return to the caller without blocking.
- It it guaranteed that the `pselect` function never modifies the time value referenced by the `timeout` argument.
- The argument `sigmask` allows the caller to install a specific signal mask for the duration of the call. When `sigmask` is a `NULL` pointer the `pselect` function behaves like `select` from this point of view.

The `poll` function is even more sophisticated and allows the caller to wait for a wider set of conditions related to a file descriptor, rather than just the three broad categories that `select` and `pselect` support:

```
int poll(struct pollfd fds[], nfds_t nfds, int timeout);
```

The array `fds` consists of `nfds` element of type `struct pollfd`. Each of these structures contain the set of conditions the caller is interested in for a certain file descriptor. More specifically:

- The `fd` field contains a file descriptor.
- The `events` field contains the bitwise or of the event flags the caller is interested in for file descriptor `fd`. Each of these event flags is represented by a symbolic constant whose name starts with `POLL`. The POSIX standard specifies about 10 different event flags.
- The `revents` field is set by `poll` to the bitwise OR of the event flags that were true for file descriptor `fd` when `poll` returned. Besides the event flags explicitly requested by the user by means of the `events` field, `poll` may also set other, unsolicited event flags associated with various error conditions.

Another difference with respect to `select` and `pselect` is the way of expressing the timeout. In this case, it is given by an `int` argument that represents a time in microseconds. The special values 0 indicates that `poll` must check whether the requested events occurred and return immediately, without waiting for them. The value −1 denotes an infinite timeout, that is, `poll` will return only when at least one of the requested events occurred or the call was interrupted.

Both synchronous I/O multiplexing and, to some extent, non-blocking I/O were introduced in the past to allow a single-threaded process to manage multiple sockets effectively, on operating systems that did not necessarily support multithreading. It can be argued whether or not these techniques are still useful on more modern systems, on which creating multiple threads and putting them in charge of one socket each, is relatively easy and quite efficient. Appropriate coordination among these threads is not a problem either, because POSIX systems nowadays offer a wide set of inter-thread communication devices presented in Chapters 7 and 9.

However, there are still several advantages in having these functions available:

- They facilitate the maintenance and the migration of legacy code to newer systems, without radically changing its design and, often, without significantly modifying its implementation.
- Polling a socket at a fixed rate instead of handling, for instance, incoming data as they arrive helps limiting the processing time dedicated to communication even if the communicating peers misbehave.

11.9 SUMMARY

This chapter described how applications can get access to network communication by means of the POSIX sockets application programming interface. In Sections 11.1–11.4, we described the main functions that applications can use to create communication endpoints, assign well-known network addresses to them when necessary, establish a connection between endpoints, and close them when they are no longer in use.

Section 11.5 highlighted the differences between connection-oriented and connectionless sockets, and provided more details about the latter kind. Next, Section 11.6 discussed data transfer functions. These functions are mostly uniform across both kinds of socket, but offer different trade-offs between complexity and power.

Some aspects of sockets behavior can be changed by setting a variety of socket options appropriately, a feature we described in Section 11.7. In the last section of the chapter, Section 11.8, we focused on two mechanisms that enable a single-threaded process to manage more than one socket at the same time, namely, non-blocking input–output and synchronous input–output multiplexing.

Part V

---

*Multicores in Real-Time
Embedded Systems*

12 Multicores in Embedded Systems

CONTENTS

This chapter provides an introduction to multicore processors in the context of embedded systems. The first part of the chapter focuses on the motivation behind the widespread diffusion of multicore processors and provides a short overview of their architecture, going from more abstract aspects down to their practical implementation on contemporary ARM processors.

The second part of this chapter is entirely devoted to summarizing the significant challenges introduced by multicore in software development for embedded systems, especially in the areas of task scheduling, schedulability algorithms and analysis, and proper inter-task communication and synchronization. The next chapters will further analyze these issues and discuss some common ways to approach and solve them.

12.1 MOTIVATION

The widespread adoption of real-time embedded systems in multiple, dissimilar application domains, combined with fast-paced, market-driven technological progress, has put a strong emphasis on continuously increasing their computing capacity.

Indeed, this additional capacity is key to accommodate the ever-growing embedded software complexity and size, and to provide end-users with appliances that are sophisticated but, at the same time, easy and convenient to use.

More recently, the demand posed on the computing capacity of embedded systems has increased further because, more often than not, they are moving towards some form of ubiquitous network connectivity, according to the so-called "Internet of Things" trend.

Considering the complexity of contemporary protocol stacks, the overall impact of network communication on the processor load of an embedded system is not at all negligible although, strictly speaking, most related activities are not subject to hard deadlines.

Also on the industrial automation front, embedded systems in charge of controlling some hardware equipment in real time are quickly evolving from a *centralized I/O* architecture, in which peripheral devices—sensors and actuators—are connected directly to the controller with dedicated, discrete buses, to a *distributed I/O* architecture. In a distributed I/O architecture, the controller and its peripherals are connected by means of a real-time communication network, for instance, the Controller Area Network [72], EtherCAT [66] or PROFIBUS [63], just to name a few.

Besides the inherent complexity of its protocol stack, which is often comparable or even exceeds the complexity of TCP/IP, distributed I/O communication is also subject to strict, hard real-time execution constraints. This is because any disruptions of communication timings are bound to hinder control algorithms accuracy and may even lead to their failure.

As a consequence, network communication can no longer be considered a straightforward add-on to an embedded system. On the contrary, its computing requirements must be taken into account right from the design phase.

For years, the most common way of increasing the computing capacity of a processor has been to increase its clock frequency by reducing chip geometries. Moreover, considerable effort has been put into improving the number of instructions per clock cycle a processor is able to complete without sacrificing the apparently sequential execution of machine instructions, that is, transparently to the software being executed.

This goal has been achieved at the expense of circuit complexity, at first by means of pipelines, and then through sophisticated execution techniques that, like the ones used in superscalar processors, dispatch multiple machine instructions for parallel execution on distinct execution units within the processor itself.

Additional hardware ensures that, although instruction execution is carried out in parallel to the maximum extent possible, execution results are still the same "as if" execution had been performed sequentially.

However, in the last decade, continuing on the path of increasing the chip clock frequency has run more and more frequently into problems of excessive power consumption and heat dissipation, along with higher manufacturing costs due to lower yield, thus leading chip makers to define sort of "barriers", that is, upper limits to the clock frequency that are unlikely to be surpassed. Currently, this symbolic barrier is set at around 5 GHz. Indeed, processors that operate beyond that frequency and, at the same time, are suitable for use in embedded systems, are uncommon.

One significant reason is that thermal issues are exacerbated in industrial embedded applications, because processors that operate in such an environment have to operate at ambient temperatures much higher than their commercial counterparts, often up to 70 °C. Moreover, even if a processor could theoretically operate at those ambient temperatures, its active thermal dissipation system—comprising, for instance, heat sinks and fans—would be too expensive, fragile, or bulky anyway.

At the same time, improvements in superscalar execution have been introduced at a slower and slower pace in recent years, after reaching the point at which further performance gains are limited and might not justify the additional circuital complexity they entail, along with the associated testing effort and risk of hardware bugs that could emerge in obscure circumstances.

12.2 MULTIPROCESSORS AND MULTICORES

As increases of clock speed and improvements of instruction execution strategy got closer to their limit, more than 20 years ago processor designers moved in another direction, that is, they started to integrate multiple processors, all connected to a common memory bus, in the same system. This gave rise to *multiprocessors*, which further evolved into *multicore* processors.

As its name says, a multicore processor embeds multiple execution cores, or just *cores* for short, in the same chip. Like in multiprocessors, all cores still have uniform access to a common, shared memory. From the hardware design point of view, this poses significant challenges that are outside the scope of this book and will be only shortly summarized in this section, mainly focusing on the effects some of them have on software design and development. More comprehensive references about multiprocessor and multicore systems include, for instance, References [57, 89].

12.2.1 BASICS OF MULTICORE ARCHITECTURES

In its most common form [21], a multicore processor consists of a number of independent cores that, at least in conventional designs, have the same instruction set architecture as legacy single-core processors, possibly extended with additional dedicated instructions for inter-core communication and synchronization. This backward compatibility has the advantage of enabling the immediate reuse of existing software development environments and toolchains when applications are migrated to a multicore processor.

The memory system of a multicore processor is considerably more complex than its single-core processor counterpart. In a single-core processor, it essentially consists of a hierarchy of on-chip *cache* memories, called *levels*. Usually, cache levels are numbered from one (the closest to the processor) to some value n (the closest to main memory).

Each cache level realizes different trade-offs between access speed and size, namely, caches become bigger but slower as their distance from the processor increases. The hierarchy has the on-chip processor memory interface at its top and ends with a bus controller, which connects to an off-chip memory, at the bottom.

Each cache level holds a subset of the contents of the level immediately below it, down to the bottom level, which holds a subset of the main memory contents. Appropriate load and eviction algorithms fill each cache level with instructions and data from the level immediately below it (or from main memory, in the case of the bottom cache), and remove data previously stored in a cache when space is needed for other instructions and data, respectively.

To improve efficiency and reduce overhead, the granularity of load and eviction operations is at the level of a cache *line* rather than individual machine words. Therefore, load and eviction algorithms treat cache lines as indivisible units when transferring instruction and data from one cache to another, or to/from main memory. Cache line sizes vary from one architecture to another, but are typically between 32 and 128 bytes.

Informally speaking, the overall goal of these algorithms is to keep as close as possible to the processor the instructions and data it is likely to need in the near future, moving them across levels as appropriate. If the algorithms work properly and the applications are sufficiently well-behaved, the net result is that, statistically, to the processor memory appears to be as large as main memory and (almost) as fast as the level-one cache.

Write operations issued by the processor may be immediately propagated through the cache hierarchy down to main memory (*write-through* policy). More often, they may be performed only on the first cache level initially, and propagated through the cache levels when an eviction takes place (*write-back* policy). In other words, to improve performance the write-back policy aims at reducing the number of cache-to-cache and cache-to-main-memory write operations by postponing them as much as possible. In this way, for instance, if the same memory location is written more than once before being evicted from a certain cache level, all write operations into the cache level below it, except the last one, are suppressed altogether.

An important consequence of the write-back policy is that there is a time window in which some caches contain fresh data written by the processor, but these data have not been written into main memory yet. This fact is completely transparent to the processor itself, because it makes access to memory exclusively through the same caches, and hence, it always reads back fresh data. On the contrary, if any other agent makes access to memory directly (or through its own, separate cache hierarchy) within this time window, it may get stale data.

On a single-core system, usually the only agents that can access memory in this way are fast devices capable of Direct Memory Access (DMA). Therefore, only device driver programmers must be aware of this peculiarity and take appropriate countermeasures when they deal, for instance, with data buffers shared between a (hardware) DMA-capable device and its (software) device driver.

When pushing data from the device driver to the device, this is done by means of special machine instructions that *flush* the whole cache or part of it—that is, ensure that cache contents are actually written back to main memory before continuing and letting the hardware access them. Symmetrically, before reading data from a buffer filled by the device, the device driver must use similar machine instructions to

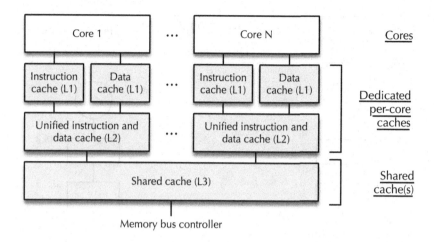

FIGURE 12.1 Example of cache hierarchy on a multicore processor.

invalidate the corresponding cache lines, to ensure that all subsequent read accesses will actually retrieve fresh data from main memory, rather than stale data from the cache.

Instead, as shown in Figure 12.1, the issue becomes considerably more complex in multicores, because each core makes access to main memory through a cache hierarchy in which at least some of the upper levels are private to that core. Namely, the figure shows an example of 3-level cache hierarchy common on contemporary, general-purpose multicore processors:

- The level-1 (L1) caches are dedicated to individual cores, hence each core has its own caches. Moreover, separate caches hold instructions and data, to take the best advantage from the differences in statistical characteristics of instruction and data access patterns.
- The level-2 (L2) caches are *unified*—that is, they hold both instructions and data without distinguishing between them—but they are still dedicated to individual cores. Hence, each core has its own unified L2 cache.
- A level-3 (L3) cache lies at the bottom of the hierarchy. This cache is shared among all on-chip cores and interfaces with the main memory bus controller that, in turn, mediates off-chip main memory accesses.

Several variations to this general scheme are possible, especially on systems with many cores. For instance, cores may be grouped into *clusters* and L2 caches may be shared among all cores in a cluster instead of being dedicated to a single core.

Regardless of the details, special care must be taken to ensure cache *coherence*, an all-important concept to be outlined in the following, to guarantee that all cores have the same image of main memory contents at any time. In turn, the cache coherence is related to the concept of *memory consistency model*, which will also be briefly discussed in the following.

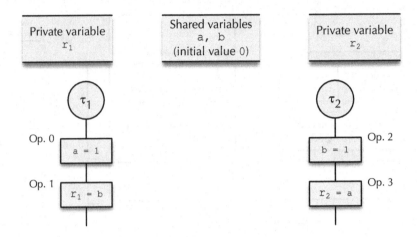

FIGURE 12.2   Sample code to illustrate the sequential consistency model.

12.2.2   MEMORY CONSISTENCY MODELS

The purpose of a *memory consistency model* is to define a set of rules that determine how memory operations must be performed by a set of agents, like cores, which operate on a shared memory. Namely, a memory consistency model stipulates if and to which extent memory operations performed by a core can be reordered during execution, and whether or not these operations must be observed by the other cores in the same order.

The problem is extremely complex and we can provide only a short introduction here, focusing on several aspects of interest to application and system programmers. For what concerns the practical implementation of memory consistency models, we will briefly describe the approach taken by ARM Cortex-A processors [14] as an example. Readers should refer to the extensive literature on the subject for more in-depth information.

Sequential consistency model

Arguably the most well-known memory consistency model is the *sequential consistency model*, defined in Lamport's seminal paper [79]. This model has the big advantage of being relatively simple and intuitive, that is, it defines systems whose behavior are easy for programmers to understand and work with. In the author's own words [79], the sequential consistency model stipulates that *"the result of any execution is the same as if the operations of all the processors were executed in some sequential order, and the operations of each individual processor appear in this sequence in the order specified by its program."*

Figure 12.2 shows two tasks τ_1 and τ_2 that share two variables a and b, whose initial value is zero, and execute on two distinct processors or cores. Each task sets one of the two variables to one, and then copies the contents of the other variable into its own private storage, for instance, into a processor register. The definition of sequential consistency model entails two distinct constraints on the order of the four operations Op. 0, ..., Op. 3 collectively performed by the two tasks:

1. The first condition of the model specifies that the four operations must take place in *some* sequential order, although it does not specify exactly *which one*. Therefore, for instance, the system is allowed to execute (Op. 0, Op. 1, Op. 2, Op. 3) or (Op. 0, Op. 2, Op. 3, Op. 1).
   The system may freely choose different sequences from one execution of the code to another provided that, for each execution, it is always possible to determine exactly which sequential order was chosen and the final system state is consistent with that order.
2. According to the second condition of the model, operations performed by a task within the sequence must take place in the same order specified by that task. For instance, since Op. 1 follows Op. 0 in τ_1, it must also follow it in the sequence.
   As a consequence, the sequence (Op. 0, Op. 2, Op. 3, Op. 1) is allowed (because Op. 1 follows Op. 0 although they are not contiguous) whereas Op. 1, Op. 0, Op. 2, Op. 3 is clearly forbidden (because Op. 1 precedes Op. 0).

The sequential consistency model avoids some counterintuitive results of the code in Figure 12.2 but, at the same time, does not make the system completely deterministic and leaves the system free to reorder operations at runtime to some extent, in order to improve performance. In this specific case, of the four possible results of the code shown in the figure:

- The results $(r_1 = 1, r_2 = 0)$, $(r_1 = 0, r_2 = 1)$, and $(r_1 = 1, r_2 = 1)$ are all possible. They correspond to the sequences (Op. 2, Op. 3, Op. 0, Op. 1), (Op. 0, Op. 1, Op. 2, Op. 3), and (Op. 0, Op. 2, Op. 1, Op. 3), respectively.
- Instead, the result $(r_1 = 0, r_2 = 0)$ is impossible, because none of the allowed sequences of operations may lead to that result.

This scenario indeed corresponds to the programmers' intuition that the read operations performed by tasks τ_1 and τ_2 (Op. 1 and Op. 3) may take place after one or both write operations (Op. 0 and Op. 2) have been performed, thus giving one of the first three results. However, they may not both be performed before both write operations, which would lead to the fourth result.

As summarized in Table 12.1, the sequential consistency model allows only 6 of the 24 possible execution sequences stemming from the code in Figure 12.2, and the allowed sequences lead to the three different results just described. Even from this simple discussion, it is evident that the definition of a memory consistency model is extremely important because it acts like a contractual boundary between the hardware architecture and software programmers. More specifically:

- The memory consistency model determines which results of a given excerpt of code are possible.
- Based on this, software programmers must be prepared to deal with all possible results and ensure their code behaves correctly in all cases.

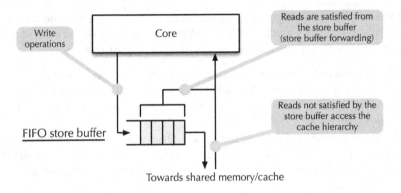

FIGURE 12.3 Simplified functional diagram of a FIFO write buffer.

Total store order (TSO)

A main issue of the sequential consistency model is its lack of efficiency because, as we just saw, it forbids many kinds of reordering. This is especially important for write operations, which are very expensive on many contemporary architectures in terms of execution time.

The underlying reason for this can be readily appreciated by referring back to Figure 12.1: In order to become visible to the other cores, a write operation must be propagated down through the cache hierarchy, starting from the core that issued the write and down to the L3 cache at least.

Moreover, looking at Figure 12.2, in order to respect the sequential consistency model, Op. 1 cannot start before Op. 0 has been completed and the write operation has become visible to all the other cores, although we can easily argue that there is no compelling reason for postponing a read of variable b until a write to the totally unrelated variable a has been completed. The same reasoning also applies to Op. 3 with respect to Op. 2.

A widespread way to alleviate this issue is to relax the memory consistency model to allow the interposition of a *store buffer* (sometimes called *write buffer*) between a core and its cache hierarchy, as shown in the simplified diagram of Figure 12.3. As its name says, the store buffer temporarily holds one or more write operations so that they can be completed at a later time, while the core continues executing other memory operations in the meantime. Informally speaking:

- Write operations are pushed into the store buffer instead of being immediately forwarded to the shared cache or memory. Buffered write operations proceed concurrently with the execution of subsequent read operations issued by the core until they are eventually committed to the shared cache or memory in first-in, first-out (FIFO) order.
- Read operations consult the store buffer to check whether it contains a write operation to the same address. In this case, the read obtains the value directly from the store buffer, with a process known as store buffer

forwarding. Otherwise, the read operation accesses the shared cache or memory through the cache hierarchy, as usual.

Due to the interposition of the store buffer, write operations issued by a core can now be postponed from the point of view of other cores although, thanks to store buffer forwarding, they always appear to be performed exactly where they are found in the code from the point of view of the core that issued them. In other words, the presence of the write buffer is transparent for the core that is connected to it.

Instead, from the point of view of other cores and of the shared memory, write operations performed by a certain core may be viewed as if they went beyond some subsequent read operations issued by that same core. On the contrary, due to the FIFO nature of the store buffer, write operations issued by a certain core cannot pass other writes issued by that same core.

Overall, although write operations can be reordered with respect to subsequent read operations issued by the same core as described previously, they are still committed to memory and observed by all other cores in some total order. For this reason, this memory model is called *total store order* (TSO). Several variants of this model are in widespread use, for instance, in modern Intel processors [107].

With respect to the simplified model illustrated here, the main complications of real models come from the fact they have to support *locked* instructions, that is, instructions that perform an indivisible read-modify-write operation on a memory cell.

Referring back to the example of Figure 12.2, we can easily note that the TSO model is more relaxed than the sequential consistency model. In particular, since the code of each task contains a write operation followed by a read operation, the hardware is free to commit to memory the write operation after the read operation.

As a consequence, all 24 possible sequences of operations listed in Table 12.1 are allowed by the TSO model and, as shown in the rightmost column of the table, the result $(r_1 = 0, r_2 = 0)$ becomes possible, although it still sounds implausible to a programmer who just looks at the code. This happens when the write buffers postpone the two write operations issued by τ_1 and τ_2 until after they have performed both read operations.

Despite its extreme simplicity, this example also illustrates how the adoption of one memory model or another is always a trade-off between:

- Improving performance, by giving the hardware more freedom to reorder memory operations.
- Accepting less intuitive behaviors and results, like the one just described.

The classical sample code listed in Figure 12.4 illustrates an execution outcome that is still forbidden by the total store order model. There are four tasks in the system, of which:

- Tasks τ_1 and τ_2 update two shared variables a and b. The update changes the value of the variables from 0, their initial value, to 1.

TABLE 12.1

Execution of the Code of Figure 12.2 Under Different Memory Models

Sequence	Sequential consistency		Total store order (TSO)	
	Allowed	Result	Allowed	Result
(Op. 0, Op. 1, Op. 2, Op. 3)	✓	$(r_1 = 0, r_2 = 1)$	✓	$(r_1 = 0, r_2 = 1)$
(Op. 0, Op. 1, Op. 3, Op. 2)	×	—	✓	$(r_1 = 0, r_2 = 1)$
(Op. 0, Op. 2, Op. 1, Op. 3)	✓	$(r_1 = 1, r_2 = 1)$	✓	$(r_1 = 1, r_2 = 1)$
(Op. 0, Op. 2, Op. 3, Op. 1)	✓	$(r_1 = 1, r_2 = 1)$	✓	$(r_1 = 1, r_2 = 1)$
(Op. 0, Op. 3, Op. 1, Op. 2)	×	—	✓	$(r_1 = 0, r_2 = 1)$
(Op. 0, Op. 3, Op. 2, Op. 1)	×	—	✓	$(r_1 = 1, r_2 = 1)$
(Op. 1, Op. 0, Op. 2, Op. 3)	×	—	✓	$(r_1 = 0, r_2 = 1)$
(Op. 1, Op. 0, Op. 3, Op. 2)	×	—	✓	$(r_1 = 0, r_2 = 1)$
(Op. 1, Op. 2, Op. 0, Op. 3)	×	—	✓	$(r_1 = 0, r_2 = 1)$
(Op. 1, Op. 2, Op. 3, Op. 0)	×	—	✓	$(r_1 = 0, r_2 = 0)$
(Op. 1, Op. 3, Op. 0, Op. 2)	×	—	✓	$(r_1 = 0, r_2 = 0)$
(Op. 1, Op. 3, Op. 2, Op. 0)	×	—	✓	$(r_1 = 0, r_2 = 0)$
(Op. 2, Op. 0, Op. 1, Op. 3)	✓	$(r_1 = 1, r_2 = 1)$	✓	$(r_1 = 1, r_2 = 1)$
(Op. 2, Op. 0, Op. 3, Op. 1)	✓	$(r_1 = 1, r_2 = 1)$	✓	$(r_1 = 1, r_2 = 1)$
(Op. 2, Op. 1, Op. 0, Op. 3)	×	—	✓	$(r_1 = 1, r_2 = 1)$
(Op. 2, Op. 1, Op. 3, Op. 0)	×	—	✓	$(r_1 = 1, r_2 = 0)$
(Op. 2, Op. 3, Op. 0, Op. 1)	✓	$(r_1 = 1, r_2 = 0)$	✓	$(r_1 = 1, r_2 = 0)$
(Op. 2, Op. 3, Op. 1, Op. 0)	×	—	✓	$(r_1 = 1, r_2 = 0)$
(Op. 3, Op. 0, Op. 1, Op. 2)	×	—	✓	$(r_1 = 0, r_2 = 0)$
(Op. 3, Op. 0, Op. 2, Op. 1)	×	—	✓	$(r_1 = 1, r_2 = 0)$
(Op. 3, Op. 1, Op. 0, Op. 2)	×	—	✓	$(r_1 = 0, r_2 = 0)$
(Op. 3, Op. 1, Op. 2, Op. 0)	×	—	✓	$(r_1 = 0, r_2 = 0)$
(Op. 3, Op. 2, Op. 0, Op. 1)	×	—	✓	$(r_1 = 1, r_2 = 0)$
(Op. 3, Op. 2, Op. 1, Op. 0)	×	—	✓	$(r_1 = 1, r_2 = 0)$

- Tasks τ_3 and τ_4 read those variables in reverse order and copy their content into local storage.

We can easily observe that the result $(r_1 = 1, r_2 = 0, r_3 = 1, r_4 = 0)$ is impossible according to the total store order model, because:

- The partial result $(r_1 = 1, r_2 = 0)$ implies that τ_3 found variable a already updated to 1 and variable b still at 0 when it performed its read operations. Therefore, from the point of view of τ_3, the write operations were performed in the order (Op. 0, Op. 1) and the read operations of τ_3 took place between the two.
- Symmetrically, The partial result $(r_3 = 1, r_4 = 0)$ implies that τ_4 observed the two write operations in the opposite order, that is, (Op. 1, Op. 0), but this contradicts the fundamental property of the total store order model.

However, it turns out that permitting this kind of behavior may be useful to further optimize the hardware. To dig deeper into the underlying reasons for this, let us

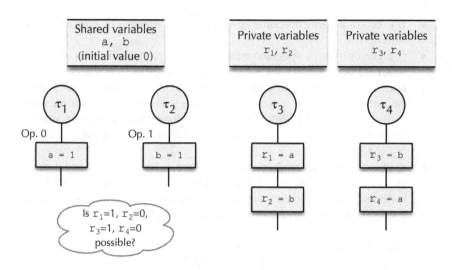

FIGURE 12.4 Sample code for more advanced forms or reordering

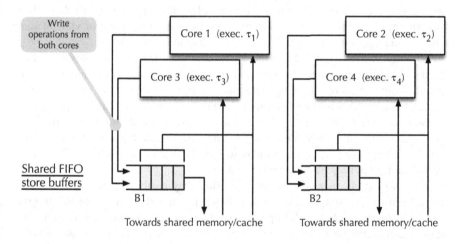

FIGURE 12.5 Hypothetical quad-core architecture with shared FIFO store buffers.

consider the hypothetical quad-core architecture depicted in Figure 12.5. With respect to the architecture shown in Figure 12.3, the main difference is that some cores (cores 1 and 3, and cores 2 and 4) are now "closer" than others because they share the same store buffer.

Let us discuss why and how executing the code of Figure 12.4 on this architecture may sometimes lead to $(r_1 = 1, r_2 = 0, r_3 = 1, r_4 = 0)$, assuming that the four tasks $\tau_1, \ldots, \tau_4$ are each executed on their own, same-numbered core. This result stems from the following execution sequence (and others):

- τ_1 executes its write operation on core 1, which temporarily goes into write buffer B1.
- τ_2 executes its write operation on core 2, which temporarily goes into write buffer B2. At this point, the value of a and b in memory is still 0.
- τ_3 executes both its read operations on core 3. Since cores 1 and 3 share the same write buffer, τ_3 gets the (updated) value of a directly from write buffer B1, whereas it reads the (original) value of b from memory since it has no visibility of the update held in write buffer B2. As a result, τ_3 obtains $(r_1 = 1, r_2 = 0)$.
- τ_4 executes both its read operations on core 4. Since cores 2 and 4 share the same write buffer, τ_4 gets the (updated) value of b directly from write buffer B2, whereas it reads the (original) value of a from memory since it has no visibility of the update held in write buffer B1. As a result, τ_4 obtains $(r_3 = 1, r_4 = 0)$.
- Eventually, the content of write buffers B1 and B2 will be flushed to main memory, but this does not affect the result.

In summary, the fact that some cores are closer than others in the hardware architecture may lead some cores to become aware of write operations sooner than others. In turn, this may alter the order of write operations from the point of view of some cores with respect to others, with side effects visible to software.

Although we may rather doubt that sharing write buffers among cores is a fruitful hardware design approach, it is worth noting that also dividing cores into groups and sharing some levels of the cache hierarchy among a group of cores may lead to the same effect, and this is indeed commonplace in contemporary multicore processors.

Some architectures, like the ARM Cortex-A [14], proceed even further along this path and specify that store buffers are not necessarily managed in first-in, first-out order [11, 13]. As a consequence, on these architectures even write operations issued by the same core may be reordered at runtime.

Unless properly managed, write reordering may easily lead to subtle software bugs, even in very simple cases. This is better illustrated by means of the program shown in Figure 12.6, which implements a one-shot synchronization between two tasks that execute on two different cores and access a common data buffer called data:

- Task τ_1 prepares the data and writes them into the shared buffer data. In order to signal that data are ready, it then sets the shared flag rdy, whose initial value is 0, to 1.
- Task τ_2 contains a busy-waiting loop in which it repeatedly checks the value of rdy. The task stays in the loop until rdy becomes 1. Afterwards, τ_2 makes use of data.

Although this code seems correct at first sight, it malfunctions if the two write operations performed by τ_1, on data and rdy, are observed in a different order by τ_2. When this happens, τ_2 may be allowed to proceed past the busy-waiting loop and access data before τ_1 has finished writing into it.

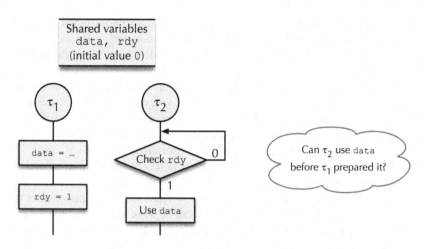

FIGURE 12.6   Simple, one-shot inter-task synchronization with busy waiting.

If the shared buffer is a complex data structure rather than just a single machine word, this is an issue not only of data *freshness* (τ_2 may access an outdated version of data), but also of data *consistency* (if the update to data requires more than one write operation by τ_1, τ_2 may get access to a mix of old and new data).

We can further generalize this concept and state that, recalling the definition of critical region entry and exit code (given in Section 7.1.3), any reordering that makes the effect of a store operation visible to other cores before the critical region entry code has been successfully executed (or, symmetrically, its effect is seen beyond the critical region exit code) undermines the very concept of mutual exclusion among tasks, which critical regions are meant to guarantee.

Memory barriers and other countermeasures

On the one hand, reordering memory operations has the indubious advantage of improving execution performance, because it allows hardware to hide memory latency in many cases. By intuition, we can also reckon that, as the approach to reordering becomes more aggressive, performance improves further and further. On the other hand, we also saw that reordering store operations may easily lead seemingly straightforward pieces of code to fail.

It is therefore important to be able to limit reordering, or disable it completely, in some crucial portions of code. On most architectures, this goal can be achieved in two possible ways, differing in the granularity of their action.

- Memory *barriers* or *fences* are special instructions that, as their name says, act as fences that cannot be crossed by memory operations. In its simplest form, a barrier instruction ensures that all memory operations that precede it in the code have been executed and are visible to other cores strictly before any of the operations that follow it starts.

Realizing the barrier's semantics usually entails a significant performance penalty because, in its most common implementation, it forces the core to wait until all memory operations that precede it have been "flushed" to memory.

For this reason, many architectures implement several different flavors of barrier, with different levels of permeability to different kinds of memory operation. For instance, besides the most restrictive form of barrier just described, they may also provide a weaker barrier, which still prevents store operations from crossing it, but lets load operations through.

- In some cases, the need to limit or disable reordering is more related to the target address of memory operations than their specific position in the code. This happens, for instance, when software must manipulate a set of device registers. More often than not, read or write operations on a device register trigger hardware-related side effects and it is usually necessary to ensure that they are performed in the exact order specified in the code. This applies not only to multicore, but also to single-core processors.

  For instance, a simple device that implements a serial interface may have a *data* register, in which the data to be transmitted must be written, and a *control* register, which starts the transmission when set appropriately. For the device to work correctly, it is clearly crucial that the processor updates the data register *before* writing into the control register.

  For similar reasons, it is also necessary that any cache interposed between the processor and the device does not hide or delay any operation on device registers from the point of view of the underlying device.

  In this case, a whole region of address space can be marked in a special way, to signal that it is used to interface with devices. This is often done during software initialization and has the effect of disabling both caching and reordering for all memory operations that target the region. This approach has an additional benefit, that is, the code that operates on the device afterward can do so without explicitly using any barrier instruction.

Both these techniques are of interest mainly to system or device driver programmers. If we focus solely on inter-task communication and synchronization, the use of operating system-provided synchronization devices, like the ones described in Chapters 7 and 8, ensures that the high-level software built upon them works properly. This is because barrier instructions are already embedded in the operating system code that implements the corresponding communication and synchronization primitives. As better described in Section 12.2.5 for the C and C++ languages, the fact that these primitives are invoked by means of a function call also prevents other kinds of undesirable reordering at the compiler level.

On the contrary, the implementation of any lock or wait-free communication algorithms, like the ones exemplified in Chapter 13, implies a judicious use of memory barriers. Again, the use of a proven synchronization library as opposed to an ad-hoc implementation of these algorithms from scratch, is valuable to solve or at least mitigate the problem.

12.2.3 CACHE COHERENCY

In Section 12.2.1 we saw that caches have become an essential component of many contemporary embedded systems, because they are meant to fill the ever-increasing speed gap between processor and memory, which has now reached one or two orders of magnitude. As the gap became wider, processor architects resorted to multiple levels of caches arranged in a hierarchical manner, like the ones depicted in Figure 12.1, in an effort to achieve the best tradeoff between access speed, capacity, additional cost, chip area and power consumption, and interconnection complexity.

In an effort to further improve performance, the rules—or *protocols*—that control cache behavior have also become more complex, especially for store operations. They went from simple *write-through* update policies, in which a store operation was immediately forwarded all the way back to main memory, to more sophisticated *write-back* policies, in which new data are temporarily held in the cache and forwarded to main memory only at a later time, for instance, when the space they occupy in the cache must be put to other use.

Informally speaking, the side effect of the write-back approach is similar to what we described for write buffers in Section 12.2.2. Namely, there are time windows in which main memory contains *stale* data, that is, its contents do not reflect the latest store operations performed by a core, when observed from another core. For write buffers, the most commonly adopted solution, briefly described in Section 12.2.2, is to use special *barrier* instructions. Essentially, those barriers prevent write buffers from postponing store operations in a way that may be harmful to the correctness of the algorithms at hand.

For what concerns caches, it is certainly possible to selectively flush and invalidate part of them when dealing with DMA-capable devices. Flushing (or *cleaning*, as some processor manuals say) the cache before letting a device read from memory ensures that the device gets access to the latest data. Symmetrically, invalidating the cache after a device wrote into memory and before letting the processor read from it ensures that the processor does not continue to use old data still present in its cache.

Given the sheer size of modern caches and the impact they have on performance, it is very important for these operations to be as *selective* as possible, that is, they must affect only the smallest possible portion of the caches. For devices, this is feasible because the location of the memory buffers shared between the processor and a device is always well-known to the device driver and virtually all cache controllers provide a way to selectively flush or invalidate cache entries based, for instance, on the physical address they refer to. Moreover, data sharing among the processor and a device takes place only in well-known points, internal to the device driver itself.

However, a similar issue also arises when data are shared among cores in which—as shown in Figure 12.1 and as is becoming more and more common in contemporary embedded processor architectures—at least the highest cache levels are *dedicated*, or *private*, to each core. This is the case of L1 and L2 caches in the figure.

Let us assume, for simplicity, that in a dual-core processor each core has only one level of dedicated, write-back cache. As shown in Figure 12.7, the following sequence of events is possible when two fragments of code running on the two cores

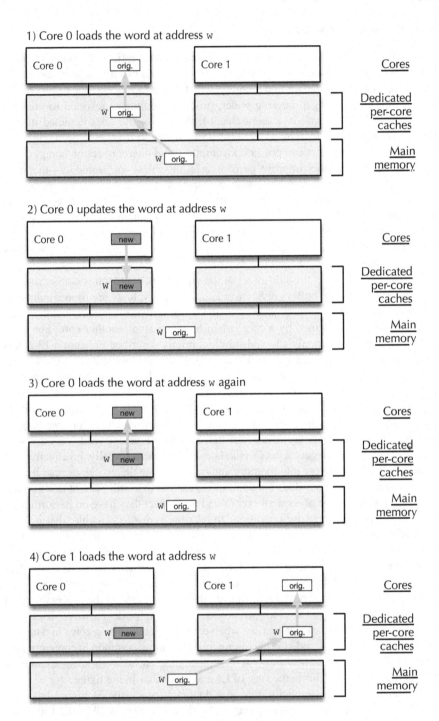

FIGURE 12.7 Sharing data between cores without cache coherency.

operate on a shared memory word at address w that was previously cached by either core:

1. Core 0 loads the word at address w. Since we are assuming that the contents of w are not cached already, a new cache line is allocated for them and filled with data from memory.
2. Core 0 updates the word at address w. Since the contents of w are now cached and the cache is write-back, the update is not propagated all the way back to main memory and the updated value stays in the cache of core 0.
3. If now core 0 reads back from w it gets the correct contents from its cache, thus confirming that the presence of a write-back cache is transparent to the corresponding core.
4. Instead, if core 1 loads the word at address w at this point, it will incur in a cache miss and retrieve the *old* contents of w from memory.

The net result of this sequence of events is that, although core 1 read from w *after* core 0 updated its contents, it still got the original, and now *stale* contents of w. In other words, the two cores no longer have a *coherent* view of memory contents. This anomaly persists until the new contents of w are eventually flushed from the cache of core 0 back into memory. In addition, if w is a complex data structure instead of just a memory word, the possible values that core 1 might get will not only be either the *old* or *new* contents, but also any mix of the two, which will inevitably lead to data corruption.

It would of course still be possible to obtain the expected behavior by properly flushing and invalidating the caches. For instance, in this particular case, it would be necessary to flush the cache of core 0 after writing into w. It would also be prudent to invalidate the cache of core 1 before reading from w, in case it already contained an obsolete copy of w's contents. However, a general use of this technique leads to two main issues:

- In a multi-threaded environment all threads belonging to the same process implicitly share the whole address space. It is therefore hard to determine which data they *actually* share in an automatic and systematic way, especially with programming languages that support aliasing and pointers. On the other hand, flushing and invalidating caches every time a thread makes access to any data that could *potentially* be shared would likely be unacceptable from the performance point of view. For these reasons, the cache maintenance burden must be left to programmers, who might well be unaware of the underlying implications of executing their code on a multicore system.
- An additional complication arises in systems that let the scheduler preempt and migrate threads from one core to another during execution, according to the *global scheduling* principle, because the software-based cache maintenance approach just mentioned may be hard to implement without interfering with the scheduling algorithm. Let us consider the following pseudo-code fragment to be executed whenever w is updated:

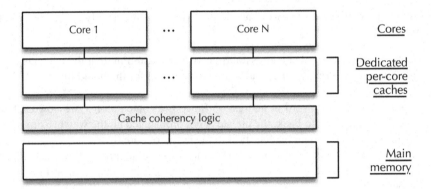

FIGURE 12.8 Cache coherency logic.

1. Store a new value into w
2. Flush the portion of cache that contains w

Although this code seems correct according to the previous discussion, it may fail if its execution starts on core 0, but the scheduler decides to migrate it to core 1 between the first and the second instruction. If this happens, the code will flush the cache of the wrong core and memory will still contain the old value of w even after both instructions have been executed. Moreover, the fact that on most architectures the code fragment can be implemented with just two assembly instructions makes the error extremely unlikely (but not at all impossible) and makes debugging even harder. Disabling thread migration during the two instructions would address the problem, but only at the expense of additional overhead.

For this reason, many architectures adopt hardware-based, rather than software-based, techniques to ensure cache coherency. As shown in Figure 12.8, this is done at the expense of some hardware complexity and additional power consumption, by interposing a *cache coherency logic* layer between the caches and main memory.

Depending on the number of cores, the cache coherency logic may span across all cores and caches, as depicted in the figure, or the system can be subdivided into multiple cache coherency *domains*, with separate and independent cache coherency logic modules. In this case, the properties guaranteed by each cache coherency logic are valid only within its domain.

As before, we assume there is only one cache level to simplify our discussion, but the principle applies equally well to multiple cache levels. In this case, distinct cache coherency logic layers must be inserted between adjacent cache levels in the hierarchy.

The cache coherency logic is responsible of maintaining cache coherency by observing all cache and memory transactions and altering the state of cache lines according to a well-defined protocol. Here, we will summarily discuss the MESI protocol, first discussed in Reference [94], because it is used in various forms and variants in several processor architectures for general-purpose and embedded computing— for instance, the MOESI protocol adopted in the ARMv8-A architecure [14] and the MESIF protocol developed by Intel [67].

In the MESI protocol each cache line is characterized by one of four possible states. It is also assumed that the cache works in write-back mode. The four states are:

- The *invalid* (*I*) state indicates that the cache line does not hold valid data.
- The *exclusive* (*E*) state indicates that the cache line contents are valid, consistent with main memory, and no other cache has the same line.
- The *shared* (*S*) state also indicates that the cache line contents are valid and consistent with main memory, but other caches may have the same line.
- The *modified* (*M*) state indicates that the cache line contents are valid but have been modified with respect to the value in main memory. No other cache has the same line.

The exclusive and shared states are often called *clean* states, because the cache line is consistent with main memory. Instead, the modified state is called *dirty* because the cache line contains data that are more recent than main memory, and hence, inconsistent with it. A key goal of the MESI protocol is to prevent other cores from reading and using outdated data from memory areas corresponding to dirty cache lines on a certain core. To this purpose, the cache coherency logic gives to a cache the ability to satisfy a read request from another core, bypassing main memory.

Cache lines transition from one state to another under the control of the cache coherency logic when two kinds of operations occur, namely:

- Local operations, which include read and write requests from the local core, that is, the core to which the cache is directly attached, as well as the cache line clean operation, applied to a non-invalid cache line before reusing it. In the following, these operations will be called L_r (local read), L_w (local write), and L_c (local clean), respectively.
- Operations initiated by other cores and *snooped* (that is, observed) through the bus that interconnects the cores and is also part of the cache coherency logic. There are three operations that a core may initiate or observe on the interconnecting bus: B_r (read), B_i (invalidate), B_{ri} (read+invalidate). They will be defined in the description that follows.

Figure 12.9 summarizes in a state/transition diagram how the MESI protocol reacts to local operations. In the figure:

- Grey circles corresponds to the four possible states of a certain cache line.
- Arrows indicate a transition of a cache line from one state to another, subject to certain conditions written in the transition label.

The transition label is composed of two parts separated by a slash:

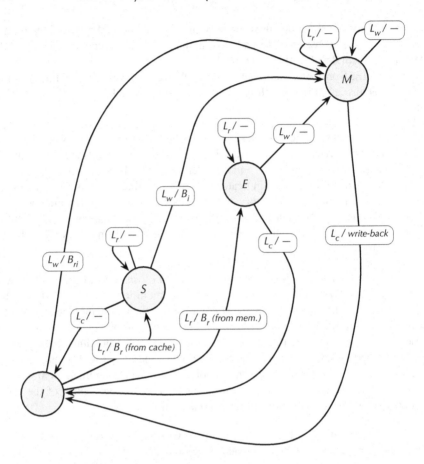

FIGURE 12.9 Main states and transitions of the MESI protocol, local operations.

- The first part is the local operation that triggers the transition when it applies to the cache line. Local read and write operations apply to the cache line that corresponds to the memory address of the operation, while the local clean operation applies to the cache line selected for reuse by the cache replacement algorithm, not covered by the MESI protocol.
- The second part lists the bus operations that are performed during the transition, if any. The notation "—" indicates that no bus operations are performed, and hence, the transition cannot be snooped by the other cores.

When a core issues a L_r (local read) operation, the following actions take place:

- The local cache is consulted to determine whether the requested data are already in the cache (cache *hit*) or not (cache *miss*).
- If there is a hit, then a cache line already holds the requested data, and it must be in the S (shared), E (exclusive), or M (modified) state. The read

operation is satisfied by the cache and the cache line keeps its current state. No bus operations are issued, and hence, the operation does not become visible to the other cores.

- If there is a miss, a new cache line must be allocated. Cache lines in the I (invalid) state are the most suitable candidates because they do not hold any useful data. If the cache is completely full, it becomes necessary to select a cache line suitable for reuse. This implies the execution of a L_c (local clean) operation on it, which will be discussed later, to possibly flush its data back to main memory and bring it to the I (invalid) state.
- Then, a B_r (bus read) operation is issued to retrieve the data and fill the new cache line. If no other caches have the requested data, the answer to the operation will come from main memory. In this case, the cache line goes to the E (exclusive) state. Otherwise, the answer will come from another cache and, accordingly, the cache line goes to the S (shared) state.

The course of action taken when a core issues a L_w (local write) operation is similar:

- As before, the local cache is consulted to determine whether there is a hit or a miss.
- If there is a hit, the cache line goes to the M (modified) state, to reflect the fact that its contents are no longer consistent with main memory. If the cache line was in the S (shared) state, other caches may hold the same, now outdated information. Therefore, a B_i (invalidate) bus operation is issued to instruct them to invalidate any matching cache lines. Instead, no bus operations are required if the cache line was in the M (modified) or E (exclusive) state, because both states guarantee that no other caches in the system hold the same information.
- If there is a miss, a new cache line must be allocated, possibly after cleaning it with a L_c (local clean) operation. Then, a B_{ri} (bus read+invalidate) operation is issued to fill the cache line. Finally, the cache line is updated according to the write operation and moves to the M (modified) state. The bus operation instructs other caches to provide the requested data, if they have it, write any updated data back to main memory, and then invalidate any matching cache lines. If no caches have the requested data, the answer to the B_{ri} request will come from main memory.

The purpose of a L_c (local clean) operation is to prepare a cache line for reuse by bringing it back to the I (invalid) state. Therefore:

- If the cache line is in the S (shared) or E (exclusive) state, its contents are consistent with main memory and are simply discarded with no further action. A side effect of this approach is that the S (shared) state is *imprecise*. As we explained previously, a cache line moves from the E (exclusive) into the S (shared) state when other caches holds the same information, but it

ﾚ

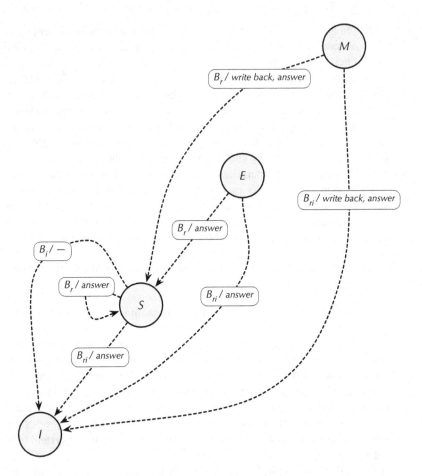

FIGURE 12.10 Main states and transitions of the MESI protocol, bus operations.

never goes back to the E (exclusive) state when this is no longer the case
due to undetectable L_c operations taking place on those other caches.

- If the cache line is in the M (modified) state, its contents are written back
  to main memory before bringing it to the I (invalid) state. Although this
  operation could potentially be snooped by other caches, the MESI protocol
  does not foresee any action they should perform in this case.

Figure 12.10 illustrates how all the other caches must react to bus operations issued by a certain cache. The reaction takes place only on *matching* cache lines, that
is, lines that are not in the I (invalid) state and correspond—by memory address—to
the cache line that gave rise to the bus operation in the first place. This address is
communicated to the other caches through the interconnecting bus along with the
kind of operation that, as stated previously, can be B_r (read), B_i (invalidate), or B_{ri}
(read+invalidate). The response to a B_r (read) operation is fairly intuitive:

- All caches with a matching cache line must reply with the requested data, regardless of the cache line state—which cannot be I (invalid). If multiple caches respond, they must by definition respond with the same data and an appropriate arbitration logic will select one of them.
- If one of the matching cache lines was in state E (exclusive), it moves to state S (shared) to reflect the fact that other caches now hold the same information. By definition, there can only be at most one matching cache line in the E (exclusive) state.
- If one of the matching cache lines was in state M (modified), its contents are written back to main memory and it is moved to the S (shared) state. As before, there can only be at most one matching cache line in the M (modified) state.

This behavior guarantees that the requesting cache always gets the most recent data available because, if there is a cache with a matching cache line in the M (modified) state, it will be the only cache to answer the request. At the same time, the most recent data is written back to main memory if necessary and the state of all matching cache lines is set to S (shared), to reflect the fact that these data are present in multiple caches (at least two). This is possible and correct because the B_r (read) operation signifies that the requesting cache needs the data it is requesting, but does not intend to modify them.

The response to a B_{ri} (read+invalidate) operation is similar, with one notable difference:

- After replying with the requested data and possibly writing back to main memory the contents of the matching cache line in the M (modified) state, if any, all caches invalidate the matching cache line, that is, they unconditionally move it to the I (invalid) instead of the S (shared) state.

The difference is due to the consideration that the B_{ri} (read+invalidate) operation indicates that the requesting cache intends to immediately modify the data it requested. Hence, it must become the sole place in which these data reside, more specifically, in a cache line in the M (modified) state.

The B_i (invalidate) bus operation is issued when a cache line in the S (shared) state moves to the M (modified) state because its contents have been updated. In response, all the other caches must invalidate any matching cache lines they may have. It is worth noting that, due to the way the MESI protocol works, all these cache lines must necessarily be in the S (shared) state.

In summary, one important property of the MESI protocol from the programmer's perspective is that it handles cache write-back and invalidation automatically, thus ensuring that all caches in the coherency domain give to the attached cores a coherent view of main memory contents.

Even more importantly, this is done transparently with respect to the application tasks that run on the cores. The memory read and write operations they perform will always operate correctly on shared data without the need of extra code or care, regardless of what the cache coherency logic is doing behind the scenes. This property stays true even if tasks are migrated from one core to another, but still within the same coherency domain, at arbitrary points of their execution.

The two main caveats are that programmers must still ensure that the hardware actually performs read and write operations in the intended order (see Section 12.2.2) and timings. As it already happened with single-core caches, the actual operations that the cache coherency logic must perform, to satisfy a read or write request from the attached core, vary depending on the state of the caches in the system. Hence, their execution time also varies accordingly and these timing variations are observable by the application code—although they do not affect the functional correctness of the code itself.

12.2.4 PRACTICAL IMPLEMENTATION ON ARM PROCESSORS

As a practical example of how the methods and techniques described in the previous sections can be implemented in practice, this section illustrates the key concepts taking the ARMv8-A architecture as a reference. The discussion will necessarily be brief and will focus only on the most essential points. More thorough information can be found in References [13, 11, 14], which address these topics with an increasing level of detail.

Memory system

The typical memory system of multicore processors that implement the ARMv8-A architecture is depicted in Figure 12.11. For what concerns the memory system, the structure is very close to the generic cache hierarchy shown in Figure 12.1. Its main elements are:

- A number of *cores*, grouped into one or more *clusters*. All cores in the same cluster are identical to each other. Instead, cores belonging to different clusters are identical in their instruction set, but may have different clock speed, internal architecture, and power requirement.
- Each core has its own instruction (I) and data (D) *L1 caches*. To attain maximum performance, the hardware does *not* enforce any kind of coherence between these two caches, because they do not normally share information except in particular cases, for instance, when executable code is loaded into memory (as data, through the D cache) and then executed (as instructions, through the I cache). Hence, software must manage these special cases through explicit flush and invalidate instructions (see Section 12.2.3).
- Each cluster has a unified *L2 cache*, shared among all cores in the cluster. This cache is equipped with a coherence control logic, often called Snoop Control Unit (SCU) in the ARM documentation, which arbitrates accesses to the L2 cache from all L1 instruction and data caches. In addition, it keeps all L1 data caches in the cluster coherent. As noted previously, L1 instruction caches are *not* part of the intra-cluster coherency domain.
- In high-performance systems a cache-coherent interconnect, which embeds a unified *L3 cache*, connects multiple clusters together and, thanks to its

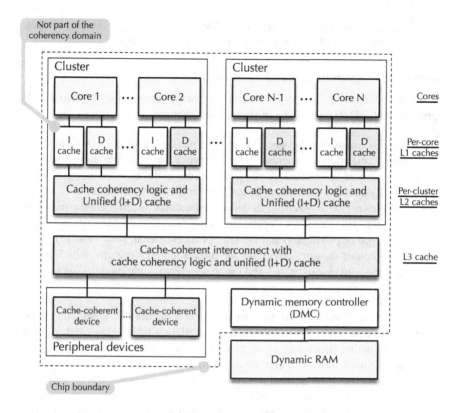

FIGURE 12.11 ARMv8-A memory system and cache coherency domain (simplified view).

own cache coherence logic, arbitrates access to the L3 cache from all L2 caches and keeps them coherent to form a single coherency domain across all clusters. Again, since L1 instruction caches were kept out of the coherency domain at a higher level in the hierarchy, they cannot be part of the coherency domain at this level.

- The cache-coherent interconnect may also be attached to some high-speed peripherals—for instance, a Graphics Processing Unit (GPU) or a network interface—that have their own Direct Memory Access (DMA) engines and internal caches. This approach enables those peripherals to access main memory through the interconnect and use the L3 cache to reduce access latency. Even more importantly, their Direct Memory Access (DMA) engines are given a coherent view of main memory contents and processor-side, software-based cache flush and invalidation before and after input-output operations becomes unnecessary.

- The cache-coherent interconnect also provides access to off-chip main memory when required. This is usually done through a Dynamic Memory Controller (DMC), which takes care of scheduling, timing and interfacing requirements of external dynamic RAM. In Figure 12.11, the entities that are part of the coherency domain are highlighted in gray.

Memory management unit and memory types

In ARMv8-A systems, memory accesses are always mediated by a Memory Management Unit (MMU), in charge of three main functions:

- *Address translation* between virtual and physical addresses. A virtual address is the address generated by a core whenever it fetches an instruction, or loads and stores data. A physical address is the corresponding address seen by memory (and often, by caches) for the same transaction.
- *Memory access protection*, to define and enforce specific access rights to certain memory regions, depending on the execution mode of the core (unprivileged or privileged) and the kind of access (fetch, load, or store).
- Definition of *memory attributes* for certain memory regions, which influence the behavior of caches and cache coherency logic modules, as well as memory operations ordering rules.

All three functions are performed by means of a memory-resident data structure called *translation table*. Up to four distinct translation tables may exist, each core uses one or another depending on its current privilege level and the range in which the virtual memory address to be translated falls.

The description of the internal structure of a translation table, and how the Memory Management Unit uses it for address translation and memory access protection, is beyond the scope of this book also because it would involve plenty of architecture-specific details that are inherently not applicable elsewhere.

Instead, we will focus on the meaning of memory attributes that—although sometimes called with different names—have a direct counterpart in other processor architectures and provide an example of how the concepts outlined in Sections 12.2.2 and 12.2.3 are embodied in real hardware.

As shown in Table 12.2, the ARM-v8A architecture defines two memory types: *normal* and *device*. These two types are mutually exclusive and all memory regions accessible to the cores must be categorized as one type or another.

Normal memory

This kind of memory provides maximum performance because it enables all memory operations reordering strategies the cores are capable of. The normal memory type should hence be used whenever possible, for instance, for regions of Flash, ROM, and RAM memory. Early versions of the architecture specification set forth very few limitations on the kind and extent of reordering cores could perform, thus making the memory order model very complex and giving a lot of freedom to implementations.

However, further limitations were introduced at a later time in order to streamline and simplify the model [101], because it was discovered that part of the freedom was never exploited by production implementation. In other words, it was determined

TABLE 12.2

Memory Types and Attributes Supported by the ARMv8-A Architecture

Type	Attribute	Description
Normal		
	Tagged [1]	Enable the tagged memory extension [14].
	Cacheable / non-cacheable [2]	Enable caching.
	Write-through / write-back [2]	Control cache behavior on write operations.
	Transient / non-transient [2]	Prefer short-term rather than long-term caching.
	Read-allocate [2]	Allocate a cache line on read miss.
	Write-allocate [2]	Allocate a cache line on a write miss.
Device		
	Gathering [3]	Group together multiple write operations.
	Reordering [3]	Reorder memory operations.
	Early write acknowledgment [3]	Continue before write operations are fully concluded.

[1] Implies inner/outer Cacheable, Write-back, Non-transient, Read-allocate, Write-allocate.
[2] These attributes can be specified separately for inner and outer caches.
[3] The negation of one of these attributes implies the negation of all the preceding ones.

that the model complexity gave programmers a burden not justified by the potential performance improvements.

More specifically, the architecture was originally *non-multicopy-atomic*. As a consequence, the model had the rather counterintuitive property that a memory write performed by a core could at first become observable only to a subset of the other cores, before becoming observable by all of them. The model was then amended to make it *multicopy-atomic*, and hence, guarantee that a memory write performed by a core becomes visible to all the *other* cores exactly at the same instant.

In the previous sentence, it is important to stress the word *other* because the visibility rule of multicopy atomicity does not include the core that issued the write operation. In other words, the write operation can still become visible to the issuing core before it becomes visible to the others, to still allow the use of the performance-critical store buffers discussed in Section 12.2.2.

Read operations issued by a core may be reordered by hardware, and are not required to be performed in program order. Write operations may also be reordered and multiple writes can be combined into a single memory write operation. Write operations can be reordered with respect to reads, too, with the only constraint of not breaking any data dependencies. For instance, it is forbidden to postpone a write operation after a read operation on the same memory address.

The core may also issue instruction fetch and data read operations from memory addresses that are not explicitly referenced in the program, as a consequence of speculative execution. For instance, a core may start fetching and executing instructions on one side of a conditional branch instruction—and also read the data they require from memory—before knowing for certain whether or not the branch will be taken, in an effort to improve performance if the speculation was correct.

Within a region of normal memory, programmers can force the cores to deviate from the default behavior and explicitly restrict reordering on a case-by-case basis by means of *barrier instructions*, to be discussed later. Barriers come in various flavors and provide different trade-offs between how strong the restriction they pose on reordering is and the consequent performance penalty.

Normal memory regions have additional attributes, also listed in Table 12.2, which determine how the caches and the cache coherence logic modules shall work and will be analyzed in the following.

Device memory

The device memory type is used to set *implicit*, rather than explicit, restrictions on reordering for a whole memory region, when using barrier instructions would be too complex or onerous. As shown in Table 12.2, the device memory type is further characterized by several attributes:

- The *gathering* attribute, when set, allows multiple memory accesses to be combined together into a single, equivalent access, to save memory bandwidth. For instance, two half-word writes at consecutive addresses can be combined into one full-word write. When not set, there is a one-to-one correspondence in number and width between memory accesses in the code and those observed by caches and main memory.
- The *reordering* attribute, when set, allows the cores to reorder accesses to addresses within the same address block within the region. The block size is implementation-dependent, but typically coincides with the part of the region allocated to a single peripheral device. When not set, accesses to non-reordering blocks are performed in program order, although they can still be reordered with respect to accesses to other regions, in which reordering is enabled.
- The *early write acknowledgment* attribute, when set, allows the cores to continue after a write operation—and assume that the write operation completed—after some intermediate bus agent within the interconnect logic provided a write acknowledgment. As a consequence, the cores may continue before the write operation has actually been observed by the target device. When this attribute is not set, the write acknowledgment must necessarily come from the target device itself.

In the previous list, attributes are ordered hierarchically, that is, the architecture forbids to set an attribute without also setting all the other attributes that follow it in the list. For example, if gathering is enabled, both reordering and early write acknowledgment must also be enabled. Instead, when gathering is disabled, reordering may or may not be enabled.

Cacheability and shareability

Two very important attributes of a normal memory region are its *cacheability* and *shareability*.

- Quite intuitively, the *cacheability* attribute controls whether or not region contents are to be cached. Somewhat counter-intuitively, there may be reasons for not caching a region, for example, when it is known in advance that its contents will be read only once, as is common for input-output buffers. In this case, caching would not bring any speed advantage upon reading (because the first and only read would not hit the cache) but would still displace other more valuable data from the cache itself.
- The *shareability* attribute specifies if region contents are private to a certain core or shared among multiple cores and other DMA-capable agents. In turn, this determines whether the coherency mechanisms described in Section 12.2.3 are to be used or not. Also in this case, unnecessarily declaring a region as shareable when it is not actually shared gives no benefits, but still brings all the overheads associated with maintaining coherency.

Since, as shown in Figure 12.11, the number of components in a memory system can easily be large, the architecture provides a way to specify the cacheability and shareability of a memory region not in absolute terms, but within and with respect to different domains, nested into each other. More specifically, the architecture defines an *inner* and an *outer* domain.

Special care must be taken when reasoning about these domains because their exact boundaries may vary from one architecture implementation to another. Even more importantly, their definition may differ for cacheability and shareability so that, for instance, the cacheability inner domain might not necessarily be the same as the shareability inner domain. For cacheability, the domains are generally defined as follows:

- The *inner cacheability* domain refers to the caches internal to a core, like the L1 caches in Figure 12.11, while the L2 caches could be either in the inner or the outer domain depending on the implementation.
- The *outer cacheability* domain comprises multiple inner cacheability domains and includes caches externally to the core. In the figure, the L3 cache would be in the outer cacheability domain.

Therefore, depending on the implementation, the contents of an inner-cacheable region will be cached only in the L1 caches (the ones closest to the core) and possibly in the L2 caches, whereas the contents of an outer-cacheable region will also be cached in the L3 cache.

When a region is marked as cacheable further attributes, specified independently for the inner and outer domains, affect the policy used for caching:

- An attribute determines whether the hardware must use the *write through* or the *write back* policy when handling a write operation.

- Unlike the previous one, the *read allocation* attribute is a hint. When set, it suggests the hardware should allocate and fill a cache line upon a cache miss during a *read* operation.
- The *write allocation* attribute is also a hint. When set, means that the hardware should allocate and fill a cache line upon a cache miss during a *write* operation.
- Optionally, a *transient* attribute may be defined. When set, it suggests that the benefit of caching is for a relatively short period and the hardware may adjust its cache allocation policy accordingly.

The shareability of a memory region specifies if, and to what extent, the contents of a memory region are shared among cores and other DMA-capable agents. Also in this case, *inner shareability* and an *outer shareability* domains are defined, giving rise to four possible settings:

- A *non-shareable* region is used by only one core, and hence, it is unnecessary to keep its contents coherent.
- A *inner shareable* region is shared among multiple cores and the hardware must keep its contents coherent for all of them. Referring back to Figure 12.11, the inner shareability domain usually includes all clusters in the system.
- The coherency scope of an *outer shareable* region is even wider. It is made of one or more inner domains and usually includes all clusters plus all of the devices attached to the cache-coherent interconnect.
- The widest possible domain is the *full system* domain, which includes all memory observers in the system and may contain multiple outer domains.

Also in this case, it is important to specify the shareability of memory regions carefully and according to how they are actually used because, generally speaking, the wider the domain is, the more complex and time-consuming guaranteeing coherency becomes. The architecture uses a MOESI-based protocol for coherency management, a variant of the MESI protocol discussed in Section 12.2.3 optimized to further decrease the number of accesses to the lower levels of the cache hierarchy or main memory.

The cache coherency protocol is enabled for normal memory, when it is also marked as shareable and write-back, write-allocate cacheable. The difference between an inner and an outer shareable region lies in how many levels in the cache hierarchy get involved and have to execute the cache coherency protocol when the region is accessed.

As in other architectures, a proper configuration of the cacheability and shareability attributes of a memory region in ARM-v8A is crucial to ensure that certain inter-core synchronization instructions to be described in Chapter 13—such as the *load exclusive* and *store conditional* instructions—work correctly within the region.

Device memory regions are always treated as non-cacheable and outer-shareable. Therefore, any operation on them does not involve any cache and write operations become immediately observable by all agents in the outer domain.

TABLE 12.3

Main Kinds of Barrier Instruction Supported by the ARMv8-A Architecture

Kind	Description
ISB	Instruction Synchronization
DMB	Data Memory [1]
DSB	Data Synchronization [1]
LDA(X)R	Load-Acquire
STL(X)R	Store-Release

[1] These instructions accept an argument that specifies the properties of the barrier, as specified below.

Property	Value
Kinds of access	Load-Load and Load-Store
	Store-Store
	Any-Any
Shareability domain	Non-shareable
	Inner shareable
	Outer shareable
	Full system

Barriers

The ARMv8-A architecture provides several barrier instructions, listed in Table 12.3, for accesses to normal memory regions. They provide programmers a way to enforce a specific order among memory operations and optionally synchronize further code execution with their completion. It is especially important to remark, as already done in more theoretical terms in Section 12.2.2, that the use of a barrier entails a significant penalty, especially on modern processors, because it negates them the opportunity to leverage the sophisticated memory access optimization techniques much of their performance depends on. It should therefore be used with care and only when necessary to ensure the functional correctness of the code.

The first barrier instruction is the Instruction Synchronization Barrier (ISB). It is peculiar with respect to the other barriers because its scope is not only limited to memory accesses like we have seen so far, but it also includes other kinds of operations of which programmers may need to ensure the completion before continuing with code execution. More specifically, the ISB instruction flushes the pipeline of the executing core and stops execution until all so-called *context-changing* instructions the core itself has previously issued have completed.

In this way, it is guaranteed that the effect of those instructions are visible to the next instruction executed after the ISB. Similarly, if there are other context-changing instructions *after* the ISB in program order, it is also guaranteed that their effect is not yet visible before the barrier itself, giving rise to a *two-way* barrier.

A notable example of context-changing instructions, beside explicit cache maintenance instructions like invalidate and flush, are the instructions that maintain the Translation Lookaside Buffer (TLB). This is a very important on-chip component that, as a cache does for memory accesses, accelerates the address translations performed by the MMU. Although the details are beyond the scope of this book, it is clear by intuition that its contents must be invalidated after modifying the translation tables in memory, to prevent the TLB from holding now-outdated translations.

The second and third instruction also establish a two-way barrier and are conceptually closer to the barriers we outlined in Section 12.2.2. They are called Data Memory and Data Synchronization Barrier (DMB and DSB, respectively). As their name say they operate on the core's data path, unlike ISB that operates on the instruction path instead. The main difference between the two is that:

- The Data Memory Barrier enforces ordering constraints on the memory access instructions that precede and follow it, but does not stop the core on which it is executed.
- The Data Synchronization Barrier enforces the same ordering constraints as the Data Memory Barrier, and has the additional effect of stopping execution of any further instruction until synchronization is complete. Moreover, it also waits until all cache, TLB, and branch predictor maintenance operations issued by the core before the barrier have completed.

In both cases, the strength of the constraints depends on the instruction argument, which affects two orthogonal properties of these barriers:

- The first property determines the *kinds of access* the barrier applies to. It has three possible settings:
  - A *load-load, load-store* barrier requires all load instrucions issued before the barrier to complete, but does not pose any requirement on store instructions. Both load and store instructions that follow the barrier in program order are constrained to wait until all loads issued before the barrier are complete. Instead, store instructions that precede the barrier may still be moved past it.
  - A *store-store* barrier requires all stores issued before the barrier to complete before any store instructions that follow the barrier complete. It is transparent to load instructions, which can therefore freely cross the barrier in both directions.
  - An *any-any* barrier requires all memory access instructions (both loads and stores) issued before the barrier to complete before any memory access instructions that follow the barrier complete.

A subtle distinction is that the barrier instruction ensures that load and/or store *instructions* complete, but not that the corresponding memory *operations* also complete. Due to the interposition of several level of caches and the early write acknowledgment feature, memory may never observe the operation (for a load instruction, if the data is already in a cache) or observe the operation only at a later time (for a store

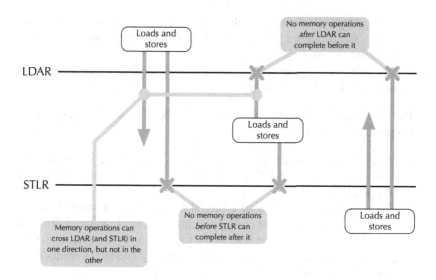

FIGURE 12.12 ARMv8-A one-way barriers.

instruction that goes through a write-back cache, only when the cache line is flushed back to memory).

This observation is also related to the second property, which defines the shareability domain to which the ordering constraints enforced by the barrier apply. Wider domains can still observe the effects of memory access instructions in a different order. Remembering the definition of shareability domain given previously, four settings are possible: not shared, inner shareable, outer shareable, and full system.

The last two barrier instructions, called Load-Acquire (LDAR) and Store-Release (STLR), differ from the previous ones in two important aspects:

- They are *one-way* rather than two-way barriers, that is, certain memory operations can cross them in one direction but not in the other.
- They combine the barrier semantics with a load or store operation in a single instruction.

Moreover, both instructions have exclusive/conditional variants (called LDAXR and STLXR) especially useful to implement wait and lock-free synchronization (see Chapter 13). The idea behind LDAR and STLR is that they must be used in pairs, as shown in Figure 12.12, to delimit a region of code and prevent memory operations from "escaping" from the region.

As shown in the figure, memory operations that precede STLR in program order cannot complete after it and memory operations that follow LDAR in program order cannot complete before it. As a result, memory operations that are between the two barriers in program order are forced to complete between them. The same figure should also make the meaning of the term *one-way* barrier clearer, because:

- Memory operations can cross the LDAR barrier only in one direction, that is, they can be moved forward and past the barrier, with respect to program order, but not backward.
- Similarly, memory operations can be moved backward, before the STLR barrier, but not forward and past it.

The advantages of these barriers, with respect to the more traditional ones presented previously, are twofold:

- Being one-way instead of two-way barriers, the ordering constraints they enforce are weaker and the performance penalty that ensues is therefore less pronounced.
- They combine a barrier with a special form of load or store instruction commonly used to delimit critical regions in multicore systems. This further enhances performance and, at the same time, ensures that critical region semantics are correct by preventing memory operations issued within it from being observed outside it by other agents.

Unlike for DMB, LDAR and STLR do not have an explicit shareability domain as argument because they implicitly infer it from the corresponding attribute of their target memory address.

12.2.5 COMPILER-LEVEL INSTRUCTION REORDERING

In the previous section, we discussed how hardware may reorder instructions, memory operations in particular, to improve efficiency. When writing code in a high-level language, programmers must be aware that the compiler may also do the same while it translates high-level language constructs into assembly code, within certain limits imposed by the language standard.

In addition, the compiler can suppress memory operations altogether if it can prove they are redundant. For instance, the compiler can omit a load operation if it already performed the same load in the past, the value is still in a processor register and, according to its knowledge, memory contents have not been modified in the meantime. The goal is the same as before: Improve performance and at the same time ensure that the code still works correctly, that is, produces correct results.

The underlying side effect of this approach is that other observers can see memory operations happening in a different order than the one stated in the high-level code, or not happening at all. On single-core processors, this is often overlooked because it is generally not an issue at the user application level. Indeed, on a single-core processor the only other observers are DMA-capable devices, and they are dealt with at the device-driver level.

Instead, the issue becomes important on multicore systems if the user application is multithreaded because, as we explained previously, all the cores are independent observers of each other's memory operations, and the unexpected reordering of operations on shared memory may lead to subtle bugs in the code. To make the matter even more complex, these bugs are often platform-dependent, because compilers targeting different processor architectures are likely to have different reordering strategies and rules.

In the C language, up to and including the C99 revision of the standard [70], reordering constraints were mainly based on the concepts of *sequence point* and *volatile object*, which we will briefly explain in the following. Sequence points are specific points in the execution sequence, defined by the standard. Examples of sequence points include function calls (after argument evaluation) and the end of a full expression (which coincides with the semicolon at its end).

At each sequence point the compiler must guarantee that, regardless of the optimizations it performed on the generated code, certain side effects of the language statements that precede the sequence point are complete, while certain side effects of any language statement that follows the sequence point have not taken place yet. Informally speaking, side effects are changes to the state of the execution environment and include, for example, load and store operations to volatile objects and store operations into non-volatile objects.

The standard specifies the least requirements that the code generated by a compiler must satisfy at sequence points, in order to be conforming to the standard. The full list can be found in the standard itself but, for the sake of this discussion, the most important one is that, at sequence points, all previous accesses to volatile objects are complete and any subsequent access to volatile objects has not yet occurred.

Volatile objects are objects whose data type is volatile-qualified, that is, the data type definition includes the `volatile` type qualifier. The underlying idea is that a volatile object may be modified in ways unknown to the compiler and accessing it may have other side effects the compiler is unaware of. This also prevents the compiler from optimizing accesses to them in other ways, for instance, by suppressing a seemingly redundant load operation according to the compiler's view. For instance, the requirements just described guarantee that, in the execution of the following fragment of code:

```
volatile int a;
volatile int b;
...
a = 0;
b = 3;
if(a == 3) ...;
```

- The store operation on a precedes the store operation on b, and the load operation on a in the if statement follows the store operation on b, because the end of each statement is a sequence point and both a and b are volatile-qualified.
- The load operation on a in the if statement is actually performed although, by looking only at the code at hand, the compiler could prove that the value of a must be zero. Indeed, a was set to zero by the first store, was not

modified by the second store, and there were no intervening function calls or other instructions that could have altered its value.

The standard puts additional constraints in place when calling a function that does not belong to the same translation unit, and when returning from it. In these cases, regardless of optimizations, the values of all externally linked objects—for instance, global variables—and all other objects the called function can access through pointers must have the expected value, according to the abstract language semantics, before the call takes place. Moreover, the value of all its parameters and all objects that can possibly be accessed through them via pointers must also have the expected value. Symmetrically, the values of all externally linked objects and all other objects that the caller can access through pointers must have the expected value before returning from a function. For instance, when translating the following fragment of code:

```
extern void f(void);
int a;
...
a = 3;
f();
```

the store operation on a cannot be moved past the call to the function f, because this function is external to the compilation unit and could access a because it is a global variable.

This is also what ensures that the critical sections discussed in Chapter 7 work correctly. They are delimited by semaphore operations, which the compiler sees as regular function calls. Hence, the previously mentioned constraints ensure that the compiler does not move any instructions in and out of them, if those instructions produce side effects observable by other threads.

In the embedded software domain, it is still important to take these aspects into consideration even in single-threaded programs, when some high-level functions are used as interrupt handlers. In this case, besides obeying all other restrictions imposed by the language and the operating system on this kind of functions, programmers must also make sure that any object they access and share with others is volatile-qualified. Otherwise, for instance, other functions might continue to use outdated object contents after they have been altered by an interrupt handler, unknowingly to the compiler.

Since version C11 of the standard [71], a more sophisticated memory model has been introduced, with weaker guarantees than its predecessor, in which explicit *fences* offer a standard, platform-independent way to impose stricter ordering constraints than the model would normally guarantee when needed, with a semantics similar to barriers. A description of the overall model would be very complex and beyond the scope of this book, but it has been the subject of a considerable amount of research work. Readers can find a suitable starting point for further reading in Reference [46].

TABLE 12.4

Parameters of the Tasks Depicted in Figure 12.13

Task	Period T_i and deadline D_i	Execution time C_i
τ_1	10	4
τ_2	11	1
τ_3	20	5
τ_4	40	4

12.3 SOFTWARE CHALLENGES INTRODUCED BY MULTICORES

As outlined in the previous section, even though the evolution from single-core to multicore systems may seem straightforward—after all, we are just duplicating the functional unit responsible of program execution to improve performance—it has deep consequences on the system as a whole. Moreover, some of its effects are definitely not transparent to application software, ranging from scheduling theory to practical software implementation.

12.3.1 LOSS OF THE CRITICAL INSTANT THEOREM

A significant challenge introduced by multiprocessor and multicore systems is that the addition of extra cores to a system may have counterintuitive, negative effects on schedulability even though the set of tasks it is supposed to execute does not change at all. More specifically, one of the most interesting outcomes is the loss of an important theorem, called the *critical instant theorem*, on which many of the results described in Chapters 3 and 4 are based.

Besides having been formally proven in 1973 [84], this theorem states a very intuitive fact. It can be informally summarized by saying that, in a fixed-priority task system that satisfies the basic task model and is executed by a single processor—as defined, for instance, in Section 3.2.3—any given task is going to have its largest possible response time when it is released simultaneously with all higher-priority tasks. This is because, in this scenario, the task suffers the worst-case amount of interference from the others.

However, let us consider a synchronous, periodic task set whose parameters are listed in Table 12.4. Without loss of generality, both the period and the execution time are expressed in terms of an arbitrary time unit. For the sake of illustration, we assume that these tasks have implicit deadlines, that is, their deadline is equal to their period. Moreover, we assume that their priority is fixed and has been set according to the Rate Monotonic priority assignment, a provably optimum algorithm on single-core systems under these circumstances.

Therefore, their priority is inversely proportional to their period and Table 12.4 lists them in order of decreasing priority. Since the tasks are executed by a priority scheduler, when the number of ready tasks exceeds the number of idle processors

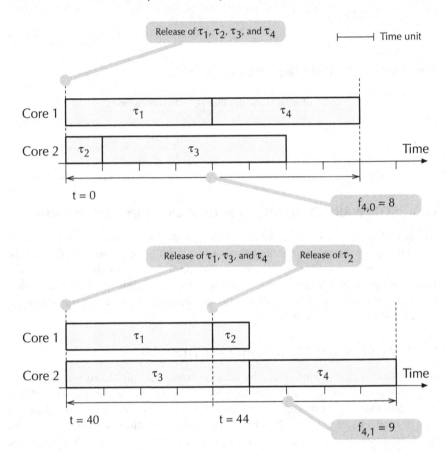

FIGURE 12.13 Counterexample of the critical instant theorem on a multicore system.

at any given time, the scheduler will pick the highest-priority tasks for execution, starting from τ_1 and down to τ_4.

Since the task set is synchronous, all tasks are released together at $t = 0$. The upper part of Figure 12.13 depicts the corresponding scheduling diagram on a two-core system. More specifically:

- At $t = 0$, all four tasks are ready for execution. Since there are only two cores available, the scheduler runs τ_1 and τ_2 concurrently, whereas τ_3 and τ_4 stay in the ready state.
- At $t = 1$, τ_2 completes its execution and the core it was executing on becomes available to another task. Both τ_3 and τ_4 are ready, but the core goes to τ_3 because it has a higher priority.
- τ_1 completes its execution at $t = 4$ and its core is assigned to τ_4, which starts executing.

- At $t = 6$, τ_3 completes its execution and its core remains idle because there are no other tasks ready at the moment.
- Eventually, the first instance of τ_4 completes its execution at $t = 8$, leading to a response time $f_{4,0} = 8$.

According to the critical instant theorem, on a single-core system the value $f_{4,0} = 8$ we just derived would be the worst-case response time of τ_4 and we could safely state that $R_4 = f_{4,0} = 8$. By analogy, to analyze schedulability on a dual-core system, we might therefore believe it would be sufficient to consider the scenario just discussed as it was done, for instance, with the Response Time Analysis (RTA) method in Section 4.1.2.

Unfortunately, this is not the case, as it becomes evident if we consider the second instance of τ_4, released at $t = 40$. Due to the harmonic relationship between the period of τ_1, τ_3, and τ_4, both τ_1 and τ_3 are released together with τ_4 at this time. On the contrary, τ_2 will be released for the fifth time at $t = 44$. At $t = 40$, τ_2 is not ready because its previous instance has already completed its execution.

The corresponding scheduling diagram is shown in the lower part of Figure 12.13. We can observe that:

- At $t = 40$, the scheduler runs τ_1 and τ_3 because they both have a priority higher than τ_4.
- At $t = 44$ two simultaneous events occur: τ_1 completes its execution and a new instance of τ_2 is released. As a consequence, τ_2 replaces τ_1 on core 1. Even though τ_4 is also ready at this time, it does not run because its priority is lower than τ_2.
- At $t = 45$, τ_3 completes its execution and the core it was running on becomes available to execute τ_4. At the same time, τ_2 completes its execution on the other core.
- Eventually, τ_4 completes its execution at $t = 49$ with a response time $f_{4,1} = 49 - 40 = 9$.
- Another option would be to execute τ_4 on core 1 instead of core 2, but this would not change its response time because, also in this case, its execution would still start at $t = 45$ as before.

In summary the diagram shows that, contrary to intuition, $f_{4,1} > f_{4,0}$ although only τ_1 and τ_3, but not τ_2, were released together with the second instance of τ_4. This result has important ramifications because, as hinted at previously, many other useful theorems and properties of the single-core scheduling algorithms described so far depend on it. For instance:

- Neither the Rate Monotonic (RM), nor the Deadline Monotonic Priority Order (DMPO), nor the Earliest Deadline First (EDF) priority assignment and scheduling algorithms are optimal on systems with more than one core.
- In the general case, it is unfeasible to determine the schedulability of a set of periodic or sporadic tasks by identifying a single worst-case release sequence of task instances and analyzing only that one. As a consequence,

TABLE 12.5

Parameters of the N Tasks Described in [41] and Depicted in Figures 12.14, 12.15, and 12.16

Task	Period T_i and deadline D_i	Execution time C_i
$\tau_1, \ldots \tau_{N-1}$	1	2ε
τ_N	$1+\varepsilon$	1

the strategy fruitfully used by the Response Time Analysis (RTA) method is inapplicable to multicore systems.

12.3.2 DHALL'S EFFECT

The loss of the critical instant theorem discussed in the previous section puts into question the optimality of Rate Monotonic (RM) and Earliest Deadline First (EDF). However, one could still rely on their excellent performance record on single-core systems and expect that their straightforward extension to multicore systems would still behave in a reasonably good way in any case.

Such an extension might operate according to the principle of *global scheduling*, that is, all M cores would be managed by a single scheduler instance, which would assign them to the M highest-priority tasks ready for execution, according to the priority assigned to them by RM or EDF.

Unfortunately, due to the *Dhall's effect* first described in [41], this is not true because some extremely problematic task sets exist, for which the use of a global RM or EDF scheduler to execute tasks on multiple cores leads to an extremely low utilization.

To better illustrate Dhall's effect, let us consider the set of $N \geq 2$ tasks presented in [41], whose parameters are listed in Table 12.5. Both the period and the execution time are expressed in terms of an arbitrary time unit, as a function of a parameter $0 < \varepsilon \ll 1$. As before, we assume that these tasks are periodic and synchronous, adhere to the basic task model, and have implicit deadlines.

The total utilization of the task set is:

$$U = \sum_{i=1}^{N} \frac{C_i}{T_i} = (N-1)\frac{2\varepsilon}{1} + \frac{1}{1+\varepsilon} = 2\varepsilon(N-1) + \frac{1}{1+\varepsilon} \qquad (12.1)$$

First of all, we note that if $N \geq 2$, then $U > 1$ for all ε. Moreover, as parameter ε becomes smaller and tends to zero, the total utilization tends to:

$$\lim_{\varepsilon \to 0^+} U = \lim_{\varepsilon \to 0^+} 2\varepsilon(N-1) + \frac{1}{1+\varepsilon} = 1 \quad \text{(from above)} \qquad (12.2)$$

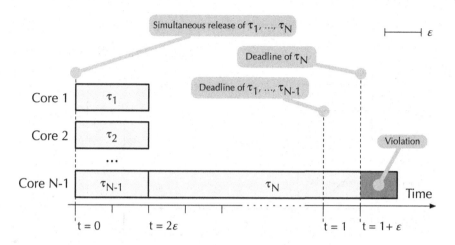

FIGURE 12.14 When executed on $N-1$ cores with global RM of EDF, the task set of Table 12.5 is not schedulable.

The fact that $U > 1$, combined with the necessary schedulability test discussed in Section 4.1.1, precludes the use of a single-core processor to execute this task set. We can then ask how many cores we need to successfully schedule it—that is, to fulfill all task deadlines—by means of RM or EDF.

Although from Section 12.3.1 we know this may or may not be the worst case for what concerns schedulability, let us focus anyway on the very first release of all tasks in the task set, which occurs at $t = 0$. In this scenario, Table 12.5 lists tasks in order of decreasing priority of their first instance, and this priority assignment is the same for both RM and EDF.

If we observe that some deadlines are violated during the analysis, we can surely conclude that the task set is not schedulable. On the other hand, if all deadlines are satisfied, we can come to the weaker conclusion that the task set *could* be schedulable in general.

Figure 12.14 shows how global EDF schedules the task set on $N-1$ cores:

- At $t = 0$ the first instance of all tasks is ready for execution. Since N task instances are ready and there are only $N-1$ cores, the scheduler picks the $N-1$ highest-priority tasks for execution, in this case $\tau_1, \ldots, \tau_{N-1}$. Task τ_N does not execute yet at this point.
- Tasks $\tau_1, \ldots, \tau_{N-1}$ complete their execution at $t = 2\varepsilon$, fulfilling their deadline. All cores become idle at this point and the execution of τ_N can start. The figure shows that the scheduler executes the task on core $N-1$, but using any other core would lead to the same result.
- The execution of τ_N continues at least until $t = 1$, when a new instance of tasks $\tau_1, \ldots, \tau_{N-1}$ becomes ready. At this point, the RM scheduler would preempt τ_N and assign its core to one of them, whereas the EDF scheduler would continue the execution of τ_N anyway, because its deadline

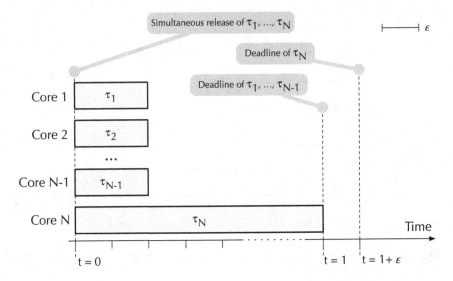

FIGURE 12.15 A successful execution of the task set of Table 12.5 with global RM of EDF requires at least N cores.

(at $t = 1 + \varepsilon$) is earlier than all the others (at $t = 2$). Figure 12.14 shows how the EDF scheduler behaves, because it is more favorable than RM for what concerns the schedulability of τ_N.

- Although the execution of τ_N continues beyond $t = 1$, it is bound to miss its deadline anyway, because the deadline expires at $t = 1 + \varepsilon$ whereas the execution of τ_N would conclude at $t = 1 + 2\varepsilon$.

It can easily be seen that using less that $N - 1$ cores makes the schedulability scenario even worse for τ_N. Instead, in order to successfully execute the task set of Table 12.5 with global RM or EDF, we need a system with at least N cores, as shown in Figure 12.15, that is, at least one core per task. Although the figure only shows that the task set is schedulable in the simultaneous-release scenario, it can easily be proven that this is true in general.

In summary, we reach this rather counterintuitive conclusion: For any system with N cores there exists at least a pathological set of N tasks, like the one just considered, which strictly requires all the cores in order to be executed successfully with global RM or EDF, because is not schedulable on $N - 1$ cores or less. Moreover, this happens even though the total utilization U of the task set can be brought as close to 1 as we desire. On the contrary, the use of a different scheduling approach may lead to much more favorable results. For instance, as Figure 12.16 shows, two cores may be sufficient to successfully execute the task set if we schedule tasks $\tau_1, \ldots, \tau_{N-1}$ on one core, and τ_N alone on the other. This is possible only as long as ε is small enough.

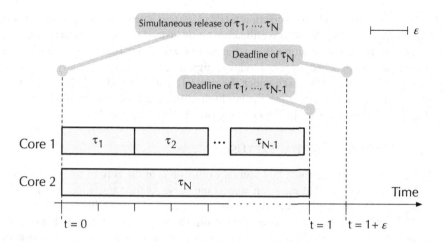

FIGURE 12.16 For small ε, two cores may be sufficient to successfully execute the task set of Table 12.5 with a different priority assignment or scheduling algorithm.

More specifically, it must be:

$$\sum_{i=1}^{N-1} C_i = (N-1)2\varepsilon \;\le\; 1, \tag{12.3}$$

because we may not exceed the total capacity of the first core. In turn, this constraint implies:

$$\varepsilon \le \frac{1}{2(N-1)}. \tag{12.4}$$

However, this is still way better than global RM or EDF, which cannot schedule the task set on less that N cores regardless of how small ε is. As before, although we verified schedulability only in the simultaneous task release scenario, it can easily be proven that this is true in general.

After publication, this seminal result influenced subsequent research, steering it towards *partitioned*, rather than global, multiprocessor and multicore scheduling algorithms. Partitioned scheduling algorithms follow the strategy outlined in the previous example and, as the name says, are algorithms that work in two stages:

1. Tasks are subdivided into N groups, one for each processor or core. The subdivision is done offline, once and for all. Therefore, tasks do not migrate from one processor to another at run time.
2. A separate scheduler is in charge of each processor or core and operates independently from the others, controlling the execution of the tasks assigned to that processor or core.

On the other hand, a closer scrutiny of the task set of the example shows that it has a peculiar characteristic. Namely, it is composed of a number of very low utilization tasks, $\tau_1, \ldots, \tau_{N-1}$, whose utilization tends to zero as ε tends to zero, plus one very high-utilization task, τ_N, whose utilization tends instead to one as ε tends to zero.

This leads to the intuition that we could obtain better results by retaining the concept of global scheduling, but treating high-utilization tasks specially, instead of putting the whole set of tasks under the control of global RM or EDF. Even more specifically, it turns out that privileging high-utilization tasks, like we did in the example, leads to better results than using global RM or EDF unaltered.

12.3.3   IMPLICIT MUTUAL EXCLUSION

In Chapter 7 we highlighted that mutual exclusion is at the core of traditional inter-task communication that relies on shared memory. At the same time, we also discussed how to implement mutual exclusion by means of a semaphore.

The semaphore-based approach works correctly regardless of the number of tasks involved and is independent of the number of processors or cores in the system. In other words, if a properly designed set of tasks makes use of semaphores for mutual exclusion, it will work correctly regardless of whether tasks are executed in a concurrent way only apparently, by time-sharing the processor among them, or for real, as it happens on a multiprocessor or multicore system.

However, on a single-core system, there are other, less orthodox ways to achieve the same result and programmers sometimes resort to them because they are more efficient than semaphores.

Task priority and mutual exclusion

Let us consider a single-core system and neglect interrupt handlers for the time being. If we focus on the highest-priority task in the system, we can easily observe that, as soon as it starts executing, it will run to completion unless it voluntarily blocks or yields the processor.

We can see this as a limited, but still useful, form of mutual exclusion because it also implies that, as long as the highest-priority task keeps executing, it cannot be preempted by any low-priority tasks and its execution will never be interleaved with them. The contrary is of course still possible, that is, any low-priority task can be preempted by the highest-priority task if it becomes ready while the low-priority task is executing.

This property can be leveraged to build the highest-priority task so that, for instance, it can still read a complex data structure d shared with one or more low-priority tasks in a controlled way, even without using semaphores or other forms of explicit mutual exclusion.

In order to do this, as shown in Figure 12.17, the low-priority tasks must apply the following protocol to update the shared data structure:

1. They coordinate among themselves to get exclusive access to the shared data. This can be accomplished by means of a mutual exclusion semaphore if there is more than one low-priority task, or be taken for granted if there is only one.

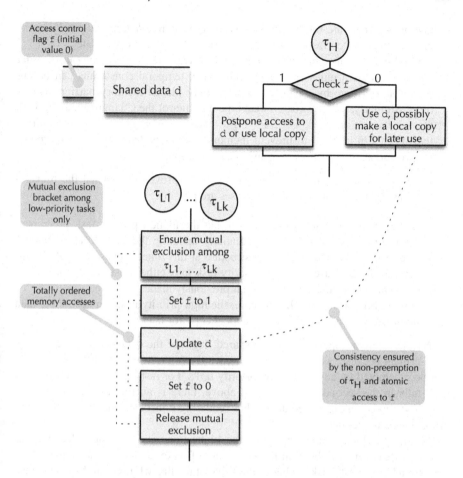

FIGURE 12.17 Implicit mutual exclusion with the highest-priority task.

It must be noted that the semaphore is used only by low-priority tasks, and not by the highest-priority task. As a consequence, the execution of the highest-priority task is not subject to any mutual exclusion constraint because of this semaphore.

2. The low-priority task that got access to the shared data sets a shared flag f with an atomic memory operation before starting to update the shared data structure. This can be accomplished in a straightforward way on most computer architectures as long as the data type of the flag is sufficiently small, typically if it fits in a machine word.

   Recent versions of the C language standard [70] define portable data types to this purpose, like `sig_atomic_t`, which is an integer type that can be accessed as an atomic entity.

3. The low-priority task then updates the shared data structure and finally resets the aforementioned flag. We assume that the memory accesses performed by the

low-priority task are totally ordered, so that terms like *before* and *finally* in the discussion above correspond to the temporal order of memory updates.

On high-performance hardware architectures that naturally perform out-of-order memory accesses unless instructed otherwise, temporal constraints can be enforced at *run time* by means of appropriately placed memory barrier instructions. Similarly, the `volatile` data type qualifier of the C language, along with the more recent `atomic_signal_fence` and `atomic_thread_fence` directives [71], can be used to prevent memory operations from being reordered or suppressed at *compile time*, during code generation.

If all these conditions are met it can easily be proven that, when the highest-priority task starts executing, only two cases are possible:

1. The shared flag `f` is *set*, which means that the highest-priority task preempted a low-priority task while it was potentially updating the shared data structure. Hence, it shall not access the data structure at this time, because it might be inconsistent. In this case, the high-priority task may either postpone the access or use a local copy of the data structure it previously made.
2. The shared flag is *not set*. In this case, the high-priority task can access the data structure at will and possibly make a local copy for later use.

Neither the read operation on the shared flag, nor the data structure access are time-critical because, as said previously, once the highest-priority task runs, it cannot be preempted by any of the low-priority tasks. For instance, if the high-priority task falls in the second case described above, there is no risk that a low-priority task starts updating the shared data—thus rendering it inconsistent—while the high-priority task is accessing it.

Similarly, which of the two cases the highest-priority task must deal with is uniquely determined right from the beginning of its execution, because there is no way for a low-priority task to change the value of the flag while the high-priority task is running.

With respect to the most obvious alternate approach—that is, using a mutual exclusion semaphore to enforce mutual exclusion among *all tasks*, including the highest-priority task—there are two main advantages:

- The only overhead that the highest-priority task has to endure is to check the shared flag `f` before accessing shared data. This operation can be performed in one or two machine instructions on most architectures and is therefore at least one order of magnitude faster than any semaphore operation.

  Given that the highest-priority periodic tasks are often also the ones with the shortest period (as stipulated, for instance, by the Rate Monotonic algorithm), reducing their overhead has a substantial positive effect on overall utilization.
- Since low-priority tasks do not share any semaphores with the high-priority task, unbounded priority inversion (defined and described in Section 8.1)

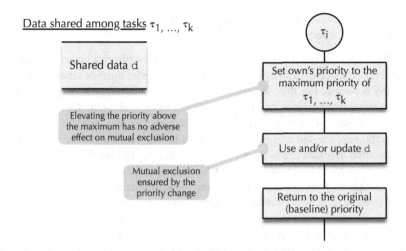

FIGURE 12.18 Implicit mutual exclusion by means of priority elevation.

cannot occur by definition between these two groups, and no special techniques are needed to avoid it.

- As a special case, if there is exactly one low-priority task, shared data consistency can be ensured without using any semaphore. This further reduces the overhead on the low-priority task to the couple of machine instructions required to set and reset the shared flag f.

  In Chapter 13, we will discuss some general methods to properly synchronize multiple tasks without semaphores or any other kind of lock, and without introducing any assumption about their relative priorities. However, they are considerably more complex than this one.

Hand-crafted priority elevation and scheduler lock

A key point of the approach just described is that, if a group of k tasks $\tau_1, \ldots, \tau_k$ shares some data structure d, the highest-priority task τ_H among them gets uninterruptible access to the data structure as soon as it starts executing, because it cannot be preempted by any of the lower-priority tasks in the group.

Even if we allow the presence of even higher-priority tasks in the system, provided they do not use any of the shared data, the basic properties of the approach are still valid. Although τ_H can now be preempted during execution, this preemption cannot alter the state of the shared data because none of the intervening tasks uses them.

On a single-core system we can then generalize and symmetrize the previous approach by having each task τ_i in the group of k tasks $\tau_1, \ldots, \tau_k$ follow the shared data access protocol depicted in Figure 12.18. More specifically, if we call $\mathscr{P}_C$ the maximum priority among tasks $\tau_1, \ldots, \tau_k$:

1. Before accessing the shared data, the task elevates its priority to $\mathscr{P}_C$.

2. It can then access and possibly update the shared data.
3. Afterwards, it returns to its original priority.

With respect to the previous approach there is now priority inversion, which is a direct consequence of priority elevation, and special care must be taken to ensure it is properly bounded. For this, we can leverage the close analogy between this method and the immediate priority ceiling protocol discussed in Section 8.1.

In particular, we can observe that the hand-crafted task priority movements in this method are the same as in the immediate priority ceiling protocol if we assume that resource accesses are not nested and the underlying semaphore foreseen by the priority ceiling protocol is always free when a task requests it. Therefore, it is easy to prove that priority inversion is bounded and the same calculations as for immediately priority ceiling can be applied to calculate the worst-case blocking time.

By itself, replacing a semaphore operation with a priority movement may or may not lead to a significant performance improvement, because on many operating systems these operations have similar complexity. However, we may push our reasoning further, and observe that the method still works correctly even if we elevate the task priority more than it is strictly necessary.

In particular, it still works if we raise the task priority to be the highest in the whole system. Although, at first sight it may seem that doing so would not affect performance at all, it must be noted that most real-time operating systems, RTEMS included, offer a way to temporarily *lock* the scheduler or, in other words, temporarily disable task preemption. More details about the interfaces provided to this purpose can be found in Section 5.7.

Such an action also temporarily gives the invoking task the highest priority in the system but, unlike a real priority movement, can be implemented in a very efficient way on a single-core system. Namely, from the operating system's perspective, it requires little more than setting/resetting an internal flag as appropriate, and checking the flag before preempting the running task.

The obvious negative side effect is that the set of tasks affected by priority inversion becomes bigger than before—because it includes all the tasks in the system instead of just $\tau_1, \ldots, \tau_k$. However, priority inversion is still bounded and, especially when critical regions are very short, the impact of blocking on the system may be acceptable or even negligible, also because it is compensated by the high efficiency of the critical region entry and exit code.

Disabling of interrupts

So far, the attention was focused on ensuring mutual exclusion among tasks, but in some cases, mutual exclusion has to be enforced with respect to interrupt handlers, too. This is typical of device drivers, in which the functions invoked from a task context must share data with interrupt handlers, and hence, must ensure their consistency in some manner.

A very low-overhead solution on single-core systems is the disabling of interrupts around critical sections. In this way, the arrival of an interrupt request and the consequent execution of the interrupt handler cannot preempt a task while it is inside a critical section. The opposite—that is, the preemption of an interrupt handler by a task—cannot happen by design, and hence, mutual exclusion is guaranteed.

The general structure of the solution is still the same as shown in Figure 12.18, but priority movements are replaced by the following actions:

1. Before accessing the shared data, interrupts are disabled after taking note of whether they were enabled or not.
2. After accessing the shared data, interrupts are re-enabled if they were enabled before step 1.

The implementation of these actions inherently depend on the underlying hardware architecture, but most operating systems provide portable interfaces for them. They are discussed in Section 5.7 in the case of RTEMS.

Recording whether interrupts were enabled or not before disabling them, and re-enabling them only if they were originally enabled, is only marginally more complex than disabling and enabling them unconditionally, but has the additional advantage of working correctly if critical regions are nested, without running the risk of re-enabling interrupts too early.

A less-invasive variant of the method, which can be used on hardware architectures that support exception priorities like the one described in Section 4.2, consists of raising the execution priority of the processor instead of disabling interrupts completely. More specifically, the execution priority must be raised to the highest priority among all exceptions whose handler may access the shared data. In this way, handlers with an even higher priority are still allowed to run and their latency is not impacted in any way.

At the same time, and without introducing further overhead, the disabling of interrupts also ensures mutual exclusion among tasks. This is because, as discussed in Section 4.2, only the acceptance and handling of an interrupt can trigger a context switch and possibly a preemption. With respect to the scheduler lock discussed previously, priority inversion is still bounded but also involves interrupt handlers, which leads to an increase of interrupt latency. As before, the actual impact on system performance can be limited if critical regions are kept sufficiently short.

Issues with multiprocessor and multicore systems

Unfortunately, none of the techniques outlined in this section works on a multiprocessor or multicore system. The main reason is that they all implicitly assume that there is at most either a single task or a single interrupt handler running at any given time. Even the description of the techniques is full of references to *the* running task (singular).

First of all, the implicit mutual exclusion methods depicted in Figures 12.17 and 12.18 are based on the assumption that the highest-priority task (in the first method) or the task that temporarily elevated its priority to be the highest (in the second) will never be preempted by other tasks unless it voluntarily blocks.

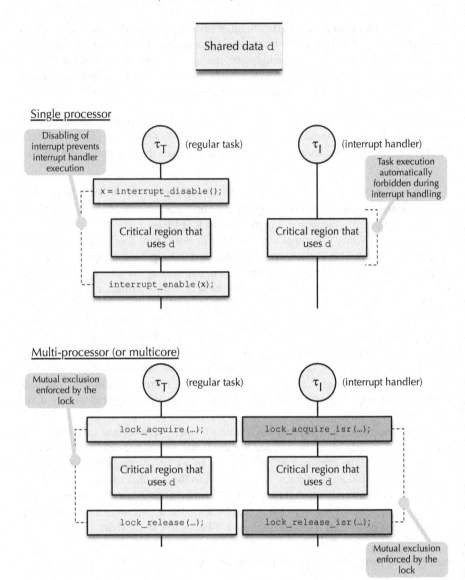

FIGURE 12.19 Mutual exclusion with respect to an interrupt handler.

Although, strictly speaking, this is still true if we focus only on a single processor or core within a multiprocessor or multicore system, nothing prevents other cores from running other tasks in the meantime. Thus, truly concurrent execution of multiple tasks, and the race condition that may ensue, are still possible even if there is no preemption at all.

As a side effect, even disabling preemption by locking the scheduler on a certain core would not prevent other cores from running other conflicting tasks. For this reason, functions of this kind are often disabled when a real-time operating system is configured for multiprocessor or multicore hardware.

Secondly, a similar issue arises with interrupt handlers. Even though the pseudocode shown on the top half of Figure 12.19 works correctly on a single-processor or single-core system, it is insufficient to ensure mutual exclusion in a multiprocessor or multicore system.

On such a system, disabling interrupt on a processor or core has no effect on the others. It is therefore still possible that the interrupt handler τ_I executes concurrently with the regular task τ_T, although τ_T brackets its critical section with functions to disable/enable (local) interrupts, if the scheduler runs them on two different cores.

Although on some multicore architectures it is indeed possible to disable interrupts *globally*, that is, on all cores, the overhead of this operation is often high, because it must be performed by working on the interrupt controller, which is a component external to the cores themselves. By contrast, local interrupts can usually be disabled in a single machine instruction.

To make the matter even more complex, since the cores and the interrupt controller may operate asynchronously, the global interrupt disable operation might not take effect instantaneously, and hence, some interrupts may still be accepted even after the processor issued the commands to disable them.

Although it is possible to circumvent these issues, as the number of cores increases, in any case the overhead of disabling *all* interrupts on *all* cores to ensure mutual exclusion within a device driver quickly becomes a bottleneck as the number of cores and devices to be managed increases.

One correct approach is therefore to use a different device for mutual exclusion, for instance, a mutual exclusion *lock*, which has the same semantics as a mutual exclusion semaphore, works correctly when used across multiple processors or cores, is more efficient than a semaphore, and can be used in an interrupt context. This entails some changes to the original code. In particular, as shown in the bottom half of Figure 12.19:

- The task code keeps the same structure, except for the operations that bracket its critical region. Instead of disabling and enabling interrupts, respectively, they acquire and release a lock.
- The critical region in the interrupt handler must now be bracketed, too, because, as explained previously, concurrent execution of a regular task and an interrupt handler is now possible. As for the task code, this is accomplished by means of lock acquire and release operations, depicted as darker rectangles in the figure.

In the figure, the actual arguments to the lock acquire and release functions have been replaced by ellipses, because the details vary across operating systems and kinds of lock. As also shown in the figure, some kinds of lock may require the use of two different sets of primitives, one on the task side and the other on the interrupt handler (ISR) side. Chapter 13 provides more specific details about the RTEMS application programming interface.

As an alternative, communication between the task and the interrupt handler can be re-designed in terms of lock or wait-free synchronization, which allow multiple agents to consistently exchange information without using any locks. An introduction to these methods will also be given in Chapter 13.

12.4   SUMMARY

The chapter first described the motivation behind the widespread adoption of multi-core processors in general-purpose computing and, in more recent times, embedded systems. Then, Section 12.2 provided an overview of multicore architectures, with practical references to contemporary ARM processors. Special attention was given to the concepts of memory model and cache coherency, because they play a much more important role in multicore system than they did in single-core ones.

Section 12.3 gave a summary of the most important challenges facing software designers and programmers when they transition to multicore systems, explaining why multicore execution has a profound—and sometimes counterintuitive—impact on several key areas of concurrent programming presented in Chapters 3–9, namely, scheduling algorithms, schedulability analysis, as well as inter-task communication and synchronization. In the following, Chapter 13 will outline some common ways to tackle those issues.

13 Multicore Concurrency: Issues and Solutions

CONTENTS

This chapter contains an introduction to the RTEMS scheduling algorithms and synchronization devices suitable for symmetric multiprocessor and multicore systems. In particular, it describes how scheduling algorithms are configured and outlines how the MrsP and OMIP semaphore protocols work to tackle unbounded priority inversion in this kind of systems, and how they can be accessed through the RTEMS application programming interface.

The central part of the chapter discusses more advanced synchronization methods, namely, lock-free and wait-free synchronization, which have not been covered in Chapters 7 and 9, along with several practical implementation examples. The last section gives a short introduction to spinlocks, a kind of synchronization device often based on lock-free or wait-free techniques. Besides being useful by themselves, especially for synchronization between tasks and interrupt handlers, spinlocks are also a main building block of more complex synchronization devices, like semaphores.

13.1 CLASSES OF MULTICORE SCHEDULING ALGORITHMS

Scheduling algorithms for multiprocessor and multicore systems have been the subject of considerable theoretical and applied research, especially in the past

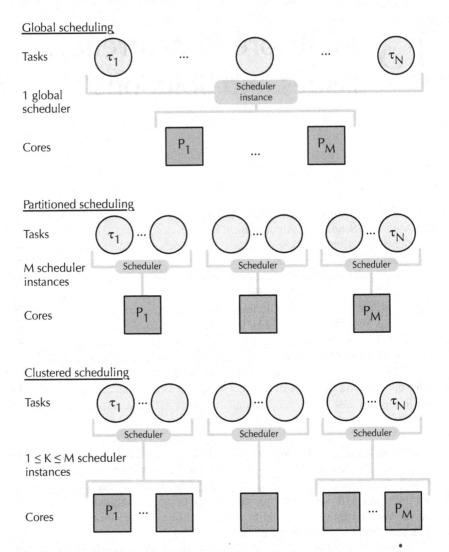

FIGURE 13.1   Global, partitioned, clustered multicore schedulers.

40 years [39]. Since the very beginning [83], it was clear to researchers that the real-time scheduling problem was much more difficult on multiprocessor and multi-core systems than single-core systems. In Section 12.3, we provided a few examples that show why this is the case and why a straightforward extension of single-core scheduling algorithms to multiprocessor and multicore systems is somewhat simplistic.

Currently, multiprocessor and multicore scheduling algorithms belong to one of three categories, depicted in Figure 13.1. At the two ends of the range lie *global* and *partitioned* scheduling algorithms:

- Global scheduling algorithms put all processors or cores in the system under the control of a single scheduler instance. In a system with M cores this scheduler instance is therefore in charge of picking, at any time, up to M ready tasks that will be executed in parallel. Although this approach is quite intuitive, it may lead to poor schedulability, at least for some peculiar sets of tasks, as described in Section 12.3.2. Moreover, algorithms that are provably optimal on single-core systems under certain hypotheses are no longer optimal when extended for global multiprocessor or multicore scheduling.
- Partitioned scheduling algorithms run one scheduler instance for each processor or core. Tasks are statically assigned to a certain core at design time and cannot migrate to other cores under normal circumstances. As a consequence, all the schedulability theorems and methods discussed in Chapter 4 can still be applied on a scheduler instance-by-instance basis. This is because, informally speaking, an M-core system works exactly like M single-core systems from the scheduling point of view, if we neglect inter-core interferences due to shared hardware components—like caches, memory buses, and main memory—outlined in Section 12.2.

  The price to be paid, with respect to global scheduling, is a generally lower processor utilization. An intuitive argument toward this conclusion is that partitioned scheduling algorithms are not *work-conserving*, that is, they may at times leave a core idle (because there are no ready tasks assigned to it) while some tasks assigned to another core are ready, but unable to execute (because there is already a higher-priority task running on that core).

The third class of algorithms are based on *clustered* scheduling. In this case, as shown at the bottom of Figure 13.1, the M cores are divided into $1 \leq K \leq M$ clusters. Each cluster consists of at least one core and all cores in a cluster are managed by one scheduler instance so that, in total, there are K scheduler instances in the system. Clustered scheduling can also be seen as a generalization of global and partitioned scheduling because these two approaches are a special case of clustered scheduling when $K = 1$ or $K = M$, respectively.

13.2 MULTICORE SCHEDULING ALGORITHMS IN RTEMS

When RTEMS is configured to run on a symmetric multiprocessor (SMP) or multicore system, several scheduling algorithms suitable for this kind of system become available. They are summarized in Table 13.1 and replace the algorithms for single-core systems listed in Table 5.1.

By default, RTEMS works according to the *global scheduling* approach depicted in Figure 13.1 and puts all available cores under the control of a single scheduler instance, whose scheduling algorithm is specified in the configuration, up to a maximum of 32 cores. However, users can instruct the operating system to use *clustered*

TABLE 13.1

Multicore Scheduling Algorithms of RTEMS

Configuration macro (CONFIGURE_SCHEDULER_...)	Default name	Description
EDF_SMP	"MEDF"	Earliest Deadline First Scheduler (default)
PRIORITY_SMP	"MPD "	Deterministic Priority Scheduler
SIMPLE_SMP	"MPS "	Simple Priority Scheduler
PRIORITY_AFFINITY_SMP	"MPA "	Arbitrary Processor Affinity Scheduler

scheduling instead, by explicitly instantiating all the scheduler instances they need in the application compile-time configuration, as described in Section 13.3. *Partitioned scheduling* is supported, too, as a special case of clustered scheduling in which each scheduler instance is in charge of exactly one core.

Multicore earliest deadline first (EDF)

If the user does not explicitly specify any scheduler in a multicore system configuration, RTEMS defaults to the multicore Earliest Deadline First (EDF) algorithm. Like its single-core counterpart introduced in Section 5.2, this scheduler uses a red-black tree [53] as its underlying data structure and divides tasks into two classes, foreground and background, depending on whether or not a deadline has been specified for them, respectively. It also enjoys the same desirable properties, namely:

- Background tasks, that is, tasks for which no deadline has been specified, are scheduled according to their priority, exactly like a fixed-priority scheduler would do. As a consequence, an application that did not specify deadlines and ran under the default fixed-priority scheduler on a single-core system will still be scheduled in the same way when ported to a multicore system, with the obvious exception that multiple tasks will run in a truly parallel way on the latter.
- All foreground tasks, that is, tasks for which a deadline has been specified by means of the Rate Monotonic Manager (see Section 5.5) have a higher priority than all of the background tasks and are scheduled based on their deadline.

In addition, on a multicore system the EDF scheduler offers limited support for affinity masks, which can be set in two possible ways:

- A *one-to-all* task-to-core association, in which the mask contains all online cores, thus giving the scheduler total freedom to choose on which core the task runs, and move it from one core to another during execution as needed. In all cases, except when semaphore sharing is involved, the cores on which a task may run is the intersection between the task's affinity mask and the set of cores managed by the home scheduler instance in charge of the task.

- A *one-to-one* association, when the mask pinpoints exactly one core. Specifying a mask in between these two cases is not flagged as an error, but RTEMS interprets it as a one-to-one association with the highest-numbered core contained in the mask. It is important to note that specifying a restrictive affinity mask does not always prevent multicore semaphore locking protocols, to be presented in Chapter 13, from running a task on cores not included in its affinity mask. This may still happen when semaphores are shared among tasks assigned to distinct scheduler instances and these protocols decide to temporarily migrate a task away from its home scheduler instance.

Besides the one-to-one task-to-core association just mentioned, which binds a task to a specific core in a relatively permanent way, that is, unless a multicore semaphore sharing protocol gets into action or the affinity mask is changed, the EDF scheduler also supports a similar, but short-term, association called *pinning*.

Pinning can be used within a task that has a liberal affinity mask to ensure that a certain region of code—delimited by calls to the RTEMS directives _Thread_Pin and _Thread_Unpin—is executed as a whole by a single core. Exactly which core is unspecified, it could be any of the cores on which the task is allowed to run according to its affinity mask and, more specifically, the one that was executing the task when _Thread_Pin was called. Regions of this kind can be nested and behave as expected, that is, the task stays pinned until it exits from the outermost region.

The use of task pinning must be limited to short regions of code because it is stronger than a one-to-one affinity mask and prevents the multicore semaphore locking protocols just mentioned from working effectively. This is because pinning prevents them from migrating tasks from one core to another when appropriate.

Multicore deterministic priority scheduler (DPS)

As shown in Table 13.1, RTEMS also offers two fixed task-level priority schedulers, whose behavior is identical to their single-core counterparts:

- the Deterministic Priority SMP scheduler, and
- the Simple Priority SMP scheduler.

Both schedulers use the same underlying data structures as their single-core counterpart, depicted in Figures 5.1 and 5.2, respectively. They support 256 priority levels by default, but the number of levels can be changed in the application compile-time configuration. These schedulers do not support affinity masks and may schedule tasks on any of the cores assigned to them, regardless of how their masks are set.

Like it happens with their single-core counterparts, although these two schedulers consistently take exactly the same scheduling decisions, they provide users with two different trade-offs between scheduler efficiency and memory footprint, namely:

- Typical scheduler operations are performed in constant time by the Deterministic Priority SMP scheduler and in linear time, with respect to the number of ready tasks, by the Simple Priority SMP scheduler. This may or may not be a shortcoming, depending on how many ready tasks there are.
- The memory footprint of the Deterministic Priority SMP scheduler is roughly linear with respect to the number of configured priority levels, because it uses one ready queue for each level. Since the Simple SMP scheduler has only one queue for all ready tasks, its memory footprint is smaller and constant, regardless of the number of configured priority levels.

Arbitrary affinity masks

The Arbitrary Processor Affinity scheduler allows users to specify arbitrary affinity masks and honors them, thus providing fine-grained control on which cores tasks are allowed to run without resorting to clustered or partitioned scheduling. It uses the same data structures as the Deterministic Priority SMP scheduler, but different algorithms to pick suitable tasks to execute on each core, considering both their priority and their affinity masks, which may prevent them from running on certain cores.

In all cases, the user-specified affinity mask of a task is further restricted to stay within the cores assigned to the scheduler in charge of the task, so the scheduler will never execute a task on a core outside of its control. An exception to this rule, which is common to all scheduling algorithms and has already been discussed previously, is the task migration put in place by the semaphore locking protocols.

This scheduler is still considered experimental at the time of this writing and may be unsuitable for production use, because its implementation is incomplete and the computational complexity of some of its operations is higher than linear with respect to the number of ready tasks. However, it highlights the flexibility of the RTEMS scheduling framework in accommodating the more and more sophisticated scheduling algorithms that, as research progresses, will be needed in future real-time multicore systems.

13.3   SCHEDULERS CONFIGURATION

For relatively simple applications, users do not have to explicitly configure RTEMS schedulers on multicore systems. By default, these systems use a global EDF scheduler across all cores, which is adequate in most circumstances. Users can also select another scheduling algorithm for the global scheduler exactly like they did in single-core systems, as described in Section 2.4.

In more complex cases, users can go through the full schedulers configuration process summarized in Figure 13.2 and opt for clustered or partitioned scheduling, possibly using different scheduling algorithms in each cluster or partition. There are four configuration steps and their logical order coincides with the order of the corresponding sections in the RTEMS configuration file.

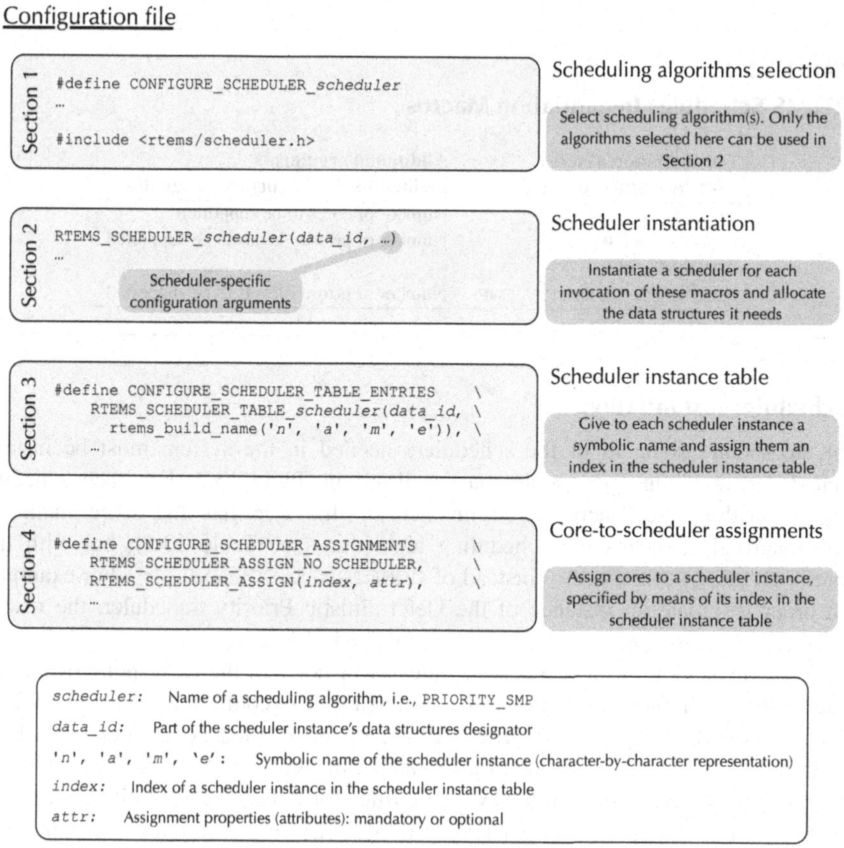

Configuration file

Section 1
```
#define CONFIGURE_SCHEDULER_scheduler
...
#include <rtems/scheduler.h>
```
Scheduling algorithms selection

Select scheduling algorithm(s). Only the algorithms selected here can be used in Section 2

Section 2
```
RTEMS_SCHEDULER_scheduler(data_id, …)
...
        Scheduler-specific
        configuration arguments
```
Scheduler instantiation

Instantiate a scheduler for each invocation of these macros and allocate the data structures it needs

Section 3
```
#define CONFIGURE_SCHEDULER_TABLE_ENTRIES    \
    RTEMS_SCHEDULER_TABLE_scheduler(data_id, \
        rtems_build_name('n', 'a', 'm', 'e')), \
    ...
```
Scheduler instance table

Give to each scheduler instance a symbolic name and assign them an index in the scheduler instance table

Section 4
```
#define CONFIGURE_SCHEDULER_ASSIGNMENTS      \
    RTEMS_SCHEDULER_ASSIGN_NO_SCHEDULER,     \
    RTEMS_SCHEDULER_ASSIGN(index, attr),     \
    ...
```
Core-to-scheduler assignments

Assign cores to a scheduler instance, specified by means of its index in the scheduler instance table

`scheduler:`	Name of a scheduling algorithm, i.e., PRIORITY_SMP
`data_id:`	Part of the scheduler instance's data structures designator
`'n', 'a', 'm', 'e'`:	Symbolic name of the scheduler instance (character-by-character representation)
`index:`	Index of a scheduler instance in the scheduler instance table
`attr:`	Assignment properties (attributes): mandatory or optional

FIGURE 13.2   Clustered scheduling configuration in RTEMS.

Scheduling algorithms selection

The first configuration step selects all the scheduling algorithms that will be used in the system, among those available. This is done by defining some of the macros listed in Table 13.1. For instance, defining the macro CONFIGURE_SCHEDULER_PRIORITY_SMP selects the Deterministic Priority Scheduler and enables it for use. Since each scheduling algorithm enabled during this step may introduce some per-task memory overhead, it is advisable to select only those scheduling algorithms that will actually be used, in order to save memory.

At the end of this configuration step, users must include the RTEMS header rtems/scheduler.h to get access to the scheduler-specific configuration macros to be used in the next steps.

TABLE 13.2

RTEMS Scheduler Instantiation Macros

Instantiation Macro (RTEMS_SCHEDULER_...)	Additional arguments besides the data structure designator
EDF_SMP	Number of cores to be supported
PRIORITY_SMP	Number of priority levels to be supported
SIMPLE_SMP	—
PRIORITY_AFFINITY_SMP	Number of priority levels to be supported

Scheduler instantiation

In the second step, all of the schedulers needed in the system must be instantiated, by invoking appropriate macros listed in Table 13.2. For each scheduling algorithm, the macro to be invoked in this step has the same name as the macro that selects the scheduling algorithm (see Table 13.1), but with the prefix RTEMS_SCHEDULER_ instead of CONFIGURE_SCHEDULER_. For example, in order to create an instance of the Deterministic Priority Scheduler, the macro RTEMS_SCHEDULER_PRIORITY_SMP must be invoked.

Each invocation of these macros creates an instance of the corresponding scheduler, which will manage a set of cores assigned to it in configuration step 4, and allocates its underlying data structures. All macros have at least one argument, which will become part of the scheduler data structure designators. For this reason, it must be unique and conforming to the syntactic rules for C-language identifiers. It is important to note that this identifier is only used in the configuration phase and is *not* the symbolic name that the application will use to refer to the scheduler. This name will be assigned to the scheduler in configuration step 3.

Some scheduler instantiation macros have additional, instance-specific arguments listed in the right column of Table 13.2. They are used to further configure each individual scheduler instance being created. It is of course possible to instantiate the same scheduling algorithm more than once, possibly with different instance-specific arguments, provided that designator names are different, and hence, unique.

Definition of the scheduler instance table

The third configuration step gathers all scheduler instances created in step 2 into a scheduler instance table. As a result, the scheduler instance table has one element for each scheduler instance in the system. This also implicitly assigns to each scheduler instance a unique index that represents its position in the table and starts from zero for the first scheduler instance. The index is important because it must be used to refer to the scheduler in the fourth and last configuration step.

Another important operation performed in this configuration step is to bind to each scheduler instance a symbolic name that the application can use to retrieve the object identifier of the scheduler instance, by means of the directive `rtems_scheduler_ident` described in Section 5.2. As shown in Figure 13.2, symbolic names are assembled by means of the `rtems_build_name` macro.

As illustrated in Figure 13.2, in order to do this, users must define the macro `CONFIGURE_SCHEDULER_TABLE_ENTRIES` to a comma-separated list of macro invocations. There must be one macro invocation for each scheduler instance created in configuration step 2 and the name of the macro must be derived from the scheduler instantiation macro by substituting the prefix `RTEMS_SCHEDULER_` with `RTEMS_SCHEDULER_TABLE_`. In other words, matching macros must consistently be used in configuration steps 2 and 3. For instance, in order to place an instance of the Deterministic Priority Scheduler in the table, the macro to be used is `RTEMS_SCHEDULER_TABLE_PRIORITY_SMP`.

Core-to-scheduler assignments

The fourth and last configuration step assigns cores to the scheduler instances created in configuration step 2. In order to do this, users must define the macro `CONFIGURE_SCHEDULER_ASSIGNMENTS` to a list of entries, one for each core configured in the system, for a total of `CONFIGURE_MAXIMUM_PROCESSORS` entries. Each entry specifies whether or not the corresponding core is assigned to a scheduler instance, and whether the assignment is mandatory or optional. More specifically, each entry can be:

RTEMS_SCHEDULER_ASSIGN_NO_SCHEDULER The core is not assigned to any scheduler instance. This does not mean that the core will sit unused, but only that RTEMS will not attempt to use it in any way. This does not preclude a hypervisor to allocate the core to another operating system or even to another RTEMS instance.

RTEMS_SCHEDULER_ASSIGN(scheduler_index, attr) The core is assigned to the scheduler instance whose index in the scheduler instance table defined in configuration step 2 is `scheduler_index`.

When a core is assigned to a scheduler instance, the `attr` argument of the assignment macro just described further specifies the assignment properties. It can assume one of the following values:

RTEMS_SCHEDULER_ASSIGN_PROCESSOR_MANDATORY The assignment is mandatory. RTEMS will raise a fatal error, which terminates the system, if one or more mandatory cores are not present in the system or cannot be started at bootstrap.

RTEMS_SCHEDULER_ASSIGN_PROCESSOR_OPTIONAL The assignment is optional. The unavailability of an optional core does not cause a fatal error, but simply prevents the scheduler instance it is assigned to from using it.

RTEMS_SCHEDULER_ASSIGN_DEFAULT This value calls for default assignment properties and is the same as `RTEMS_SCHEDULER_ASSIGN_PROCESSOR_OPTIONAL` at the time of this writing.

Besides the static core-to-scheduler assignments just described, which are specified at compile-time, it is also possible to dynamically add/remove cores to/from scheduler instances by means of the directives `rtems_scheduler_add_processor` and `rtems_scheduler_remove_processor`, respectively. Both were described in Section 5.2.

In a multicore system one of the cores, often called *boot core*, is generally responsible of initializing the operating system before the other cores are started and the system enters multitasking mode. This core must necessarily be assigned to a scheduler instance in the compile-time configuration. A fatal error results if this requirement is not fulfilled.

13.4   MULTICORE SYNCHRONIZATION DEVICES

Like scheduling algorithms, also the protocols against unbounded priority inversion behind lock-based synchronization devices must be redesigned to work properly and effectively on multicore systems. When configured for a multicore system, RTEMS supports two state-of-the-art protocols for binary semaphores with priority queuing:

1. The multiprocessor resource sharing protocol (MrsP) by Burns and Wellings [28], derived from the priority ceiling protocol.
2. The $O(m)$ independence-preserving protocol (OMIP) by Brandenburg [23], derived from the priority inheritance protocol.

Besides being of practical interest because they can be readily implemented in a real-time operating system, both protocols come accompanied by significant theoretical results. In the first case, a method to extend response time analysis (RTA) to consider MrsP-induced blocking time exists, while in the case of OMIP the asymptotic optimality of its priority-inversion induced blocking time has been proven.

13.4.1   MULTIPROCESSOR RESOURCE SHARING PROTOCOL

The multiprocessor resource sharing protocol (MrsP) [28] is an evolution of the immediate priority ceiling protocol described in Chapter 8 and was designed with schedulability analysis in mind. More specifically, one important design goal was to be able to analyze the schedulability of a system that makes use of MrsP by means of the response time analysis (RTA) method discussed in Chapter 4.

Like the immediate priority ceiling protocol, MrsP requires programmers to specify, for each semaphore, what is the priority of the highest-priority task that can ever acquire that semaphore, for each scheduler instance. As a result, each MrsP semaphore has multiple priority ceiling values, one for each scheduler instance in the system, rather than just one system-wide value as for the immediate priority ceiling protocol.

Without going deep into MrsP implementation, there are nevertheless three aspects that are important from the programmer's point of view:

- When a task successfully acquires a semaphore, its priority is temporarily elevated to the ceiling priority of the semaphore for the scheduler instance the task is normally assigned to. In RTEMS, this scheduler instance is also called the *home* scheduler instance of the task.

   The priority elevation lasts until the task releases the semaphore. From this point of view, MrsP works in the same way as the immediate priority ceiling protocol.

- Tasks that are waiting to acquire a semaphore *spin*, that is, perform an active wait and remain running from the scheduler's point of view, instead of waiting passively by moving to the blocked state of the task state diagram. Informally speaking, this prevents the spinning task from suffering further priority inversion-induced blocking after it eventually acquires the semaphore, due to lower-priority tasks that might have been run and have had their priority elevated by MrsP in the meantime.

- In some cases, a *helping* protocol among scheduler instances may temporarily migrate a task from the set of cores it is normally assigned to, according to the partitioned or clustered scheduling approach, onto other cores.

   In the most basic case, this happens when the task is preempted by a higher-priority task while it holds a semaphore, and on the other core there is a task waiting for the same semaphore. In other words, this also means that a task may temporarily execute on cores that are not managed by its own home scheduler instance, but belong to other scheduler instances where there is at least another task that shares a semaphore with it.

MrsP usage in RTEMS

The RTEMS operating system implements the MrsP protocol on symmetric multiprocessor (SMP) and multicore systems for local binary semaphores with priority-based queues, which is arguably the kind of semaphore most suitable for mutual exclusion. This protocol is selected by specifying the RTEMS_MULTIPROCESSOR_RESOURCE_SHARING attribute (together with the RTEMS_LOCAL, RTEMS_BINARY_SEMAPHORE, and RTEMS_PRIORITY attributes) upon semaphore creation.

Initially, all priority ceiling values of a new semaphore are set to the priority specified upon creation, which is hardly adequate in most practical cases. A specific directive exists, rtems_semaphore_set_priority, to set a more appropriate value for each scheduler instance afterwards. Moreover, MrsP semaphores cannot be created locked, and hence, their initial value must necessarily be 1. Any attempt to create a semaphore with an initial value of 0 results in the RTEMS_INVALID_NUMBER status code. Since MrsP semaphores are used only to delimit critical regions for mutual exclusion, this is not a practical issue in most cases.

An MrsP semaphore can be used like any other mutual exclusion semaphore for the most part, bearing in mind the following restrictions:

- Tasks must release MrsP semaphores in the reverse order with respect to the sequence in which they were acquired, as it comes natural if critical regions are properly nested. Attempts to deviate from the prescribed order are detected and reported with the RTEMS_INCORRECT_STATE status code upon semaphore release.

- MrsP semaphores cannot be acquired recursively. Any attempt to do so results in the RTEMS_UNSATISFIED status code being returned by rtems_semaphore_obtain.

- Besides self-deadlocks that would result from recursive semaphore acquisition, the system also detects more complex kinds of deadlock that would be created by acquiring a semaphore. In this case, the rtems_semaphore_obtain directive fails and returns the status code RTEMS_UNSATISFIED.

- Finally, but this is a restriction that does not concern only MrsP semaphores, an MrsP semaphore cannot be taken from an interrupt context or whenever thread dispatching is disabled. Any attempt to do so results in an internal RTEMS error or undefined behavior.

Scheduling analysis

Let us recall the main recurrence relationship of response time analysis (RTA) [16, 17], that is, Equation (4.5) of Section 4.1.2:

$$w_i^{(h+1)} = C_i + \sum_{j \in \text{hp}(i)} \left\lceil \frac{w_i^{(h)}}{T_j} \right\rceil C_j. \tag{13.1}$$

It has been proven that, if we let $w_i^{(0)} = C_i$ and the succession $w_i^{(0)}, w_i^{(1)}, \dots$ converges, it converges to R_i, the worst-case response time of τ_i. If we consider a *partitioned* multicore system, that is, a system in which tasks are statically assigned to a specific core for execution and cannot autonomously migrate elsewhere, this recurrence formula is still valid to calculate the worst-case execution time of task τ_i if we take into account only the interference it may suffer from other tasks being executed on the same core.

This is reasonable because a task running on a certain core does not suffer interference from tasks running in parallel on other cores, if we neglect the indirect interference caused, among other things, by cache and memory access contention outlined in Chapter 12. In this scenario, the set hp(i) must be replaced by another set hpl(i), defined as the set of indices of the tasks with a priority higher than τ_i and local to τ_i, that is, assigned to the same core as τ_i:

$$w_i^{(h+1)} = C_i + \sum_{j \in \text{hpl}(i)} \left\lceil \frac{w_i^{(h)}}{T_j} \right\rceil C_j. \tag{13.2}$$

The basic RTA succession (4.5) was then extended to consider blocking time, that is, the time tasks spend waiting on semaphores for shared resources to become available for use. This led to a new recurrence relation, Equation (4.6). It can be proved that the same relation still holds for MrsP-induced blocking, by replacing $\text{hp}(i)$ with $\text{hpl}(i)$ as before:

$$w_i^{(h+1)} = C_i + B_i + \sum_{j \in \text{hpl}(i)} \left\lceil \frac{w_i^{(h)}}{T_j} \right\rceil C_j. \tag{13.3}$$

in which the term B_i represents the worst-case blocking time endured by τ_i. In single-core priority ceiling, the worst-case B_i is bounded (somewhat pessimistically) by B_i^{PC}, calculated as in Equation (4.8):

$$B_i^{\text{PC}} = \max_{k=1}^{K} \{\text{usage}(k,i)C(k)\}, \tag{13.4}$$

where:

- K is the number of semaphores in the system.
- $\text{usage}(k,i)$ is a function that returns 1 if semaphore S_k is used by (at least) one task with a priority less than the priority of τ_i, and also by (at least) one task with a priority higher than or equal to the priority of τ_i, *including τ_i itself*. Otherwise, $\text{usage}(k,i)$ returns 0.
- $C(k)$ is the worst-case execution time among all critical regions associated with, or guarded by, semaphore S_k.

To proceed further, if we restrict our attention to non-nested critical regions for the time being, we can express the worst-case execution time C_i of τ_i as:

$$C_i = W_i + \sum_{k \in \text{usedby}(i)} n_{k,i}C(k), \tag{13.5}$$

- W_i is the worst-case execution time of τ_i, excluding the time spent within critical regions.
- $\text{usedby}(i)$ is the set of indices of the semaphores that τ_i uses at least once.
- $n_{k,i}$ is the number of times that τ_i "uses" semaphore S_k, that is, enters a critical region controlled by S_k.

It can then be proven that the worst-case response time of τ_i when using the MrsP protocol is given by value to which recurrence relation (13.3) converges when substituting (13.4) and (13.5) into it, and finally replacing $C(k)$ with the quantity:

$$E(k) = |\text{cores}(\text{uses}(k))| C(k), \tag{13.6}$$

where:

- $\text{uses}(k)$ represents the set of indexes of the tasks that use semaphore S_k at least once.
- $\text{cores}(\cdot)$, given a set of task indexes, returns the set of cores these tasks are assigned to.

- $|\cdot|$ gives the cardinality (number of elements) of a set.

Finally, we obtain:

$$
w_i^{(h+1)} = W_i + \overbrace{\sum_{k \in usedby(i)} n_{k,i} E(k)}^{C_i}
$$

$$
+ \overbrace{\max_{k=1}^{K} \{usage(k,i) E(k)\}}^{B_i}
\tag{13.7}
$$

$$
+ \sum_{j \in hpl(i)} \left\lceil \frac{w_i^{(h)}}{T_j} \right\rceil \overbrace{\left(W_j + \sum_{k \in usedby(j)} n_{k,j} E(k) \right)}^{C_j}.
$$

The formal proof of (13.6) is given in [28] and is outside the scope of this book. However, the intuition behind the formula is that, when there can be parallel acquisition requests for a semaphore S_k from up to m distinct cores with FIFO queueing, it is reasonable to expect that, in the worst case, a task has to wait for $(m-1)C(k)$ to acquire the semaphore and then spends $C(k)$ within the corresponding critical region. As a consequence, the total worst-case cost of executing the critical region is:

$$
(m-1)C(k) + C(k) = mC(k)
$$

and the correct value of m to be used for S_k is indeed given by $|cores(uses(k))|$.

The original MrsP paper [28] suggested that nested critical regions can easily be accommodated in the analysis with a different and only slightly more complex definition of $E(k)$:

$$
E(k) = (|nested(k)| + |cores(uses(k))|)\,C(k),
\tag{13.8}
$$

where $nested(k)$ is the set of semaphores that access semaphore S_k in a nested way. In other words, a semaphore $S_h \in nested(k)$ if and only if there is at least one critical region controlled by S_k nested within a critical region controlled by S_h.

The informal reasoning behind this formula is that, in the worst case, semaphore S_k can be accessed directly from up to $|cores(uses(k))|$ contending sources like in the non-nested case, and indirectly from up to $|nested(k)|$ critical regions controlled by a different semaphore. In the second case, the outer critical region serializes accesses to the semaphore, so these sources must be counted only once regardless of from how many sources the semaphore that controls the outer critical region is accessed.

Therefore, from the point of view of a task that wants to acquire the semaphore and execute the corresponding critical region, each of those sources introduces an execution overhead of up to $C(k)$, including the time the task needs to execute its own critical region. The same reasoning also provides the starting point to further extend the analysis and consider clustered scheduling. As a final remark, in all the above formulae we neglected operating system-induced blocking due to critical regions needed to implement, for instance, task switching and dispatching, like we did in Chapter 4. Further insights on nested critical regions in MrsP, along with a more comprehensive and refined analysis, were given in [50].

13.4.2 O(M) INDEPENDENCE-PRESERVING PROTOCOL

Like MrsP does, also the $O(m)$ independence-preserving protocol (OMIP) [23] extends a protocol originally devised for single-core systems—the priority inheritance protocol described in Chapter 8—to multiprocessor and multicore systems by means of a helping mechanism that enables temporary task migration from the home scheduler instance to another.

However, unlike MrsP, OMIP is not based on spinning and only relies on passive wait to regulate critical region access. Even more importantly, OMIP does not require any user-provided prior information about which tasks contend for which semaphores and their priorities. For this reasons, in RTEMS the OMIP protocol is a drop-in replacement for priority inheritance on symmetric multiprocessor and multicore systems. In other words, when users select the priority inheritance protocol on such systems, they automatically get OMIP instead.

Independence-preserving protocols

Another OMIP design goal of practical importance, also in common with MrsP, is the fact of being an *independence-preserving* protocol, as its name says. Informally speaking, the concept of independence preservation captures a very desirable property of a locking mechanism, which can be expressed in a very simple way: The execution of high-priority tasks must be independent of, and shielded from, any interference coming from accesses to unrelated critical regions by lower-priority tasks.

Last, but not the least, it can be proved that the blocking time due to priority inversion in OMIP is asymptotically optimal, that is, it grows like $O(m)$ as m, the number of cores, grows. This result is valid under suspension-oblivious analysis, that is, a kind of schedulability analysis in which task blocking is considered in an implicit way, as part of the execution time of the task itself. Moreover, the proof does not take nested critical regions into account and considers clustered scheduling in which the m cores are uniformly divided among clusters.

A short discussion of how the priority inheritance protocol falls short of expectations on a multicore system and how temporary task migration solves the issue, along with an example of a different approach that is not independence-preserving, will help us understand the underlying strategy of the OMIP protocol. Most of the reasoning to be carried out in the following is applicable to task migration in MrsP as well.

A further, interesting result proved in [23] is that, except in the case of global scheduling, it is impossible for a protocol to simultaneously fulfill the following three properties, at least when using a scheduling algorithm in which task priorities may

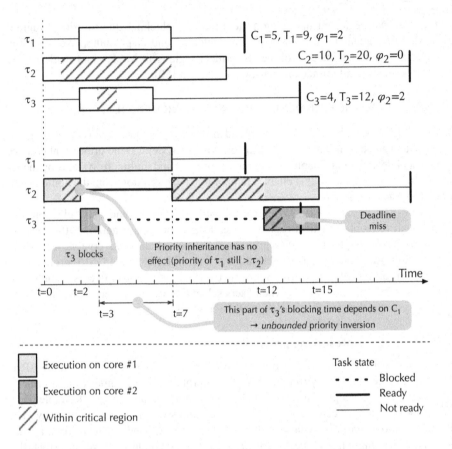

FIGURE 13.3 Priority inheritance and unbounded priority inversion in multicores.

be dynamic, but job priorities are fixed (like Rate Monotonic or Earliest Deadline First):

1. Prevent unbounded priority inversion
2. Be independence-preserving
3. Not use task migration

Priority inheritance on multicores

Let us consider, as shown in Figure 13.3, three periodic tasks to be executed on two different cores under partitioned Rate Monotonic scheduling. The top half of the figure depicts task characteristics. Their execution time is represented by a rectangle in which diagonal stripes highlight two critical regions in τ_2 and τ_3, controlled by the same semaphore. More specifically:

- Task τ_1 has an execution time $C_1 = 5$, a period $T_1 = 9$, and an initial phase $\varphi_1 = 2$. All these quantities are times expressed in an arbitrary, but consistent, time unit. The task does not contain any critical region and, being the task with the shortest period, has the highest priority in the system.
- Task τ_2 has an execution time $C_2 = 10$, a period $T_2 = 20$, and an initial phase $\varphi_2 = 0$. It enters a critical region controlled by a semaphore S after executing for 1 time unit and executes within the region for 6 time units. This task has the longest period, and hence, the lowest priority.
- Task τ_3 has an execution time $C_3 = 4$, a period $T_3 = 12$, and an initial phase $\varphi_3 = 2$. It contains a critical region controlled by the same semaphore S that starts after 1 time unit since the beginning of its execution and lasts for 1 time unit. According to the Rate Monotonic priority assignment, its priority is lower than the priority of τ_1 and higher than the priority of τ_2.

Tasks τ_1 and τ_2 are statically assigned to core 1 for execution, while τ_3 is assigned to core 2. In the figure, execution on core 1 is represented by a light gray color, while a darker gray indicates execution on core 2. Thin horizontal lines denote that a task is not ready for execution, because it is waiting for its next activation, thick lines indicate that a task is ready, and dotted thick lines indicate that a task is blocked, that is, it is waiting to acquire semaphore S.

The bottom half of Figure 13.3 shows how the system behaves when using the priority inheritance protocol:

- At $t = 0$ task τ_2 becomes ready and start executing on core 1. The other two tasks, τ_1 and τ_3, are not ready for execution yet. Since $\varphi_1 = \varphi_3 = 2$, these tasks are still waiting for activation.
- At $t = 2$ both τ_1 and τ_3 become ready for execution. Both start executing immediately, because τ_3 is alone on core 2 and τ_1 preempts τ_2 on core 1 because it has a higher priority. It is worth noting that τ_2 is preempted *after* it acquired S and entered its critical region.
- At $t = 3$ task τ_3 tries to acquire S as well, and blocks because τ_2 did not release the semaphore yet. The priority inheritance protocol elevates the priority of τ_2 to the priority of τ_3.
- However, τ_1 keeps executing, because its priority is even higher. As a consequence, τ_2 resumes execution only when τ_1 completes at $t = 7$ and waits for its next activation.
- Task τ_2 eventually exits from its critical region at $t = 12$ and unblocks τ_3.
- Task τ_3 would finish at $t = 15$, but it misses its deadline, which is set at $t = \varphi_3 + T_3 = 14$.

Overall, we can conclude that priority inheritance not only does not make τ_3 meet its deadline, but it also does not even allow us to determine an upper bound in its worst-case blocking time. This is because, by looking at Figure 13.3, we can see that part of the blocking time depends on C_1, the execution time of a totally unrelated task that does not even compete for the same semaphore. This is exactly the same

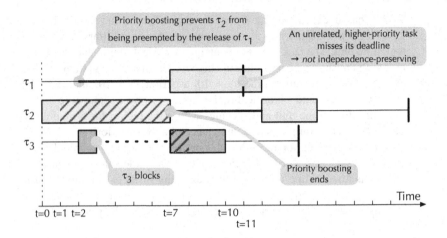

FIGURE 13.4 Priority boosting is not independence-preserving.

scenario that led us to the concept of unbounded priority inversion in Section 8.1 and Figure 8.1.

Priority boosting

The observation that one reason of the unbounded blocking time just described is the preemption of τ_2 by τ_1 may lead to the introduction of *priority boosting* techniques, based on a concept not unlike priority ceiling. More specifically, a task that enters a critical region may temporarily be granted a boost that makes its priority higher than any other normal priority. Many multiprocessor and multicore semaphore protocols actually employ various variations on this basic theme. This approach prevents tasks within a critical region from being preempted by tasks that do not hold any lock but, as illustrated in Figure 13.4, it has other side effects.

If we consider again the same set of tasks analyzed previously, introducing priority boosting leads to the following behavior:

- As before, task τ_2 becomes ready at $t = 0$ and start executing on core 1.
- At $t = 1$, it enters its critical region and gets a priority boost, which has no immediate consequences.
- At $t = 2$, both τ_1 and τ_3 become ready for execution. However, only τ_3 starts executing immediately, because it is alone on core 2. Instead, τ_1 does not preempt τ_2 on core 1 because of the priority boost just discussed.
- As a consequence, τ_2 continues executing until it leaves the critical section at $t = 7$ and the priority boost ceases. At this point τ_1 preempts τ_2. In addition, task τ_3 is unblocked and resumes execution in its own critical section.
- Task τ_3 successfully concludes its execution at $t = 10$.
- However, τ_1 misses its deadline, which is set at $t = \varphi_1 + T_1 = 11$.

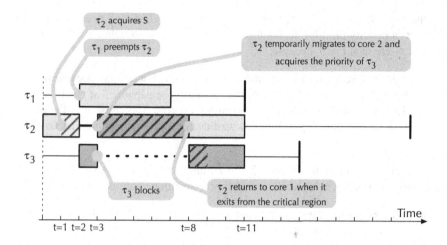

FIGURE 13.5 Temporary task migration in OMIP.

Overall, priority boosting prevented τ_3 from missing its deadline, at the expense of τ_1 though. This example shows that priority boosting is not independence-preserving because the execution of τ_1—the highest-priority task in the system—was not at all independent of the accesses to critical regions by τ_2 and τ_3, although these critical regions are completely unrelated to τ_1 itself.

How deeply the lack of independence preservation affects real-world systems of course depends on how long critical regions are because, in general, the shorter critical sections are, the less they disrupt unrelated task timings. However, worst-case critical regions length may be hard to estimate in many cases of practical relevance, which further increases the interest of independence-preserving protocols. For instance:

- Complex applications that make use of proprietary, third-party libraries where the exact nature of the shared data structures they manipulate and the algorithms they use may be unknown or hard to assess.
- Presence of shared data structures whose size is variable and hard to bound at design time, even if they are only shared among low-priority tasks.
- Open systems, in which the actual task mix may be little known at design time and may comprise unrelated tasks of different timing criticality.

Task migration

With the help of Figure 13.5 we can now informally see how task migration solves the issue. A more complete, formal proof of why this strategy works can be found in [23], and also in [28] for what concerns task migration in MrsP. More specifically, OMIP defines a so-called *migratory priority inheritance* rule. The rule states that whenever a task τ_i is ready for execution, but not running, and there is another task τ_j that would be eligible for execution on its cluster, but is blocked waiting for one of the semaphores held by τ_i, then τ_i migrates to the cluster of τ_j and temporarily takes τ_j's priority until it releases the semaphore. This is exactly the situation tasks τ_2 and τ_3 are in at $t = 3$, because:

- Task τ_2 is ready for execution on core 1, but it is not running because it has been preempted by τ_1 at $t = 2$, after acquiring semaphore S at $t = 1$.
- Task τ_3 would be eligible for execution on core 2, because it is the only task on that core, but is blocked waiting for S, the semaphore that task τ_2 holds.

Therefore, τ_2 migrates to core 2 and assumes τ_3's priority so that, informally speaking, it "runs on τ_3's place" on that core. Execution continues as follows:

- Thanks to the migration, τ_2 continues its execution on core 2 until it eventually releases semaphore S at $t = 8$.
- At this point, τ_2 returns to core 1 and reverts back to its original priority. It keeps executing on core 1, though, because in the meantime τ_1 successfully concluded within its deadline.
- Also at $t = 8$, τ_3 is given semaphore S and unblocked. Being the only ready task on core 2, it immediately starts execution.
- Both τ_2 and τ_3 conclude their execution at $t = 11$, thus meeting their deadlines.

OMIP queuing strategy

The last aspect to be discussed is how tasks are queued when they are blocked on a semaphore. In the case of priority inheritance, a single, priority-based queue for each semaphore is adequate and sufficient for blocking time analysis, as described in Chapter 8. In the case of OMIP, Reference [23] advocates a more complex queuing strategy, depicted in Figure 13.6, to optimize blocking time and reach $O(m)$ asymptotic optimality. From right (closer to the semaphore) to left (closer to tasks), the queues associated to a certain semaphore S are organized in three stages:

1. A global first-in, first-out (FIFO) queue $\mathscr{G}$ that can hold up to one task for each cluster in the system. The task at the head of this queue is the one that holds the semaphore, when the queue is empty it means that no tasks hold the semaphore.
2. A per-cluster FIFO queue $\mathscr{F}$, whose maximum length is the number of cores in the cluster.
3. A per-cluster priority queue $\mathscr{P}$, of unlimited length.

The three queues are used according to the following rules:

- When a task τ_i tries to acquire S and queue $\mathscr{F}$ is empty, it is enqueued in both $\mathscr{F}$ and $\mathscr{G}$. If $\mathscr{G}$ was also empty, τ_i obtains the semaphore immediately, otherwise it is blocked.
- Otherwise, if queue $\mathscr{F}$ is not full, τ_i is enqueued in $\mathscr{F}$, but not in $\mathscr{G}$, and blocked.

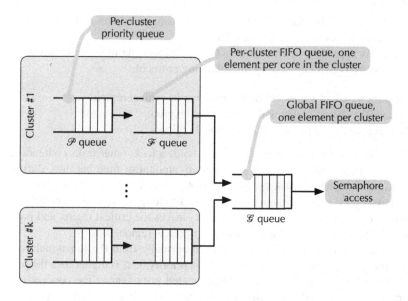

FIGURE 13.6 OMIP waiting queues for critical region access.

- If queue $\mathscr{F}$ is full, τ_i is enqueued only in $\mathscr{P}$ and blocked. Queue $\mathscr{P}$ has unlimited length, so it cannot be full as well.
- When a task τ_i releases semaphore S, it is removed from both $\mathscr{G}$ and $\mathscr{F}$.
- Since now there is at least one empty element in $\mathscr{G}$ and $\mathscr{F}$, the system fills it according to the queue hierarchy. Namely, if $\mathscr{P}$ is not empty, its head is *removed* and enqueued in $\mathscr{F}$. Moreover, if $\mathscr{F}$ is not empty, its head is *duplicated* and enqueued in $\mathscr{G}$.
- If $\mathscr{G}$ is not empty after these updates, the task at its head is unblocked and given semaphore S.

13.5   LOCK-FREE AND WAIT-FREE COMMUNICATION

From the previous chapters, Chapter 7 in particular, we know that a shared object is a data structure that can be accessed and modified by means of a fixed, predefined set of operations by a number of concurrent tasks. Uncontrolled access to a shared object may lead to a race condition, which usually entails the unrecoverable corruption of object contents.

For this reason, object access is traditionally controlled by means of critical sections or regions. A mutual exclusion mechanism, for example, a semaphore or a monitor, governs the access to critical regions and makes tasks wait if necessary, so that only one of them at a time is allowed to access or modify the object. Although critical sections are not apparent in inter-task communication based on message passing, they are still needed within message-passing primitives if they rely on some form of shared memory for data transfer.

With this approach, often called *lock-based* object sharing, a task that wants to access a shared object must obey the following protocol:

1. Acquire a lock of some sort before entering its critical region.
2. Access the shared object within the critical region.
3. Release the lock when exiting the critical region.

A crucial point is that, as long as a task holds a lock, other tasks contending for the same lock are forced to wait for a certain amount of time because, during the lock acquisition phase, a task τ_1 blocks if another task τ_2 is currently within a critical region associated with the same lock. The block takes place regardless of the relative priorities of the tasks. It lasts at least until τ_2 leaves the critical region and possibly more, if other tasks are waiting to enter their critical region, too.

Even though both τ_1 and τ_2 still proceed normally from the functional point of view, and hence produce correct results, if the priority of τ_1 is higher than the priority of τ_2, the way mutual exclusion is implemented goes against the concept of task priority, because a higher-priority task is forced to wait until a lower-priority task has completed part of its activities. In Chapter 8, we saw that special techniques are needed to solve this issue, and still be able to bound and calculate the worst-case blocking time suffered by higher-priority tasks.

Even so, it is still true that in modern processors multiple hardware components may introduce execution delays that are inherently hard to predict precisely, like the caches we described in Chapter 12. This implies a certain uncertainty on how much time a given task will actually spend within a critical region in the worst case and may lead to an overestimation of this time. In turn, the uncertainty is reflected back into worst-case blocking time computation.

As shown in Figure 13.7, bringing this reasoning to the extreme we can also observe that if τ_2 halts for any reason while within its critical region, τ_1 and all of the other tasks willing to enter a critical region associated with the same lock will be blocked forever and will be unable to make any further progress. Although in most operating systems it is possible to specify a timeout for lock acquisition (see Chapter 7) at the very least the shared object cannot be safely used again, because it may have been left in an inconsistent state by a partial update performed by τ_2 before halting.

Even though a full discussion of the topic is beyond the scope of this book, this section contains a short introduction to a radically different approach to object sharing, known as *lock-free* and *wait-free* communication. Unlike lock-based object sharing, this approach is able to guarantee the consistency of a shared object without ever forcing any task to wait for another. For this reason, lock-free and wait-free communication is unaffected by random task halts, although it is still sensitive to soft failures, which lead a task to sporadically perform incorrect operations and continue rather than halting completely.

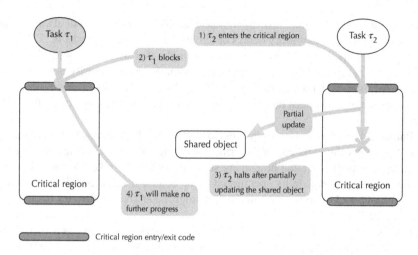

Task τ_1

2) τ_1 blocks

1) τ_2 enters the critical region

Task τ_2

Partial update

Shared object

Critical region

3) τ_2 halts after partially updating the shared object

4) τ_1 will make no further progress

Critical region

Critical region entry/exit code

FIGURE 13.7 Issues with lock-based synchronization when a task halts.

After giving some basic definitions, we will first look at two very specific algorithms, one of which is lock-free while the other is wait-free. These examples will give readers a general feeling of what these algorithms look like. Then we will discuss a method that solves the problem in general terms and allows objects of any kind to be shared in a lock-free way. From these examples, it will become clear that the inner workings of this kind of algorithms are considerably more complex than, for instance, semaphores. For this reason, although lock-free and wait-free algorithms are already in use for real-world applications, they are still an active research topic, above all for what concerns their actual implementation.

The development and widespread availability of open-source libraries containing a collection of lock-free data structures such as, for example, the Concurrent Data Structures library (libcds) [74] is encouraging. In this way, more and more programmers will be exposed to them in the near future, and they will likely bring what today is still considered an advanced topic into mainstream programming. Interested readers may refer to References [4, 5, 6, 56, 57, 58, 78] for a more detailed introduction to this subject and its applications to real-time embedded systems.

13.5.1 BASIC PRINCIPLES AND DEFINITIONS

Given a system of tasks $\mathscr{T} = \{\tau_1, \ldots, \tau_n\}$ wishing to access a shared object O, the implementation of O is *lock-free* if *some* task $\tau_i \in \mathscr{T}$ must complete an operation on O after *the system $\mathscr{T}$ takes* a finite number of steps.

In this definition, a step means an elementary execution step of a task, such as the execution of one machine instruction. The concept of execution step is not tightly related to time, but it is assumed that each execution step requires a finite amount of time to be accomplished. In other words, the definition guarantees that, in a finite amount of time, some task τ_i—not necessarily all of them—will always make progress regardless of arbitrary delays or halting failures of other tasks in $\mathscr{T}$.

It is easy to show, by means of a counterexample, that sharing an object by introducing critical regions leads to an implementation that is necessarily not lock-free. That is, let us consider a shared object accessed by a set of tasks $\mathscr{T} = \{\tau_1, \ldots, \tau_n\}$ and let us assume that each task contains a critical section, all protected by the same lock.

Pursuing the same reasoning presented at the very beginning of this section, if task τ_i acquires the lock, enters the critical section, and halts without releasing the lock, regardless of what and how many steps the system $\mathscr{T}$ takes as a whole, none of the tasks will complete its operation. On the contrary, all tasks will wait forever to acquire the lock, without making further progress, as soon as they try to enter the critical section.

Similarly, given a set of tasks $\mathscr{T} = \{\tau_1, \ldots, \tau_n\}$ wishing to access a shared object O, the implementation of O is *wait-free* if *each* task $\tau_i \in \mathscr{T}$ must complete an operation after *taking* a finite number of steps, provided it is not halted.

The main difference is therefore that a wait-free implementation guarantees that *all* non-halted tasks—not just *some* of them as for a lock-free implementation—will make progress, regardless of the execution speed of other tasks. For this reason, the wait-free property of an implementation is strictly stronger than the lock-free property.

Both lock-free and wait-free objects have a lot of interesting and useful properties. First of all, they are not (and cannot be) based on critical sections and locks. Therefore, lock-free objects are typically implemented using *retry loops*, in which a task repeatedly attempts to operate on the shared object until it succeeds.

Those retry loops are potentially unbounded and theoretically might give rise to starvation for some tasks, that is, they might be unable to make progress although the scheduler allows them to spend some time executing. However, the probability p that a task has to retry is usually extremely small and the probability of retrying k times can be approximated by p^{-k}. Being a negative exponential, this quantity quickly decreases toward zero as the exponent k increases. Wait-free methods go one step further and preclude all waiting dependencies among tasks, including retry loops. Individual wait-free operations are therefore necessarily starvation-free, another important property in a real-time system.

13.5.2 LOCK-FREE MULTI-WORD COUNTER READ

As an example of ad-hoc lock-free communication technique we will examine a programming problem that often arises in practice, that is, how to consistently *read* a counter, which is being concurrently updated by another agent, called *writer*.

To better define the problem, let us specify that the counter we would like to use is composed of multiple (two or more) *digits*, according to the nomenclature chosen in a seminal reference paper about this subject [78]. It is assumed that hardware makes it possible for the processor to read *individual* digits atomically, but not the whole counter. Instead, it is possible for another agent to write *all* digits in an atomic way.

Although no assumptions are needed on how wide a digit is, in the following we will assume that digits coincide with machine words for simplicity.

This is realistic because in most processors—including multi-core ones—atomic access to individual machine words is easily implemented in hardware, typically by means of a bus arbiter, and always guaranteed. The real issue may be to guarantee that memory or device access operations are actually performed in the intended order, avoiding unwelcome optimization performed either by the compiler or by the hardware itself but, as we saw in Chapter 12, valid methods do exist to address this issue.

On the writer's side, the assumption that atomic updates to multiple digits are possible is often satisfied, for instance, if all digits are actually registers of the same hardware device, which is then able to make their new values visible to the processor all at once. When needed, it is also possible to relax the assumption about the atomic update of the whole counter, but this requires the use of algorithms more sophisticated than the one being described here. Interested readers can refer to [78] for more information about this point, as well as an example of those algorithms.

The most important consequence of atomic access is that, if a digit is read while it is being updated by the writer, the reader gets either the old or the new value, but no other cases are possible. However, since atomic read access from software is not possible for the whole counter, the only way to access it is to read its digits in a non-atomic *sequence*. By intuition, this may lead to inconsistent readings unless appropriate countermeasures are taken.

To better highlight the practical application of this method, let us consider an example. Many microcontrollers embed a Real Time Clock (RTC) able to keep track of wall-clock time autonomously. Often, thanks to an auxiliary power source, the RTC also works when the processor is inactive and the rest of the microcontroller is powered off. For convenience, the RTC hardware often makes available different parts of the wall-clock time in distinct registers. For instance, the LPC1768 RTC [90] has 8 registers for seconds, minutes, hours, day of month, and so on. Atomic access to individual registers is generally possible, but not to all of them together. Instead, all registers are updated atomically on the RTC side.

As a running example, in the following we consider a very simplified RTC that provides only two wall-clock time registers: SECS, holding seconds, and MINS, holding minutes, although the method we are going to describe can be easily extended to handle more than two registers. Going back to the original nomenclature, each register can be considered as a digit of a counter, in base 60. The RTC hardware is responsible of updating these digits every second.

As shown in the left part of Figure 13.8, the writer updates both registers in an atomic way, bringing them from an old value (in light gray) to a new one (dark gray). Instead, as shown in the right part of the figure, the reader must perform a minimum of two independent read operations to read registers individually. Although the figure shows a single reader, the reasoning that follows stays true for any number of concurrent readers. Read operations may be carried out in two different orders:

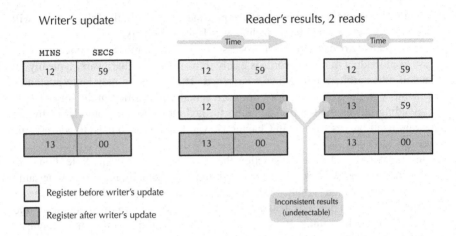

FIGURE 13.8 Possible outcomes of non-atomic RTC reads with a concurrent update.

1. From *left to right*. Assuming that most significant digits (and registers) are written on the left of a number, this order corresponds to reading the most significant digit MINS first, and then SECS.
2. From *right to left*, that is, least significant digit first or, in other words, SECS and then MINS.

Considering that read operations take place concurrently with the update, it is evident that neither order is sufficient to guarantee that the reader gets correct results—that is, either 12:59 or 13:00. Indeed, as shown in the figure, in both cases it is possible that the reader obtains a wrong result, 12:00 or 13:59. What is more, the reader has no way to *detect* this condition, and hence, no opportunity to correct it, for instance, by retrying the operation.

From this observation we can readily draw the conclusion that *two* read operations, regardless of the order in which they are performed, are not enough to consistently read a two-digit counter under our assumptions. As a side note, it is not a coincidence that the figure depicts the scenario in which there is carry propagation from one digit to the next, because this is the only scenario in which inconsistent results may be obtained.

However, by looking more carefully at Figure 13.8, we can also notice that *three* read operations are able to give a reader both the value of the whole counter—if it was able to read it correctly—or an indication to retry. The order of read operations is actually a combination of the left-to-right and then right-to-left orders shown in the figure, namely:

1. The reader first reads MINS;
2. Then, it reads SECS (in left-to-right order);
3. Last, it reads MINS again (in right-to-left order).

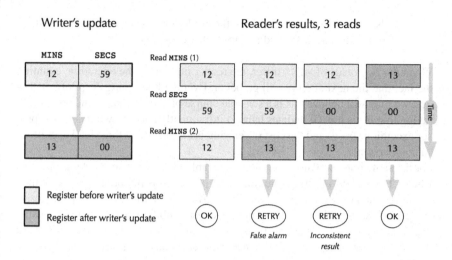

FIGURE 13.9 Possible outcomes of a sequence of 3 non-atomic RTC reads.

For the code to work as intended, besides ensuring that read operations are actually forwarded to the memory or I/O bus in the same order as they appear in the code, it is also important that the second read of MINS is performed "for real", that is, the data it returns actually come from the bus and not from somewhere else. Ensuring this requires a twofold approach, which works on both the hardware and software side:

- On the hardware side, individual register accesses or the address range in which the register address resides must be marked non-cacheable. As recalled in Section 12.2.4, on the ARMv8-A architecture, this is done implicitly when an address range is classified as device memory.
- On the C-language software side, the data type used to access the register must be qualified as volatile, as indicated in Section 12.2.5, to prevent the compiler from reusing an outdated value that might still be present in some processor register.

Figure 13.9 shows the four possible outcomes of the sequence of read operations just described. In the figure, time flows vertically and outcomes are shown besides each other horizontally. Light gray and dark gray rectangles still denote old and new values, respectively.

- In two cases (the leftmost and the rightmost) the reader gets the *same value* from the first and second read of MINS. This ensures that SECS was not updated between the two reads of MINS, and hence, the counter value obtained by the reader is correct.
- In two other cases, the reader gets *different values* from the first and second read of MINS. This outcome may correspond to a genuine inconsistent result (in which the reader obtained 12:00 from the counter). Otherwise, it

may also be a false alarm, in which the reader obtained a consistent result, 12:59, but the check spotted an inconsistency anyway.

When the two reads of MINS differ, the reader must retry the operation, counting on the fact that the next attempt will go better. Even without going deep into mathematics, we may still believe this is a sensible course of action because the probability of having to retry is low (if updates happen infrequently enough with respect to processor execution speed) and the cost of retrying is also low (just three read operations and a comparison).

For the same reason, it also becomes acceptable to retry for no good reasons, that is, when a false alarm occurs. From this point of view it is important to remark that, although the method we are describing may give false alarms, or false positives, it never gives false negatives, thus ruling out any possibility of obtaining and using inconsistent readings. In other words, we are willing to accept this method because a false positive only affects its performance, whereas a false negative would affect its correctness.

13.5.3 FOUR-SLOT ASYNCHRONOUS COMMUNICATION

A seminal work by Simpson [113] analyzed and solved the all-important problem of the asynchronous communication of an arbitrary memory-resident data structure between a single writer and a single reader. In the context of the work, the meaning of asynchronous is the absence of any kind of timing interference between the tasks involved, which corresponds to the definition of wait-free implementation given in Section 13.5.1. Briefly discussing this solution gives us the opportunity to explore wait-free communication methods and highlight some of the possible pitfalls in the quest for a fully correct solution. Besides asynchronism, the two additional, quite intuitive requirements posed on the solution are:

- *Coherency:* data produced by the writer must become available to the reader as an atomic unit, although they consist of multiple memory words that cannot be read or written atomically as a whole. In other words, the reader must never obtain a mix of partial data produced by multiple write operations because this would likely lead to data corruption.
- *Freshness:* the reader must obtain the most recent data fully produced by the writer at the time it starts the read operation, rather than some older version. This property is especially important for real-time systems because it guarantees that communication will not only be correct, but also timely.

The main assumption of the work—to avoid an obvious chicken-and-egg issue—is that the underlying hardware architecture supports asynchronous, coherent, and fresh access to one-bit variables. As discussed in the previous chapters, this assumption is amply satisfied on virtually all contemporary multicore architectures, which support atomic access to data at least as big as a memory word.

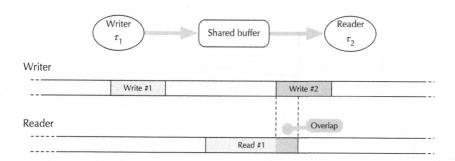

FIGURE 13.10 Incoherent read in single-buffer communication.

Strictly speaking, hardware-based atomic access does not fully satisfy the asynchronism assumption because the memory system most often implements it by serializing memory accesses, and hence, there is indeed a certain amount of timing interference among competing tasks. However, given that the time needed to perform a single memory access—or wait until the access initiated by another task completes—is usually orders of magnitude smaller than the time scale on which tasks operate, this interference can be neglected in most practical cases.

Since memory-based communication is accomplished by means of a certain number of memory buffers shared among the tasks involved, an interesting question is what is the minimum number of buffers needed to obtain a solution that satisfies the asynchronism, coherency, and freshness properties just mentioned.

As shown in Figure 13.10, a single shared buffer is inadequate because the coherency property is no longer satisfied as soon as a read operation overlaps with a write operation. In the example shown in the figure, if read operation #1 overlaps with write operation #2, it will get part of the "old" (light gray) data pertaining to the previous write operation, write #1, and part of the "new" (dark gray) data presently begin written. Instead, the freshness property is satisfied because, as also shown in the figure, a read operation always gets access to the most recent data.

In the previous chapters, notably Chapter 7, we learned to call this issue a *race condition* and solve it by establishing and enforcing some timing constraints between the reader and the writer—for instance, with the help of a mutual exclusion mechanism—to prevent unwelcome overlaps from ever happening. However, this approach invalidates the asynchronism property and is hence unsuitable in this case. Solutions of this kind, which satisfy the asynchronism property only with the help of additional timing constraints external to the solution itself, are called *conditionally asynchronous* in [113].

The two-buffer approach is similar to the previous one, and still conditionally asynchronous, but allows for more relaxed timing constraints. It was called the *swung buffer* method in [113] and is commonly referred to as *double buffering* nowadays. As depicted in Figure 13.11, a shared index L identifies which of the two buffers contains the latest data, that is, the most recent data that has been *completely* written by the writer. The reader and the writer operate as follows:

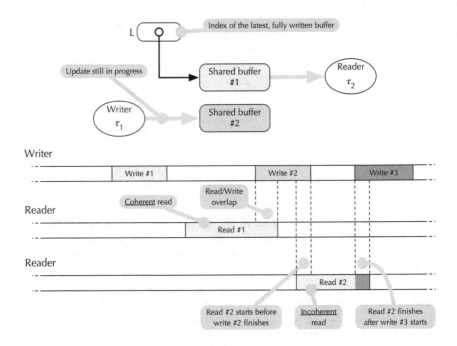

FIGURE 13.11   Incoherent read in double-buffer communication.

- The reader retrieves L immediately before starting an operation and performs the whole operation reading from the buffer indicated by it, even if the value of L changes during the operation itself.
- The writer consults L immediately before starting a write operation to identify the "other" buffer, that is, the buffer not referenced by L itself. The writer fills the other buffer and, at the end of the write operation, updates L so that it refers to it.

In other words, the writer alternates between the two buffers for writing and marks a buffer as containing the latest data, by switching L to it, only after completing a write operation. This ensures that the writer does not corrupt the buffer referenced by L until new data have been completely written, and hence, solves the simple race condition due to overlapping operations we identified in the one-buffer approach.

Referring back to Figure 13.11, if read #1 starts after write #1 has been completed, it will read from the buffer last written by the writer (light gray in the figure), thus gaining access to the latest data at the moment and fulfilling the freshness property. Any new write operation, like write #2, which starts while read #1 is still in progress does not corrupt the data being read by the reader, because the writer operates on the other buffer (darker gray in the figure). The asynchronism property is implicitly satisfied, too, because the algorithm just described does not introduce by itself any form of timing dependency between tasks.

Although this gives the reader more timing freedom with respect to the one-buffer approach, it still does not address all the possible race conditions that may arise. The worst case from the timing point of view happens when a read operation starts while a write operation is in progress and partially overlaps with it, as it happens to read #2 with respect to write #2 in the figure.

Since write #2 has not finished—and L has not yet been updated to refer to the new data being written—read #2 still operates on the buffer written by write #1. When write #2 eventually finishes, the writer switches L to the dark gray buffer, which does not directly affect the reader as we saw previously. However, if write operation #3 starts before read #2 finishes, it will write into the light gray buffer again, thus corrupting the data being read.

Therefore, the solution just described fulfills the coherency property only if the interval between the end of a write and the beginning of the next is guaranteed to be longer than the time it takes to fully perform a read operation. Even though this is still not completely satisfactory in general, the condition holds in a variety of situations of practical interest, which makes the solution useful especially when combined with methods that allow the lack of coherency to be detected at least, if not avoided.

A very simple detection method, which will also be introduced in a slightly different form in Section 13.5.4, is to use an index L larger than strictly needed to identify which buffer contains the latest data. More specifically, the least-significant bit could be used to identify the buffer, while the n most-significant bits could be reserved for consistency checks.

Then, when the writer wants to switch L from one buffer to the other, it *increments* L instead of simply toggling its least-significant bit. The reader reads L before starting a read operation (as it already did), and then again after finishing it. It is easy to see that the read operation may have retrieved corrupt data only if the two values of L before and after the operation differ by more than one.

The detection method may actually lead to a false negative if there were so many intervening write operations while the read operation was in progress that L wrapped around completely. However, the probability of this unfortunate event may be made negligible with respect to other causes of failure by making n sufficiently big.

An obvious shortcoming of the double-buffer approach is that in some circumstances the writer is forced to choose between two undesirable outcomes when it initiates a new write operation:

- It may reuse the buffer currently being read by the reader, thus sacrificing the coherency property. This is what happens in write #3 of Figure 13.11.
- Otherwise, it may reuse the same buffer as in the previous write operation, thus overwriting the data it previously wrote and violating the freshness property while the write is in progress.

Based on this reasoning, we might think that offering the writer a third option, in the form of a third buffer to write into, should completely solve the problem. However, pursuing this approach for a while will give us the opportunity to show that intuition, based on sequential programming concepts, does not always work as

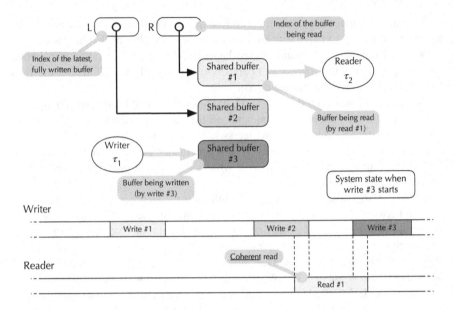

FIGURE 13.12 Triple-buffer asynchronous communication.

expected when designing a lock-free or wait-free algorithms and leads to other, subtler issues.

As shown in Figure 13.12, an apparently correct solution may be based upon:

- Three buffers, one that stores the latest fully written data, another that keeps data being read, and the third to accommodate an in-progress write operation.
- A shared index L that identifies the buffer with the latest fully written data, as before.
- An additional shared index R, which refers to the buffer currently being read.

Together, L and R give the writer the means to identify the third buffer, which is the buffer it can use for a new write operation.

To perform a write operation, the writer operates as follows:

- It reads L and R to determine which buffer it can write into, that is, the buffer that is referenced by neither L nor R.
- It writes into the buffer.
- It updates L to refer to the buffer just written, to indicate to the reader where the latest data now are.

Symmetrically, the reader:

- Sets R to the value of L, to indicate to the writer that it is about to start reading from that buffer.
- It reads from the buffer.

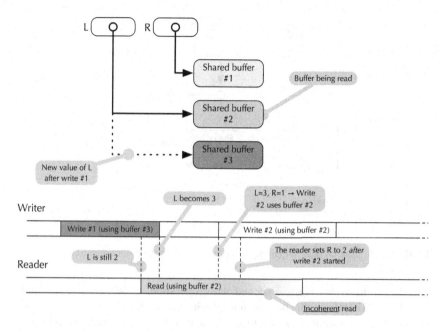

FIGURE 13.13 Race condition in triple-buffer asynchronous communication.

Even if we assume that L and R consist of a single memory word and can *individually* be read and written in an atomic way—as it is reasonable to do on most contemporary processor architectures—this solution can still malfunction if an unfortunate interleaving between read and write operations takes place. However, we can ascertain the issue only if we do not stop at a too-high level of abstraction when looking at the operations discussed previously, but go down to examine how the processor really implements them.

A key point is to notice that the first step to be performed by the reader, *set R to the value of L*, cannot be performed atomically because it consists of a *sequence* of two atomic memory operations, that is, the load of R into a processor register, followed by the store of that register into L.

With this aspect in mind, it becomes clear that the sequence of steps illustrated in Figure 13.13 may indeed occur when a read and two write operations overlap:

- The writer starts a write operation. For the sake of the example, let us assume that $L = 2$ and $R = 1$, as shown in Figure 13.13, so the writer is going to write into buffer #3.
- The reader starts a read operation, reads the value of L, and gets 2.
- Before the reader has the chance to update R and set it to 2, the writer completes the write operation and sets L to 3.

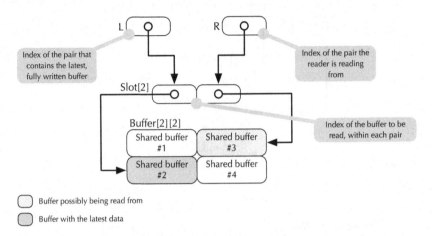

FIGURE 13.14   Shared data in quadruple-buffer asynchronous communication.

- Then, the writer starts a second write operation, which finds $L = 3$ and $R = 1$. Therefore, this operation will write into buffer #2.
- The reader eventually sets R to 2 to indicate it is about to start reading from that buffer.

At this point, both the reader and the writer continue concurrently, both using buffer #2. The reader can therefore retrieve corrupt data that have been partially overwritten by the writer.

The next step will be to show that adding a fourth shared buffer can solve this issue and lead to a full solution of the problem. Although the discussion will be kept at an informal level in this book, Reference [113] includes a full correctness proof.

As depicted in Figure 13.14, the four buffers are arranged in a matrix with two rows and two columns, named *Buffer*, which gives rise to two logical pairs of buffers, one on the left and one on the right in the figure. Three additional, shared control variables regulate how the reader and the writer use them. In particular:

- The one-bit index variable L has a similar meaning as in the triple-buffer solution, but now indicates which *pair*, rather than which buffer, contains the latest fully written buffer.
- Also the one-bit index variable R mostly keeps the meaning it had previously, and indicates from which pair the reader is possibly reading at the moment.
- The array of two one-bit indexes *Slot* has one element for each pair. For the pair referenced by L, it indicates which buffer in the pair contains the latest data, and hence, should be read from. For the pair referenced by R, it indicated which buffer the reader is possibly being reading from.

It is possible that L and R reference the same pair, when the reader is reading from the pair that contains the latest fully written buffer. This does not lead to a

Shared data

d1.   `int L`
d2.   `int R`
d3.   `int Slot[2]`
d4.   `data Buffer[2][2]`

Read operation	*Write operation*
r1. `p = L`	w1. `p = 1 - R`
r2. `R = p`	w2. `i = 1 - Slot[p]`
r3. `i = Slot[p]`	w3. Write into `Buffer[p][i]`
r4. Read from `Buffer[p][i]`	w4. `Slot[p] = i`
	w5. `L = p`

FIGURE 13.15 Quadruple-buffer asynchronous communication (from [113]).

contradiction because, in this case, the *Slot* index for the pair has both the meanings just described, while the *Slot* index for the other pair is unused and holds no specific meaning.

The reader and the writer perform their operations according to the pseudo-code listed in Figure 13.15, which also summarizes the shared data presented so far at the top. For simplicity, variables *L*, *R*, and *Slot* are integers although, as said previously, a one-bit index would suffice. It is also assumed that the abstract data type `data` represents a buffer. Shared variables are capitalized, while local variables are lowercase. In order to perform a write operation, the writer:

- Chooses the pair of buffers *not* currently used by the reader at the moment, by consulting *R* and calculating $1 - R$, then stores its index into local variable p (step w1 in Figure 13.15).
- Locates the buffer *not* referenced by the *Slot* index of the pair with the same technique, and stores its index into local variable i (step w2).
- Writes into the buffer (step w3).
- Indicates that the buffer it just used now contains the latest fully written data, by setting the *Slot* index of the pair to the value of i and *L* to the value of p (steps w4 and w5).

Symmetrically, the reader:

- Determines which pair contains the latest fully written data by consulting *L* and declares it is about to start reading from that pair by setting *R* to the value of *L*. The pair index is now in local variable p (*non-atomic* sequence of steps r1 and r2 in Figure 13.15).
- Within that pair, uses the *Slot* index to locate the index i of the buffer it must read from (step r3).
- Reads from the buffer (step r4).

The way the reader and the writer operate on L and R is very similar to what they did in the triple-buffer algorithm and may hence lead to the same race condition, due to the non-atomicity of reader's steps r1 and r2. As a consequence, it is possible that the reader and writer use the *same pair* for a concurrent read and write operation.

However, the fact that L and R now work at the pair (rather than at the buffer) level together with the use of the *Slot* index within the pair prevents them from clashing on the same buffer. Namely, if multiple write operations occur while a read is in progress:

- The first write might use the same pair p as the read, but it will for sure write into a different slot, because the read operation uses the $Slot[p]$ index of the pair, whereas the write uses $1 - Slot[p]$.
- The next writes will certainly use the other pair because the reader steered R away from the pair it was going to use before it started reading data from the buffer.

In summary, the algorithm just presented does not contain any conditional statement or retry loop. Therefore, it implements a completely wait-free data transfer between the writer and the reader, and also guarantees transfer coherency and freshness.

13.5.4   UNIVERSAL CONSTRUCTION OF LOCK-FREE OBJECTS

Ad-hoc techniques, such as the ones described in Sections 13.5.2 and 13.5.3, have successfully been used as standalone solutions or as building blocks to realize more complex lock-free objects. Although the algorithms we examined, especially the fully asynchronous communication mechanism of Section 13.5.3, are applicable in a reasonable large number of practical scenarios, they still do not solve the problem in completely general terms and may be difficult to understand and prove correct. Even more importantly, they must be proved correct again, possibly from scratch, whenever any aspect of the algorithms change.

Both issues may hinder the adoption of those methods for real-world applications and go against consolidated software design concepts, like design patterns [49]. This stimulated other researchers to look for a more general and practical methodology. As summarized by Herlihy [56], the underlying goal is that *"A practical methodology should permit a programmer to design, say, a correct lock-free priority queue without ending up with a publishable result."*

According to Herlihy's proposal [56], an arbitrary lock-free object can be designed in two steps:

1. First, the programmer designs and implements the object as a stylized, *sequential* program with no explicit synchronization, following certain simple conventions.
2. Then, the sequential implementation is transformed into a lock-free or wait-free implementation by surrounding its code with a general synchronization and memory management layer that is independent on the inner workings of the sequential implementation.

It can be proven that, if the sequential implementation is correct, the transformed implementation is correct as well. Crucially, the proof still holds even if the sequential implementation changes radically. Therefore, the lock-free or wait-free transformation can be seen as a well-formed universal design pattern.

Besides the usual atomic read and write operations on shared variables already mentioned for digit access in Section 13.5.2 Herlihy's approach requires hardware support for two additional operations:

- A *load linked* or *load exclusive* operation, which atomically reads a shared variable and copies it into a local task variable.
- A *store conditional* operation, which must follow a *load exclusive* at the same memory address. It stores back a new value into the shared variable, but *only if* no other task has modified the same variable in the meantime.
- Otherwise, *store conditional* does not do anything and, in particular, does not modify the shared variable.
- In both cases *store conditional* returns a success/failure indication, so that the task that executed it is made aware of its outcome.
- To provide room for an easier and more efficient implementation, *store conditional* is permitted to fail, with a low probability, even if the variable has not been modified at all.

Even if these requirements may seem exotic at first sight, most processors nowadays provide such instructions, or similar ones, due to their extreme usefulness. For example, starting from version V6 of the ARM processor architecture [7], the following two instructions are available:

1. LDREX loads a register from memory. In addition, if the address belongs to a shared memory region, it marks the physical address as exclusive access for the executing core.
2. STREX performs a conditional store to memory. The store only occurs if the executing core has exclusive access to the memory addressed.

Moreover, STREX puts into a destination register a status value that represents its outcome. The value returned is 0 if the operation updated memory, or 1 if the operation failed to update memory because the core did not have exclusive access anymore. As said before, exclusive access is revoked when another core modifies the same memory word.

Moreover, in Section 12.2.4 we also remarked that, in the more recent V8-A architecture [14], even more sophisticated instructions exists—called LDAXR and STLXR—which conveniently combine the semantics of LDREX and STREX with a one-way barrier.

At the same time, the practical implementation of these instructions is still reasonably straightforward and efficient when there is an underlying cache coherency protocol like, for instance, MESI. The fine details are beyond the scope of this book but, by intuition, it can be seen that:

- If a core issues a *load exclusive* instruction on a cacheable memory location shared with other cores, its contents will be cached and the corresponding cache line will be in the *exclusive* (E) state.
- If other cores also issue a *load exclusive* instruction on a memory location whose address hits the same cache line, the cache line will move to the *shared* (S) state.
- Assuming that no cache lines have been evicted in the meantime, the first core that issues a *store conditional* instruction will find that the store hits a cache line in either the E or the S state. Based on this information, it will be able to conclude that no other cores have modified the same location in the meantime—or another location whose address has hit the same cache line—so it will actually perform the store and succeed.
- As a consequence of the store, the MESI protocol will invalidate all the cache lines corresponding to the same location in the other cores.
- At this point, the execution of a *store conditional* instruction on other cores will result in a cache miss. From this, the cores will conclude that the same location—or another location whose address has hit the same cache line—has been modified by another core and report that the *store conditional* failed.

The fact that cache lines are wider than the memory words on which *load exclusive* and *store conditional* operate makes *store conditional* fail for no good reason with low probability. Moreover, it is usually necessary to reset the per-core logic that couples a *load exclusive* with the subsequent *store conditional*, thus forcing *store conditional* to fail, when switching a core from a task to another or when handling an exception (including interrupts). On most architectures this is done automatically in whole or in part, and can also be manually forced by means of a dedicated instruction, for instance, the CLREX instruction in the ARM architecture.

As we did previously, we define a *concurrent system* as a collection of N tasks. As usual, tasks communicate through shared objects and are sequential, that is, all of the operations they invoke on the shared objects are performed in a well-known fixed sequence, possibly with the help of barriers. Moreover we allow tasks to *halt* at arbitrary points of their execution and exhibit arbitrary variations in speed.

Shared objects are typed. The type of an object defines the set of its possible values, as well as the operations that can be carried out on it. Objects are assumed to have a *sequential specification* that defines how the object behaves when its operations are invoked sequentially by a single task.

As was done in Chapter 3, when dealing with a concurrent system it is necessary to give a meaning to interleaved operation execution. According to the definition given in Reference [58], an object is *linearizable* if each operation on the object appears to have taken place instantaneously at some point between the invocation of the operation and its conclusion.

This property implies that tasks operating on a linearizable object appear to be interleaved at the granularity of complete operations. Moreover, the *order* of

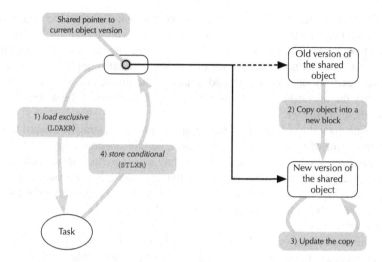

FIGURE 13.16 Basic technique to transform a sequential object implementation into a lock-free implementation.

non-overlapping operations is preserved. Linearizability is the basic correctness condition of Herlihy's concurrent objects.

Since the universal construction we are going to discuss relies on making a copy of shared objects before operating on them, it is assumed that those objects are small enough to be copied efficiently. Other universal constructions have been proposed for large objects, for instance Reference [4].

Moreover, it is understood that a sequential object occupies a fixed-size, contiguous area of memory called a *block*. Any sequential operation invoked on an object cannot have any side effect other than modifying the block occupied by the object itself. Sequential operations must also be *total*, that is, they must have a well-defined behavior for any legal or *consistent state* of the object they are invoked on. For example, a dequeue operation invoked on an empty queue is allowed to return an error indication but is not allowed to trigger an exception because it tried to execute an illegal instruction or follow an invalid pointer.

Figure 13.16 depicts the basic technique to transform a sequential implementation into a lock-free implementation. All tasks share a pointer that references the current version of the shared object and can be manipulated by means of the *load linked* and *store conditional* instructions. Each task operates on the object by means of the following steps:

1. It reads the pointer using *load exclusive*.
2. It copies the version it references into another block.
3. It performs the sequential operation on the copy.
4. It tries to move the pointer from the old version to the new by means of a *store conditional*.

If the store at step 4 fails, it means that another task disturbed the shared object somewhere between the *load exclusive* and the *store conditional* itself, that is, between steps 1 and 4. In this particular scenario—barring the sporadic spontaneous failure of *store conditional* mentioned previously—this also implies that some other tasks successfully updated the shared object in the meantime. Therefore, the task must retry the operation from the beginning. Each iteration of these steps is sometimes called an *attempt*.

The linearizability of the concurrent implementation is straightforward to prove. From the point of view of the other tasks, a certain operation appears to happen instantaneously, exactly when the corresponding *store conditional* succeeds, because the underlying memory system guarantees that the pointer update takes place atomically. Moreover, the order in which operations appear to happen is total and well-defined, and is the same as the order of the final, successful execution of *store conditional* operations.

If *store conditional* cannot fail spontaneously, even if all the N tasks in the system try to perform an operation on the shared object concurrently, at least one out of every N attempts to execute *store conditional* must succeed, the one that is performed first. Hence, the implementation is indeed lock-free according to the definition given in Section 13.5.1.

However, we saw that in cache-based implementations, *store conditional* may fail, with low probability, even if the shared pointer has not been touched. The probability of this event, which does not affect the correctness of the method just described but is detrimental to its practical efficiency because it leads to an unnecessary retry, depends on the length of the time window between the *load exclusive* and the corresponding *store conditional*.

One way of narrowing the window by leaving out the object copy operation is to perform a regular load instead of a *load exclusive* in the first step and then read the pointer again, this time with a *load exclusive*, immediately before the *store conditional*. The task then retries the operation if the pointer has changed between the first load and the second. Otherwise, it proceeds with the *store conditional* as before.

An important issue still to be discussed is memory management, that is, how the memory blocks that hold distinct object versions should be allocated to individual tasks and released as appropriate. From the practical point of view, it is also important to keep in mind that only a finite number of such blocks are available, and they must therefore be reused over and over again.

Apparently, if we refer back to Figure 13.16, $N + 1$ blocks of memory should suffice, provided they are used according to the following straightforward rules:

- At each instant, each of the N tasks owns one block of memory.
- The $N + 1$-th block holds the current version of the object and is not owned by any tasks.
- In step 2 of the algorithm, the task copies the object's current version into its own block.

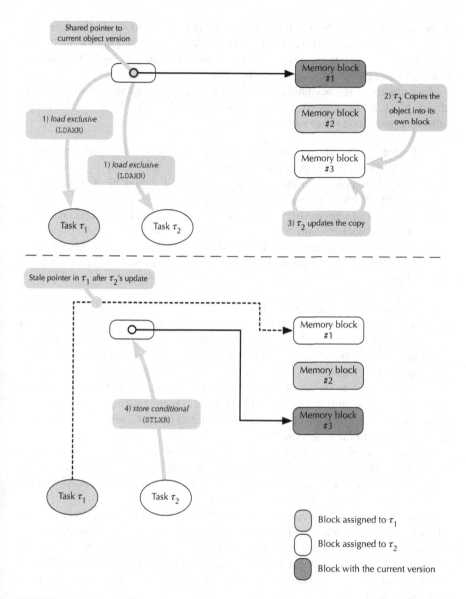

FIGURE 13.17 A possible race condition in object access.

- When a task performs a successful *store conditional*, it becomes the owner of the block that held the object's previous version and releases the block it owned previously because it now contains the current object version.

However, this approach is incorrect because it may lead to a subtle race condition in object access. Let us assume, as shown in Figure 13.17, that there are two tasks in the system, τ_1 and τ_2, and three blocks of memory, #1 to #3. In the figure, the allocation of each block is indicated by the block's color: dark gray means that the block holds the current version of the shared object, light gray and white indicate that the block belongs to τ_1 and τ_2, respectively.

Accordingly, at the beginning block #1 contains the current version of the shared object, block #2 belongs to τ_1, and block #3 belongs to τ_2. In this scenario, the following interleaving may take place:

- Both τ_1 and τ_2 retrieve a pointer to the current object version, held in block #1, as shown in the upper half of Figure 13.17.
- One of the task, τ_2 in the example, continues. It copies block #1 into its own block, block #3, and updates the object.
- The *store conditional* of τ_2 succeeds. The new version of the object is now stored in block #3, and τ_2 is now the owner of block #1.
- However, as depicted in the lower half of the figure, τ_1 still holds a stale pointer to block #1, which contains what is now an obsolete version of the object.

If τ_2 begins a new operation at this point, it will retrieve a pointer to the current object version, now in block #3, will copy its contents into its own block #1, and then operate on it. Since τ_1 is still engaged in its operation, and the next action it is due to perform is to copy the object from block #1 to block #2, the two copies performed by τ_1 and τ_2 may overlap.

In other words, as shown in Figure 13.18, τ_1 may read from block #1 while τ_2 is overwriting it with the contents of block #3, and then updating the object it contains. As a result, τ_1's copy of the shared object into block #2 might not represent a consistent state of the object itself.

This race condition is subtle because it does not harm the consistency of the shared object itself. Indeed, when τ_1 will eventually performs its *store conditional*, this operation will certainly fail because τ_2's first *store conditional* preceded it. As a consequence, τ_1 will carry out a new update attempt.

Nevertheless, it still poses significant issues from the software engineering point of view. Although it is relatively easy to ensure that any operation invoked on a consistent object will not do anything nasty (execution of an illegal instruction, division by zero, and so on), this property may be extremely difficult to guarantee in practice when the operation is invoked on an arbitrary bit pattern, like the one τ_1 may be confronted with.

The issue can be addressed by inserting a consistency check between steps 2 and 3 of the algorithm, that is, between the copy and the execution of the sequential operation. If the consistency check fails, the copy might be inconsistent and the task must retry the operation from step 1, without acting on the copy in any way. On some architectures, the consistency check is assisted by hardware and is built upon a *validate* instruction. This instruction works like *store conditional* and checks whether a variable previously read with *load exclusive* has been modified or not, but does not store any new value into it.

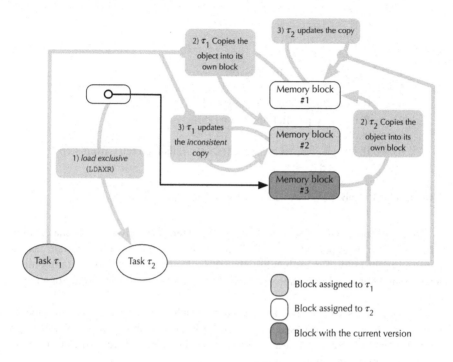

FIGURE 13.18 A task may operate on an inconsistent copy of the object.

If the underlying architecture does not provide this kind of support, like the ARM V6 processor architecture [7], a software-based consistency check can be used as a replacement. In this case, two counters, c_0 and c_1, complement each object version. It is usually assumed that counters can be read and written atomically, although in Section 13.5.2 we saw that techniques to ensure consistent readings of, and updates to, multi-word counters do exist [78]. Both counters start from the same value and are used according to the following rules:

- Before starting to modify an object, a task increments c_0. After finishing, it also increments c_1.
- A task reads c_1 before starting to copy an object and reads c_0 after finishing the copy, respectively.
- The consistency check consists of comparing these values: if they match, the copied object is definitely consistent. Otherwise, it might be inconsistent and must be discarded as described previously.

In [56], both counters are also assumed to be unbounded—that is, they never overflow or wrap around—but for practical purposes it is sufficient that they are large enough to make the probability of false matches negligible. Since the consistency check may succeed incorrectly only when a counter cycles all the way around, this probability can be made arbitrarily small by enlarging the counters.

```
1.    old = load_exclusive(Q)
2.    N->C₀ = N->C₁+1
3.    O₁ = old->C₁
4.    Copy Q->obj into N->obj
5.    O₀ = old->C₀
6.    If O₀ ≠ O₁ then retry from step 1
7.    Perform sequential operation on N->obj
8.    N->C₁++
9.    If store_conditional(Q, N) fails then retry from step 1
10.   N = old
```

FIGURE 13.19 Universal construction of a lock-free object (from [56]).

As an example, 64-bit unsigned integer counters may give a false match only when they are incremented $2^{64} \simeq 10^{19}$ times by other tasks while a certain task is performing the copy. The probability of such an occurrence is almost certainly much lower than other equally disruptive events, for instance, a hardware failure or a memory error.

We can now discuss the full algorithm to build a lock-free object from a sequential one in a universal way proposed in Reference [56], shown in Figure 13.19. Although we will keep the discussion at an informal level here, Reference [56] also formally proved its correctness. In the listing:

- Q is a shared variable that points to the current version of the shared object and corresponds to the pointer at the top left of Figures 13.16–13.18.
- N is a per-task local variable that points to the block owned by the task according to the memory management scheme described previously.
- Shared objects are assumed to be structures containing three fields: the obj field holds the object contents, while C_0 and C_1 are the two counters used for consistency checks.
- The obj field does not incorporate the counters. So, when the obj field is copied from one structure to another, the counters of the destination structure are not overwritten and retain their previous value.
- It is assumed that the code includes appropriate barriers to ensure that memory operations are performed in program order when necessary, although they are not shown for clarity.

The algorithm proceeds as follows:

- It starts by getting a local pointer, called old, to the current object version, using a *load exclusive* operation (step 1 of Figure 13.19). This corresponds to step 1 of the abstract algorithm described earlier in this section.
- The block owned by the executing process, pointed by N, is then marked as invalid. This is done by incrementing its C_0 counter using C_1 as a reference. This step is very important because, as discussed previously, other processes may still hold stale pointers to it (step 2).

- The counters associated to the object version pointed by `old` are copied into local variables before and after copying its contents into the block pointed by N (steps 3–5).
- After the copy, the values obtained for the two counters are compared. If they do not match, another process must have worked on the object in the meantime and the whole operation must be retried (step 6).
- If the consistency check is successful, then the block pointed by N contains a consistent copy of the shared object, and the executing process can perform the intended operation on it (step 7). This step is not detailed in the listing because it only depends on the sequential object implementation, and not on the lock-free wrapper.
- After finishing the update, the object version held in the block pointed by N is marked as consistent by incrementing its C_1 counter (step 8). The increment brings C_1 to the same value as C_0, which was incremented in step 2.
- It is now time to publish the updated object by performing a *store conditional* of the local pointer N into the shared object pointer Q (step 9). The conditional store will fail if Q was modified by some other process since it was loaded by the executing process in step 1. In this case, the whole operation must be retried.
- If the conditional store was successful, the executing process lost ownership of the block that contains the new object version, but can claim ownership of the block that contains the now-outdated version. The local pointer N is therefore updated accordingly (step 10).

It is worth noting that, although here we only discussed the construction of a *lock-free* object starting from a sequential object, a transformation that makes the object *wait-free* is also possible, for instance, by means of a general technique known as operation combining presented in Reference [56].

13.6   SPINLOCKS AND INTERRUPT HANDLING SYNCHRONIZATION

Section 12.3.3 highlighted the need of switching from implicit to explicit mutual exclusion between tasks and interrupt handlers when using a multicore processor. RTEMS provides two synchronization devices specifically tailored to this purpose, one accessible through the Classic Interrupt Manager API, the other by means of the POSIX API.

Both are *spinlocks*, that is, synchronization devices in which tasks are blocked by trapping them in an active spin loop rather than moving them to the blocked state of the task state diagram. Their implementation, not covered in this book, often relies on lock-free and wait-free synchronization. Being very efficient, these devices are often used within operating system kernels, RTEMS included, to protect critical data structures and as building blocks of more complex and user-friendly synchronization devices. For instance, spinlocks can be used to ensure that semaphore primitives are atomic in a multicore system.

TABLE 13.3

RTEMS Interrupt Manager API, Spinlocks

Function	Purpose
lock_initialize [1]	Initialize an interrupt service routine (ISR) lock
lock_destroy	Destroy a lock
lock_acquire	Acquire an ISR lock from a task
lock_release	Release an ISR lock from a task
lock_acquire_isr	Acquire an ISR lock from an ISR
lock_release_isr	Release an ISR lock from an ISR

[1] All Interrupt Manager functions start with the rtems_interrupt_ prefix.

The portion of Classic Interrupt Manager API that applies to single-core systems has been presented in Section 5.7. The same manager also makes available the directives listed in Table 13.3. They implement an efficient, low-overhead kind of lock called *interrupt lock*, built as a thin API layer on top of the ticket locks, first proposed in [87], which the RTEMS core uses internally for low-level synchronization. Interrupt locks must be allocated by the user and initialized before use. The directive:

```
void rtems_interrupt_lock_initialize(
    rtems_interrupt_lock *lock,
    const char *name
);
```

initializes the interrupt lock referenced by lock and gives it a symbolic name, used for debugging purposes. The directive does not return any status code because it always succeeds. However, as for many other initialization functions that operate on synchronization devices, users must ensure that locks are initialized sequentially, that is, multiple concurrent calls to rtems_interrupt_lock_initialize referencing the same lock do not occur, because this would lead to unpredictable results. In addition, since the string pointed by name could be referenced at any time during the lifetime of the lock, users must guarantee that it will remain valid throughout it.

The directive:

```
void rtems_interrupt_lock_destroy(
    rtems_interrupt_lock *lock,
);
```

destroys the lock referenced by lock. As for lock initialization, concurrent lock destruction is unsupported and leads to unpredictable results.

On the tasks side, an interrupt lock referenced by lock shall be acquired and released by means of the directives:

```
void rtems_interrupt_lock_acquire(
    rtems_interrupt_lock *lock,
    rtems_interrupt_lock_context *lock_context
);
```

```
void rtems_interrupt_lock_release(
    rtems_interrupt_lock *lock,
    rtems_interrupt_lock_context *lock_context
);
```

Besides a pointer to the lock, both directives take as second argument a pointer `lock_context` to another user-allocated data structure that represents the lock context. A separate lock context must be provided for each pair of acquire/release directive calls that can possibly be executed concurrently. Lock contexts do not need to be shared and can be stored on the stack in automatic variables.

Invoking `rtems_interrupt_lock_acquire` disables interrupts on the calling core and acquires the lock referenced by `lock`. Any wait that could be necessary to acquire the lock is implemented by means of a busy loop and does not lead to the preemption of the calling task. Symmetrically, `rtems_interrupt_lock_release` releases the lock and restores interrupts as they were when the matching `rtems_interrupt_lock_acquire` was called. The lock context passed to `rtems_interrupt_lock_release` must be the same given to the matching `rtems_interrupt_lock_acquire`.

Interrupt locks do *not* support recursive lock acquisition. Any attempt to acquire a lock within the critical region delimited by an acquire/release pair of the same lock results in an infinite busy loop with interrupts disabled on the calling core.

These two directives can also be called from an interrupt handler and work as intended, with the side effect that all mask-able interrupts—including the ones with a higher priority than the interrupt handler being executed—will be disabled within the critical region they delimit. When this is undesirable, specific variants are available for use from within an interrupt handler, which only acquire/release the lock without changing the interrupt level:

```
void rtems_interrupt_lock_acquire_isr(
    rtems_interrupt_lock *lock,
    rtems_interrupt_lock_context *lock_context
);
```

```
void rtems_interrupt_lock_release_isr(
    rtems_interrupt_lock *lock,
    rtems_interrupt_lock_context *lock_context
);
```

When using these variants, it is especially important to consider that the interrupt handler might still be preempted while within the critical region by a higher-priority interrupt handler, if the underlying architecture supports interrupt nesting and the

TABLE 13.4

RTEMS Interrupt Manager Spinlocks on a Single-Core System

Function	Behavior on a single-core system
lock_initialize	Do nothing
lock_destroy	Do nothing
lock_acquire	Disable all mask-able interrupts
lock_release	Restore interrupt level as it was before lock_acquire
lock_acquire_isr	Do nothing
lock_release_isr	Do nothing

interrupt request is accepted by the same core. If this happens and the nested handler also tries to acquire the same interrupt lock, results are unpredictable.

An interesting and useful aspect of these primitives is that, on a single-core system, they automatically assume the behavior listed in Table 13.4. Accordingly, the data types `rtems_interrupt_lock` and `rtems_interrupt_lock_context` are defined differently, and become smaller, on these systems. This enables the same code to also compile and run on a single-core system and, at the same time, keeps lock semantics unchanged.

POSIX spinlocks represent an alternative to the Interrupt Manager spinlocks just discussed. A POSIX spinlock is a user-allocated object of type `pthread_spinlock_t` used with the functions listed in Table 13.5. A comparison between Tables 13.3 and 13.5 reveals that the interfaces in the two cases are very similar. In particular, the function:

```
int pthread_spin_init(
    pthread_spinlock_t *lock,
    int pshared
);
```

initializes the POSIX spinlock referenced by `lock`. The second argument `pshared`, if non-zero, allows the spinlock to be shared among multiple processes. Unlike other POSIX objects, spinlocks do not have a static initializer and must therefore always be initialized at runtime, by means of this function. The `pshared` argument is ignored in RTEMS, since it implements multiple concurrent threads within a POSIX single-user, single-process (SUSP) execution environment.

Unlike its Interrupt Manager counterpart, this function returns a status code to the caller, which is zero upon successful completion. Otherwise, it shall return one of the following error numbers:

EAGAIN There are insufficient resources, other than memory, to initialize the lock.

ENOMEM There is not enough memory to initialize the lock.

Moreover, the standard recommends that the function detects and reports one additional error condition:

TABLE 13.5
RTEMS POSIX Spinlocks

Function	Purpose
pthread_spin_init	Initialize a spinlock
pthread_spin_destroy	Destroy a spinlock
pthread_spin_lock	Lock (acquire) a spinlock
pthread_spin_trylock	Lock a spinlock without waiting if it is already locked
pthread_spin_unlock	Unlock (release) a spinlock

EBUSY The lock referenced by lock is already initialized.

In the case of RTEMS, neither EAGAIN nor ENOMEM may occur, because the spinlock is implemented as a completely self-contained object and no additional resources besides the memory allocated to the pthread_spinlock_t object are needed to initialize and use it. The error number EBUSY may not occur either, because at the time of this writing RTEMS does not check for the corresponding, optional error condition. The function:

```
int pthread_spin_destroy(
    pthread_spinlock_t *lock
);
```

destroys the spinlock referenced by lock. The function returns to the caller a status code that reflects the outcome of the operation, zero indicates successful completion. In general, destroying a locked spinlock or a spinlock that has not been initialized leads to undefined behavior, but the standard recommends to check for this condition and return one of the following error codes instead:

EBUSY The lock referenced by lock is locked.
EINVAL The lock referenced by lock is not initialized.

At the time of this writing, RTEMS does not perform these optional checks. Therefore, pthread_spin_destroy always returns zero.

Spinlocks are acquired and released—*locked* and *unlocked* in POSIX terms—by means of the following functions:

```
int pthread_spin_lock(
    pthread_spinlock_t *lock
);
```

```
int pthread_spin_unlock(
    pthread_spinlock_t *lock
);
```

Both function take a pointer to an initialized lock as argument and return to the caller a status code, which is zero if they completed successfully. They can be called from both a task and an interrupt context. Passing an uninitialized lock to either function or attempting a recursive lock leads to undefined behavior. Optionally, implementation may check for these conditions and return one of the status codes listed below. Currently RTEMS does not perform these checks and both functions always return zero.

EINVAL The lock referenced by `lock` is not initialized.
EDEADLK The lock referenced by `lock` is already locked by the caller or, more generally, locking the lock would lead to a deadlock.

It is worth noting that, although tasks are not allowed to acquire the same spinlock or interrupt lock recursively, they may still acquire multiple, distinct locks, giving rise to nested critical regions. In this case, programmers must ensure that this behavior does not lead to a deadlock, by means of one of the techniques described in Section 8.2. Considering that in most practical cases the use of these locks aims at maximum execution efficiency, deadlock prevention strategies that involve zero runtime overhead (like total lock order) are especially suitable to this purpose.

The POSIX standard also defines a non-blocking variant of `pthread_spin_lock`, which immediately returns to the caller with an `EBUSY` indication if the spinlock referenced by `lock` is already locked:

```
int pthread_spin_trylock(
    pthread_spinlock_t *lock
);
```

In the current RTEMS implementation, this function is defined as an alias of `pthread_spin_lock`. It always succeeds and returns zero to the caller.

Like interrupt locks do, POSIX spinlocks also change their behavior automatically on single-core systems. In this case, they implement mutual exclusion by completely disabling mask-able interrupts.

13.7 SUMMARY

This chapter briefly described several scheduling algorithms and synchronization technique suitable for symmetric multiprocessor and multicore systems. In particular, after giving a general outline of scheduling policies and algorithms in Section 13.1, it described the multicore scheduling algorithms available in RTEMS and how they are configured in Sections 13.2 and 13.3, respectively.

Section 13.4 discussed the MrsP and OMIP semaphore protocols, both available in RTEMS. These protocols, especially the ones that support nested, fine-grained locking [50, 123], are still an active area of research even though some proposals, like the RNLP protocol [124], achieved significant theoretical and practical results. Interested reader may refer to the recent survey by Brandenburg [24] for comprehensive information about the evolution and current status of these protocols.

Then, Section 13.5 provided some information on lock-free and wait-free synchronization, together with several simple example to outline how it works in practice and the principle it is based upon. At the end of the chapter, Section 13.6 gave an introduction to spinlocks, a very efficient and specialized mutual exclusion lock used to synchronize tasks and interrupt handlers in multicore systems, and as a building block of more complex and sophisticated synchronization devices.

References

1. L. Abeni and G. Buttazzo. Integrating multimedia applications in hard real-time systems. In *Proceedings of the 19th IEEE Real-Time Systems Symposium (RTSS)*, pp. 4–13, December 1998.

2. Aeronautical Radio, Inc. *Avionics Application Software Standard Interface: ARINC Specification 653*, 2010.

3. Alfred V. Aho, Monica S. Lam, Ravi Sethi, and Jeffrey D. Ullman. *Compilers: Principles, Techniques, and Tools*. Pearson Education Ltd., Harlow, England, 2nd edition, September 2006.

4. James H. Anderson and Mark Moir. Universal constructions for large objects. *IEEE Transactions on Parallel and Distributed Systems*, 10(12):1317–1332, 1999.

5. James H. Anderson and Srikanth Ramamurthy. A framework for implementing objects and scheduling tasks in lock-free real-time systems. In *Proc. 17th IEEE Real-Time Systems Symposium*, pp. 94–105, December 1996.

6. James H. Anderson, Srikanth Ramamurthy, and Kevin Jeffay. Real-time computing with lock-free shared objects. In *Proc. 16th IEEE Real-Time Systems Symposium*, pp. 28–37, December 1995.

7. ARM Ltd. *ARM Architecture Reference Manual*, July 2005. DDI 0100I.

8. ARM Ltd. *ARMv6-M Architecture Reference Manual*, September 2010. DDI 0419C.

9. ARM Ltd. *ARMv7-M Architecture Reference Manual*, February 2010. DDI 0403D.

10. ARM Ltd. *Cortex-M4(F) Lazy Stacking and Context Switching — Application Note 298*, March 2012. DAI 0298A.

11. ARM Ltd. *ARM® Cortex®-A Series — Programmer's Guide for ARMv8-A*, March 2015. DEN 0024A.

12. ARM Ltd. *Procedure Call Standard for the ARM® Architecture*, November 2015. IHI 0042F.

13. ARM Ltd. *ARMv8-A Memory Systems*, February 2017. 100941_0100_en.

14. ARM Ltd. *ARM Architecture Reference Manual — ARMv8, for ARMv8-A architecture profile*, July 2019. DDI 0487E.a.

15. N.C. Audsley. On priority assignment in fixed priority scheduling. *Information Processing Letters*, 79(1):39–44, 2001.

16. Neil C. Audsley, Alan Burns, Mike Richardson, and Andy J. Wellings. Hard real-time scheduling: The deadline monotonic approach. In *Proc. 8th IEEE Workshop on Real-Time Operating Systems and Software*, pp. 127–132, 1991.

17. Neil C. Audsley, Alan Burns, and Andy J. Wellings. Deadline monotonic scheduling theory and application. *Control Engineering Practice*, 1(1):71–78, 1993.

18. Theodore P. Baker and Alan Shaw. The cyclic executive model and Ada. In *Proc. IEEE Real-Time Systems Symposium*, pp. 120–129, December 1988.

19. J. H. Baldwin. Locking in the multithreaded FreeBSD kernel. In *BSDC'02: Proceedings of the BSD Conference 2002*, Berkeley, CA, 2002. USENIX Association.

20. E. Bini, G. C. Buttazzo, and G. M. Buttazzo. Rate monotonic analysis: the hyperbolic bound. *IEEE Transactions on Computers*, 52(7):933–942, July 2003.

21. G. Blake, R. G. Dreslinski, and T. Mudge. A survey of multicore processors. *IEEE Signal Processing Magazine*, 26(6):26–37, November 2009.

22. R. Braden, L. Zhang, S. Berson, S. Herzog, and S. Jamin. *Resource ReSerVation Protocol (RSVP) – Version 1 Functional Specification, RFC 2205.* September 1997.

23. B. B. Brandenburg. A fully preemptive multiprocessor semaphore protocol for latency-sensitive real-time applications. In *Proceedings of the 25th Euromicro Conference on Real-Time Systems (ECRTS)*, pp. 292–302, July 2013.

24. Björn B. Brandenburg. Multiprocessor real-time locking protocols: A systematic review, September 2019. arXiv preprint number 1909.09600.

25. P. Brinch Hansen. Structured multiprogramming. *Communications of the ACM*, 15(7): 574–578, 1972.

26. P. Brinch Hansen. *Operating System Principles.* Prentice-Hall, Englewood Cliffs, NJ, 1973.

27. P. Brinch Hansen, editor. *The Origin of Concurrent Programming: from Semaphores to Remote Procedure Calls.* Springer-Verlag, New York, 2002.

28. A. Burns and A. J. Wellings. A schedulability compatible multiprocessor resource sharing protocol – MrsP. In *Proceedings of the 25th Euromicro Conference on Real-Time Systems (ECRTS)*, pp. 282–291, July 2013.

29. Alan Burns and Andy Wellings. *Real-Time Systems and Programming Languages.* Pearson Education, Harlow, England, 3rd edition, 2001.

30. G. Buttazzo and G. Lipari. Ptask: An educational C library for programming real-time systems on Linux. In *Proc. 18th Conference on Emerging Technologies Factory Automation (ETFA)*, pp. 1–8, September 2013.

31. Giorgio C. Buttazzo. *Hard Real-Time Computing Systems. Predictable Scheduling Algorithms and Applications.* Springer-Verlag, Santa Clara, CA, 2nd edition, 2005.

32. Brad Cain, Steve Deering, Isidor Kouvelas, Bill Fenner, and Ajit S. Thyagarajan. *Internet Group Management Protocol, Version 3, RFC 3376.* October 2002.

33. Steve Chamberlain and Cygnus Support. *Libbfd — The Binary File Descriptor Library.* Free Software Foundation, Inc., 2008.

34. Steve Chamberlain and Ian Lance Taylor. *The GNU linker ld (GNU binutils) Version 2.20.* Free Software Foundation, Inc., 2009.

35. Kuan-Hsun Chen, Georg von der Brüggen, and Jian-Jia Chen. Overrun handling for mixed-criticality support in RTEMS. In *Proc. Workshop on Mixed-Criticality Systems (WCS)*, Porto, Portugal, November 2016.

36. Ivan Cibrario Bertolotti and Gabriele Manduchi. *Real-Time Embedded Systems: Open-Source Operating Systems Perspective.* CRC Press, Taylor & Francis Group, Boca Raton, FL, 1st edition, January 2012.

37. E. G. Coffman, M. Elphick, and A. Shoshani. System deadlocks. *ACM Computing Surveys*, 3(2):67–78, 1971.

38. P. J. Courtois, F. Heymans, and D. L. Parnas. Concurrent control with "readers" and "writers". *Communications of the ACM*, 14(10):667–668, October 1971.

39. Robert I. Davis and Alan Burns. A survey of hard real-time scheduling for multiprocessor systems. *ACM Comput. Surv.*, 43(4):35:1–35:44, October 2011.

40. Raymond Devillers and Joël Goossens. Liu and Layland's schedulability test revisited. *Information Processing Letters*, 73(5-6):157–161, 2000.

41. Sudarshan K. Dhall and C. L. Liu. On a real-time scheduling problem. *Operations Research*, 26(1):127–140, February 1978.

42. Edsger W. Dijkstra. Cooperating sequential processes. Technical Report EWD-123, Eindhoven University of Technology, 1965. Published as [44].

43. Edsger W. Dijkstra. The multiprogramming system for the EL X8 THE. Technical Report EWD-126, Eindhoven University of Technology, June 1965.

44. Edsger W. Dijkstra. Cooperating sequential processes. In F. Genuys, editor, *Programming Languages: NATO Advanced Study Institute*, pp. 43–112. Academic Press, Villard de Lans, France, 1968.

45. Edsger W. Dijkstra. The structure of the "THE"-multiprogramming system. *Communications of the ACM*, 11(5):341–346, 1968.

46. Marko Doko and Viktor Vafeiadis. A program logic for C11 memory fences. In Barbara Jobstmann and K. Rustan M. Leino, editors, *Proc. 17th International Conference on Verification, Model Checking, and Abstract Interpretation (VMCAI)*, pp. 413–430, Berlin, Heidelberg, 2016. Springer Berlin Heidelberg.

47. Free Software Foundation, Inc. *GNU Make*, 2014. Available online, at http://www.gnu.org/software/make/.

48. Edgar Gabriel, Graham E. Fagg, George Bosilca, Thara Angskun, Jack J. Dongarra, Jeffrey M. Squyres, Vishal Sahay, Prabhanjan Kambadur, Brian Barrett, Andrew Lumsdaine, Ralph H. Castain, David J. Daniel, Richard L. Graham, and Timothy S. Woodall. Open MPI: Goals, concept, and design of a next generation MPI implementation. In *Proceedings of the 11th European PVM/MPI Users' Group Meeting*, pp. 97–104, Budapest, Hungary, September 2004.

49. Erich Gamma, Richard Helm, Ralph Johnson, and John Vlissides. *Design Patterns: Elements of Reusable Object-oriented Software*. Addison-Wesley Longman Publishing Co., Inc., Boston, MA, USA, 1995.

50. Jorge Garrido, Shuai Zhao, Alan Burns, and Andy Wellings. Supporting nested resources in MrsP. In Johann Blieberger and Markus Bader, editors, *Reliable Software Technologies – Ada-Europe 2017*, pp. 73–86, Cham, 2017. Springer International Publishing.

51. Charles M. Geschke, James H. Morris, Jr., and Edwin H. Satterthwaite. Early experience with mesa. *Communications of the ACM ACM*, 20(8):540–553, August 1977.

52. M. D. Godfrey and D. F. Hendry. The computer as von Neumann planned it. *IEEE Annals of the History of Computing*, 15(1):11–21, 1993.

53. L. J. Guibas and R. Sedgewick. A dichromatic framework for balanced trees. In *Proceedings of the 19th Annual Symposium on Foundations of Computer Science (SFCS)*, pp. 8–21, Oct 1978.

54. A. N. Habermann. Prevention of system deadlocks. *Communications of the ACM*, 12(7):373–377, 1969.

55. J. W. Havender. Avoiding deadlock in multitasking systems. *IBM Systems Journal*, 7(2):74–84, 1968.

56. Maurice P. Herlihy. A methodology for implementing highly concurrent data objects. *ACM Trans. on Programming Languages and Systems*, 15(5):745–770, November 1993.

57. Maurice P. Herlihy and Nir Shavit. *The Art of Multiprocessor Programming, Revised Reprint*. Morgan Kaufmann Publishers Inc., San Francisco, CA, USA, 1st edition, 2012.

58. Maurice P. Herlihy and Jeannette M. Wing. Axioms for concurrent objects. In *Proc. 14th ACM SIGACT-SIGPLAN Symposium on Principles of Programming Languages*, pp. 13–26, New York, 1987.

59. C. A. R. Hoare. Towards a theory of parallel programming. In *Proc. International Seminar on Operating System Techniques*, pp. 61–71, 1971. Reprinted in [27].

60. C. A. R. Hoare. Monitors: An operating system structuring concept. *Communications of the ACM*, 17(10):549–557, 1974.

61. Richard C. Holt. Some deadlock properties of computer systems. *ACM Computing Surveys*, 4(3):179–196, 1972.

62. Gerard J. Holzmann. *The Spin Model Checker: Primer and Reference Manual*. Pearson Education, Boston, MA, 2003.

63. IEC. *Industrial Communication Networks—Fieldbus specifications—Part 3-3: Data-Link Layer Service Definition—Part 4-3: Data-link layer protocol specification—Type 3 elements*, December 2007. Ed 1.0, IEC 61158-3/4-3.

64. *IEEE Std 1003.13™-2003, IEEE Standard for Information Technology—Standardized Application Environment Profile (AEP)—POSIX® Realtime and Embedded Application Support*. IEEE, 2003.

65. *IEEE Std 1003.1™-2008, Standard for Information Technology—Portable Operating System Interface (POSIX®) Base Specifications, Issue 7*. IEEE and The Open Group, 2008.

66. *Industrial Communication Networks—Fieldbus specifications—Part 3-12: Data-Link Layer Service Definition—Part 4-12: Data-link layer protocol specification—Type 12 elements*. IEC, December 2007. Ed 1.0, IEC 61158-3/4-12.

67. Intel Corp. *Intel® 64 and IA-32 Architectures Software Developer's Manual*, 2007.

68. *International Standard ISO/IEC/IEEE 9945, Information Technology—Portable Operating System Interface (POSIX®) Base Specifications, Issue 7*. IEEE and The Open Group, 2009.

69. International Organization for Standardization and International Electrotechnical Commission. *ISO/IEC 9899, Programming Languages — C*, 1st edition, December 1990.

70. International Organization for Standardization and International Electrotechnical Commission. *ISO/IEC 9899, Programming Languages — C*, 2nd edition, December 1999.

71. International Organization for Standardization and International Electrotechnical Commission. *ISO/IEC 9899, Programming Languages — C*, 3rd edition, December 2011.

72. *ISO 11898-1—Road vehicles—Controller area network (CAN)—Part 1: Data link layer and physical signalling*. International Organization for Standardization, 2003.

73. *ISO 17356-1 — Road vehicles — Open interface for embedded automotive applications—Part 1: General structure and terms, definitions and abbreviated terms*. International Organization for Standardization, January 2005.

74. Max Khiszinsky. *CDS: Concurrent Data Structures library*. Available online, at http://libcds.sourceforge.net/.

75. Marc Kleine-Budde. SocketCAN — the official CAN API of the Linux kernel. In *Proc. Intl. CAN Conference (iCC)*, pp. 5-17–5-22, March 2012.

76. Sambasiva Rao Kosaraju. Limitations of Dijkstra's semaphore primitives and Petri nets. *Operating Systems Review*, 7(4):122–126, January 1973.

77. Karthik Lakshmanan and Ragunathan Rajkumar. Scheduling self-suspending real-time tasks with rate-monotonic priorities. In *Proc. 16th IEEE Real-Time and Embedded Technology and Applications Symposium*, pp. 3–12, April 2010.

78. Leslie Lamport. Concurrent reading and writing. *Communications of the ACM*, 20(11):806–811, November 1977.

79. Leslie Lamport. How to make a multiprocessor computer that correctly executes multiprocess programs. *IEEE Transactions on Computers*, 9:690–691, September 1979.

80. Leslie Lamport. The mutual exclusion problem: part I—a theory of interprocess communication. *Journal of the ACM*, 33(2):313–326, 1986.

81. Leslie Lamport. The mutual exclusion problem: part II—statement and solutions. *Journal of the ACM*, 33(2):327–348, 1986.

82. J. Y.-T. Leung and J. Whitehead. On the complexity of fixed-priority scheduling of periodic, real-time tasks. *Performance Evaluation*, 2(4):237–250, 1982.

83. Chung L. Liu. Scheduling algorithms for multiprocessors in a hard real-time environment. In *JPL Space Programs Summary 37-60*, volume 2. Jet Propulsion Laboratory (JPL), November 1969.

84. Chung L. Liu and James W. Layland. Scheduling algorithms for multiprogramming in a hard-real-time environment. *Journal of the ACM*, 20(1):46–61, 1973.

85. Jane W. S. Liu. *Real-Time Systems*. Prentice Hall, Upper Saddle River, NJ, 2000.

86. Marshall Kirk McKusick, Keith Bostic, Michael J. Karels, and John S. Quarterman. *The Design and Implementation of the 4.4BSD Operating System*. Addison-Wesley, Reading, MA, 1996.

87. John M. Mellor-Crummey and Michael L. Scott. Algorithms for scalable synchronization on shared-memory multiprocessors. *ACM Trans. Comput. Syst.*, 9(1):21–65, February 1991.

88. Motorola Microcomputer Division and Software Components Group. *Real Time Executive Interface Definition*, January 1988. Draft 2.1.

89. Bryon Moyer. *Real World Multicore Embedded Systems*. Newnes, May 2013.

90. NXP B.V. *LPC17xx User manual, UM10360 rev. 2*, August 2010. Available online, at http://www.nxp.com/.

91. On-line Applications Research Corp. *RTEMS Documentation*, December 2011. Available online, at http://www.rtems.com/.

92. The ORKID Working Group Software Subcommittee of VITA. *ORKID — Open Real-Time Kernel Interface Definition*, August 1990. Draft 2.1.

93. OSEK/VDX. *OSEK/VDX Operating System Specification*. Available online, at http://www.osek-vdx.org/.

94. Mark S. Papamarcos and Janak H. Patel. A low-overhead coherence solution for multiprocessors with private cache memories. *SIGARCH Comput. Archit. News*, 12(3):348–354, January 1984.

95. C. Perkins. *IP Encapsulation within IP, RFC 2003*. October 1996.

96. Roland H. Pesch, Jeffrey M. Osier, and Cygnus Support. *The GNU Binary Utilities (GNU binutils) Version 2.20*. Free Software Foundation, Inc., October 2009.

97. Jon Postel. *User Datagram Protocol, RFC 768*. Information Sciences Institute (ISI), August 1980.

98. Jon Postel. *Internet Control Message Protocol—DARPA Internet Program Protocol Specification, RFC 792*. Information Sciences Institute (ISI), September 1981.

99. Jon Postel, editor. *Internet Protocol—DARPA Internet Program Protocol Specification, RFC 791*. USC/Information Sciences Institute (ISI), September 1981.

100. Jon Postel, editor. *Transmission Control Protocol—DARPA Internet Program Protocol Specification, RFC 793*. USC/Information Sciences Institute (ISI), September 1981.

101. Christopher Pulte, Shaked Flur, Will Deacon, Jon French, Susmit Sarkar, and Peter Sewell. Simplifying ARM concurrency: Multicopy-atomic axiomatic and operational models for ARMv8. *Proceedings of the ACM on Programming Languages*, 2(POPL):19.1–19.29, January 2018.

102. Ragunathan Rajkumar, L. Sha, and John P. Lehoczky. Real-time synchronization protocols for multiprocessors. In *Proc. 9th IEEE Real-Time Systems Symposium*, pp. 259–269, December 1988.

103. Red Hat Inc. *eCos User Guide*, 2013. Available online, at http://ecos.sourceware.org/.

104. W. Richard Stevens and Gary R. Wright. *TCP/IP Illustrated (3 Volume Set)*. Addison-Wesley Professional, Boston, MA, USA, November 2001.

105. RTEMS Project. *RTEMS C User Documentation — Release 4.11.3*, February 2018.

106. RTEMS Project. *RTEMS Source Builder — Release 4.11.3*, February 2018.

107. Peter Sewell, Susmit Sarkar, Scott Owens, Francesco Zappa Nardelli, and Magnus O. Myreen. x86-TSO: A rigorous and usable programmer's model for x86 multiprocessors. *Communications of the ACM*, 53(7):89–97, July 2010.

108. Lui Sha, Tarek Abdelzaher, Karl-Erik Årzén, Anton Cervin, Theodore P. Baker, Alan Burns, Giorgio C. Buttazzo, Marco Caccamo, John Lehoczky, and Aloysius K. Mok. Real time scheduling theory: A historical perspective. *Real-Time Systems*, 28(2):101–155, 2004.

109. Lui Sha, Mark H. Klein, and John B. Goodenough. Rate monotonic analysis for real-time systems. Technical Report CMU/SEI-91-TR-006, Software Engineering Institute, Carnegie Mellon University, 1991.

110. Lui Sha, Ragunathan Rajkumar, and John P. Lehoczky. Priority inheritance protocols: an approach to real-time synchronization. *IEEE Transactions on Computers*, 39(9):1175–1185, September 1990.

111. A. Shoshani and E. G. Coffman. Prevention, detection, and recovery from system deadlocks. In *Proc. 4th Princeton Conference on Information Sciences and Systems*, March 1970.

112. A. Silberschatz, P. B. Galvin, and G. Gagne. *Operating System Concepts*. John Wiley & Sons, New York, 7th edition, 2005.

113. H. R. Simpson. Four-slot fully asynchronous communication mechanism. *IEE Proceedings E (Computers and Digital Techniques)*, 137(1):17–30, January 1990.

114. Socket-CAN. The Socket-CAN project. Available online, at `https://github.com/linux-can/`, 2019.

115. Brinkley Sprunt, Lui Sha, and John Lehoczky. Scheduling sporadic and aperiodic events in a hard real-time system. Technical Report CMU/SEI-89-TR-011, Software Engineering Institute, Carnegie Mellon University, 1989.

116. Richard M. Stallman, Roland McGrath, and Paul D. Smith. *GNU Make — A Program for Directing Recompilation, for GNU make Version 4.0*. Free Software Foundation, Inc., October 2013.

117. Richard M. Stallman and the GCC Developer Community. *GNU Compiler Collection Internals, for GCC Version 4.3.4*. Free Software Foundation, Inc., 2007.

118. Richard M. Stallman and the GCC Developer Community. *Using the GNU Compiler Collection, for GCC Version 4.3.4*. Free Software Foundation, Inc., 2008.

119. A. S. Tanenbaum and A. S. Woodhull. *Operating Systems Design and Implementation*. Pearson Education, Upper Saddle River, NJ, 3rd edition, 2006.

120. L. R. Turner and J. H. Rawlings. Realization of randomly timed computer input and output by means of an interrupt feature. *IRE Transactions on Electronic Computers*, EC-7(2):141–149, June 1958.

121. William Von Hagen. *The definitive guide to GCC*. Apress, Berkeley, CA, 2006.

122. J. von Neumann. First draft of a report on the EDVAC. *IEEE Annals of the History of Computing*, 15(4):27–75, 1993. Reprint of the original typescript circulated in 1945.

123. Bryan C. Ward and James H. Anderson. Supporting nested locking in multiprocessor real-time systems. In *Proceedings of the 24th Euromicro Conference on Real-Time Systems*, ECRTS'12, pp. 223–232, USA, 2012. IEEE Computer Society.

124. Bryan C. Ward and James H. Anderson. Fine-grained multiprocessor real-time locking with improved blocking. In *Proceedings of the 21st International Conference on Real-Time Networks and Systems*, RTNS'13, pp. 67–76, New York, NY, USA, 2013. Association for Computing Machinery.

125. G. Bloom and J. Sherrill. Scheduling and thread management with RTEMS. *ACM SIGBED Review*, 11(1): 20–25, February 2014.

126. S. Gadia, C. Artho, and G. Bloom. Verifying Nested Lock Priority Inheritance in RTEMS with Java Pathfinder. In Ogata K., Lawford M., Liu S. (eds) *Formal Methods and Software Engineering, ICFEM 2016*. Lecture Notes in Computer Science, vol. 10009. Springer, Cham, October 2016.

Index

Printed in the United States
By Bookmasters